Kausch
Advanced Thermoplastic Composites

Advanced Thermoplastic Composites

Characterization and Processing

Edited by Hans-Henning Kausch
in Collaboration with Roger Legras

With 209 Illustrations

Hanser Publishers, Munich Vienna New York Barcelona

The Editor: Professor Dr. H.H. Kausch, EPFL, Laboratoire de Polyméres, MX-D-Ecublens, CH-1015 Lausanne, Switzerland

The Collaborator:
Professor Roger Legras, Université de Louvain la Neuve, Physique et Chimie des Hauts Polymères, 1 Place de la Croix du Sud, B-1348 Louvain-La-Neuve, Belgium

The use of general descriptive names, trademarks, etc., in this publication, even if the former are not especially identified, is not to be taken as a sign that such names, as understood by the Trade Marks and Merchandise Marks Act, may accordingly be used freely by anyone.

While the advice and information in this book are believed to be true and accurate at the date of going to press, neither the authors nor the editors nor the publisher can accept any legal responsibility for any errors or omissions that may be made. The publisher makes no warranty, express or implied, with respect to the material contained herein.

Library of Congress Cataloging-in-Publication Data
Advanced thermoplastic composites : characterization and processing /
  H.H. Kausch.
    p. cm.
  Includes bibliography, references and index.
  ISBN 3-446-17028-6
  1. Thermoplastic composites. I. Kausch, H.H.
TA18.9.C6A2835 1993
620.1'923--dc20   92-26914

Die Deutsche Bibliothek - CIP-Einheitsaufnahme
Advanced thermoplastic composites: characterization and processing / ed. by Hans-Henning Kausch in collab. with Roger Legras. - Munich ; Vienna ; New York ; Barcelona : Hanser, 1993
  ISBN 3-446-17028-6
NE: Kausch, Hans-Henning (Hrsg.)

All rights reserved. No part of this book may be reproduced or transmitted in any form or by any means, electronic or mechanical, including photocopying or by any information storage and retrieval system, without permission from the publisher.

© 1993 Carl Hanser Verlag, Munich Vienna New York Barcelona
Typesetting by J.W. Arrowsmith, Bristol, Great Britain
Printed and bound in Germany by Passavia Druckerei GmbH, Passau

Preface

Advanced thermoplastic materials are noted for their excellent thermal stability, ease of processing, and attractive mechanical properties at high temperatures. Retention of mechanical properties at high temperature is particularly important when the thermoplastic forms the matrix of (fibre-reinforced) composites. The generally stiff molecular backbones, high processing temperatures, and presence of reinforcing fibres, however, give rise to problems which have been the subject of intensive research—and of numerous publications—in the last decade.

This volume owes its inception to the fruitful discussions between several groups working on major complementary research projects in the field of advanced thermoplastic materials—in particular, to the interactions between the polymer laboratories in Louvain-la-Neuve and Lausanne. The work of these groups has resulted in new methods of molecular and structural analysis. The results from these and other methods have a strong bearing on processing, design and (long term) performance of advanced thermoplastic matrices and their composites. In the editor's opinion, these developments form an essential complement to the other major work of reference in this field, F. N. Cogswell's *"Thermoplastic aromatic polymer composites"*.

In Louvain-la-Neuve innovative methods of molecular characterization have been developed, which by way of chemical substitution, overcome a serious handicap to such a characterization, the insolubility of PEEK in most solvents—normally considered to be one of its biggest advantages. For the same reason, chemical reactions such as degradation or reticulation could not be easily identified previously—although they have an obvious bearing on mechanical properties, and proper precautions should be taken during processing so as to avoid such modifications. A new approach is also proposed concerning the assessment of crystallinity using DSC, IR and gravitometric techniques (Chapters 1 to 3).

In Lausanne the mechanical properties of several advanced thermoplastics—including liquid crystalline polymers and their composites—have been analyzed in terms of their morphological changes. Such studies form the essential bridge between processing and optimization of properties, and are the basis of all reliable service life predictions (Chapters 4 to 6 and 8).

The long term properties of continuous fibre composites are treated by D. R. Moore (ICI, Chapter 7).

A brief overview of processing methods is given by Manson in Chapter 9 (a more comprehensive treatment of processing—mostly restricted to that of PEEK, however—has just become available in the form of Cogswell's book). One particularly interesting aspect of processing has been dealt with in Chapter 10 by J. Müller who—in another major research project—has evaluated the economical and technical concepts for design and processing of composite structures for the machine tool industry.

This book closes with a discussion of one of the major advantages of thermoplastic matrix composites which is their potential for bonding and repair. It is strongly hoped that the discussion of important materials science aspects of advanced thermoplastic materials from the molecular to the macroscopic level as presented in this book will help the researcher engaged in development as well as the design engineer and the processing manager.

The Editor would like to thank wholeheartedly his colleague, Prof. R. Legras, Louvain-la-Neuve, for his enthusiastic and unqualified support in the editing of this book, all authors for their valuable and efficient collaboration, and Carl Hanser Verlag for the speedy production.

Lausanne, July 1992   Henning Kausch

Contents

Preface	V
Part I: The Matrix	1

1 Poly(Ether Ether Ketone) Characterization 3
D. Daoust, J. Devaux, P. Godard, A. Jonas, and R. Legras

Abstract	3
1 Introduction	3
2 Synthesis, Solubility, and Sulphonation of PEEK	5
3 Spectroscopic Characterization of PEEK and Sulphonated PEEK	15
4 Molecular Mass and Distribution Determination	35
Conclusion	52
Acknowledgments	53
References	54

2 PEEK Degradation and Its Influence on Crystallization Kinetics 57
A. Jonas and R. Legras

Abstract	57
1 Introduction	57
2 Thermal Stability of Molten PEEK in an Oxidative Atmosphere	58
3 Thermal Stability of Molten PEEK in Inert Atmosphere	63
Conclusion	79
Acknowledgments	79
References	80

3 Assessing the Crystallinity of PEEK 83
A. Jonas and R. Legras

Abstract	83
1 Introduction	83
2 Density Determination of the Degree of Crystallinity of PEEK	84
3 Differential Scanning Calorimetry Determination of the PEEK Degree of Crystallinity	87
4 Infrared Absorption Spectroscopy Determination of the Degree of Crystallinity of PEEK	94
5 Limitations of the Two-Phase Crystallinity Concept for PEEK	99

Conclusion ... 106
Acknowledgments ... 107
References ... 107

4 The Physical Structure and Mechanical Properties of Poly(Ether Ether Ketone) ... 111
P.-Y. Jar and Ch.J.G. Plummer

Abstract ... 111
1 Introduction ... 111
2 The Microstructure of PEEK ... 112
3 The Mechanical Properties of Bulk Samples ... 121
4 Microdeformation Behaviour in Thin Films ... 129
Conclusions ... 137
Acknowledgments ... 138
References ... 138

5 Structure and Mechanical Properties of Other Advanced Thermoplastic Matrices and Their Composites ... 141
P. Davies and Ch.J.G. Plummer

Abstract ... 141
1 Introduction ... 141
2 Case Studies ... 146
Conclusions ... 166
Acknowledgments ... 166
References ... 166

Part II: Evaluation of the Composites ... 171

6 The Short-Term Properties of Carbon Fibre PEEK Composites ... 173
W.J. Cantwell and P. Davies

Abstract ... 173
1 Introduction ... 173
2 Comparison of the Mechanical Properties of Carbon Fibre PEEK with Other High Performance Composites ... 174
3 A More Detailed Characterization of the Mechanical Properties of Carbon Fibre PEEK ... 180
References ... 190

7 Long-Term Properties of Aromatic Thermoplastic Continous Fibre Composites: Creep and Fatigue 193
D.R. Moore

Abstract 193
1 Introduction 193
2 Materials 194
3 Creep 196
4 Fatigue 208
5 Other Thoughts on Long-Term Properties 224
Acknowledgments 225
References 225

8 Structure and Mechanical Properties of Filled and Unfilled Thermotropic Liquid Crystalline Polymers 227
Ch.J.G. Plummer

Abstract 227
1 Introduction 227
2 Injection Mouldings 231
3 Mechanical Properties 238
4 Discussion 258
Conclusions 264
Acknowledgments 267
References 267

Part III: Processing and Joining 271

9 Processing of Thermoplastic-Based Advanced Composites 273
J.-A. E. Månson

Abstract 273
1 Introduction 273
2 Aspects of Material Processability and Properties 275
3 Basic Product Forms 277
4 Forming Mechanisms 281
5 Shaping and Consolidation Techniques 282
6 Process-Structure-Property Relations 290
7 Processing Window 296
Acknowledgments 297
References 297

10 Conception and Processing of Advanced Thermoplastic Composite Structures for the Machine Industry ... 303
J. Müller

1. General Situation for Advanced Composites in the Machine Industry ... 303
2. Requirements of the Machine Industry for Advanced Composites ... 303
3. Significance of Advanced Thermoplastic Composites for the Machine Industry ... 304
4. Applicability and Limitations, Guidelines for Application ... 306
5. Design Procedure ... 306
6. Manufacturing ... 325
7. Quality Assurance ... 332
8. Cost Effectiveness ... 333
9. Service Experiences ... 333
10. Future Aspects, Requirements for Material ... 334

References ... 335

11 Bonding and Repair of Thermoplastic Composites ... 337
P. Davies and W.J. Cantwell

1. Introduction ... 337
2. Mechanical Fastening ... 338
3. Adhesive Bonding ... 339
4. Fusion Bonding ... 347
5. Comparison of Methods ... 358
6. Repair ... 360

Conclusions ... 362
References ... 365

Index ... 367

Part I
The Matrix

Chapter 1
Poly(Ether Ether Ketone) Characterization

D. Daoust, J. Devaux, P. Godard, A. Jonas, and R. Legras

Abstract

Different poly(ether ether ketone) (PEEK) synthesis procedures are described. A review of PEEK solvents for UV-visible, ^{13}C and ^{19}F NMR spectroscopies, light scattering (LS), viscometry and SEC is presented.

Upon dissolution in H_2SO_4 aqueous solution, PEEK is chemically modified by sulphonation of phenyl rings. Sulphonated PEEK is soluble at room temperature in various solvents. This derivatization method leads to the development of a room temperature SEC (RTSEC) of PEEK.

Neat PEEK and sulphonated PEEK are characterized by UV-visible, ^{13}C and ^{19}F NMR and IR spectroscopies. Fluoroarylketone and hydroxyaryl chain ends are identified and quantitatively evaluated by ^{19}F NMR and IR spectroscopies.

PEEK average molecular masses and molecular mass distribution are also obtained by high temperature SEC and RTSEC. The reliability of the derivatization method is proven. The agreement between results obtained by both methods is very satisfactory.

Finally, a PEEK-carbon fibre composite is characterized.

1 Introduction

The remarkable solvent resistance of poly(oxy-1,4-phenylene-oxy-1,4-phenylene-carbonyl-1,4-phenylene) (PEEK) has notably delayed its characterization, especially the determination of molecular masses and molecular mass distribution which are most important parameters controlling the end-use properties of the polymer. The structural unit of the polymer is represented on Scheme 1.

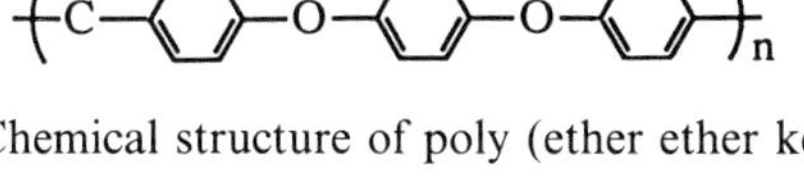

Scheme 1 Chemical structure of poly (ether ether ketone) (PEEK)

Daniel Daoust, Jacques Devaux, and Roger Legras, Laboratoire de Chimie et de Physique des Hauts Polymères, Université Catholique de Louvain, Place Croix du Sud, 1, B-1348 Louvain-la-Neuve, Belgium; Pierre Godard, Research Associate of the National Fund for Scientific Research, Belgium; Alain Jonas, Research Assistant of the National Fund for Scientific Research, Belgium.

PEEK being mostly used as an advanced fibre-reinforced composite matrix, such a product often necessitates the development of separation methods prior to the analysis of the matrix itself.

These reasons explain why chemical structure and molecular mass characterization of PEEK is still an active research area. Identification of the nature of the chain ends, conformation in dilute solution, new mixtures of solvents, chemical modification are among other topics of recent works and publications.

This chapter is devoted to an overview of some characterization methods of PEEK.

After a brief review of the different synthesis procedures, an important part concerns the solubility of PEEK. The use of concentrated sulphuric acid as solvent and the analytical possibilities offered by the sulphonation explain the important place reserved to PEEK sulphonation throughout this chapter. The improvement of solubility of the sulphonated polymer is a positive factor allowing the development of a room temperature size exclusion chromatography (RTSEC) of PEEK. The nature of the sulphonating species, the mechanism, the chemical modification induced, the kinetics, the operating conditions to achieve a 100% sulphonation level and the solubility of the sulphonated PEEK in organic solvents are the various subjects that will complete this first section.

The second section of the chapter is dedicated to the spectroscopic characterization of neat and sulphonated PEEK. The UV-visible, nuclear magnetic resonance (RMN) and infrared (IR) spectroscopic analyses of the main chain are described and attempts are made to determine the nature and the amount of the various chain ends in experimental and commercial PEEK grades. This latter study is of some interest, since chain ends will be shown in chapter 3 to play a crucial role in the PEEK thermal stability and crystallization properties. Furthermore, a quantitative evaluation of the number of chain ends allows the determination of the number average molecular mass.

The third section covers the molecular mass characterization of PEEK with the help of established techniques like light scattering, viscometry and more particularly size exclusion chromatography (SEC).

SEC is an appreciated tool because it provides absolute molecular masses and molecular mass distribution which are important parameters to understand the physical, rheological and mechanical properties and to control the polymerization kinetics. However SEC is intrinsically a relative method and must be calibrated with PEEK standards. As PEEK standards were not commercially available, it was necessary to synthesize them and to determine directly their weight average molecular masses by light scattering experiments.

A detailed description of two SEC methods suitable for the determination of the molecular characteristics of PEEK are given in this third section. The first method is high temperature SEC which cannot be easily used as a routine procedure. The second one based on the so-called derivatization concept has the advantage to be performed at room temperature.

This derivatization must fulfil some specific criteria.

1. The molecular mass distribution of the original polymer must absolutely be

preserved: no chain scission or rebuilding can occur during the derivatization process.
2. The chemical modification must be reproducible.
3. The ratio between the structural unit molecular mass of the polymer and the derivative polymer must be known.
4. The derivative polymer must be chemically stable and soluble in a classical room temperature SEC solvent without any degradation.
5. The molecular characteristics of the original polymer can be found from the analytical results describing the derivative polymer.

It will be shown in this third section that sulphonation fulfills the previous criteria.
Finally, the derivatization procedure will be applied to the characterization of an experimental PEEK-carbon fibre composite.

2 Synthesis, Solubility, and Sulphonation of PEEK

The first part of this section describes two main synthesis procedures of poly(ether ether ketone): the acylation way and the displacement way based on a nucleophilic substitution mechanism. Both methods allow the preparation of polymers characterized by weight average molecular masses below ~ 60000 and polydispersity going from ~ 2.0 to ~ 2.5. Narrow molecular mass distribution fractions of PEEK have also been prepared by hydrolysis of poly(aryl ether ether ketimine) prepolymers. This latter method is very interesting from an analytical point of view: higher molecular mass and isomolecular species can be obtained and used as standards for viscometric and chromatographic techniques.

The second part deals with the solubility of PEEK. The remarkable chemical resistance of PEEK has set an important problem for the chemical structure and molecular mass characterization. Indeed, in this field, the dissolution conditions are particularly critical, viz. the nature of the solvent, the temperature, the concentration limit. A review of the different analytical solvents discovered up to now is given.

The subject of the third part draws the attention to a derivatization method of PEEK which leads to the preparation of a modified polymer with improved solubility properties.

2.1 Synthesis

It is not the objective of this paragraph to provide an exhaustive compilation of the existing literature on PEEK synthesis. For more details, the reader is invited to refer to recent reviews, for instance the paper from *Mullins* [1] on which most of the informations reported below are based. Moreover, in the present summary, mainly synthesis methods explicitely referring to PEEK will be selected.

There are two main synthesis procedures for PEEK, a first one by linking aromatic ether species through ketone groups, and a second one by linking aromatic ketone species by an ether bond. The former is sometimes referred to as the acylation way and is based on an electrophilic reaction of *Friedel–Crafts* type. The latter, sometimes called displacement way, is a nucleophilic substitution.

An original synthesis procedure, involving alternation of both kinds of methods and leading to well-defined oligomers of increasing lengths was recently described [2].

2.1.1 Acylation Way

Because of its inherent solvent resistance and its propensity to reach high crystallinity level, PEEK cannot successfully be synthesized in common organic solvents. Early attempts in methylene chloride or in nitrobenzene, for instance, only led to low molecular mass products, due to premature crystallization [3]. As emphasized by *Colquhoun* [4], strong acids are virtually the only low temperature solvents suitable for synthesis work.

Although the unique combination of anhydrous hydrogen fluoride/boron trifluoride was discovered more than twenty years ago [5], the failure of this system to use phosgene as reagents explains probably the lack of reference dealing with this PEEK synthesis until the paper of *Jansons* [6]. This author reports successful syntheses of PEEK using various alkylthiochloroformates.

Rose [7] discovered that trifluoromethanesulphonic acid can be used as solvent and as catalyst of polyacylation reactions leading to polyetherketones. *Colquhoun* [4] described later in details the *Friedel–Crafts* type polycondensation of 4-(4'-phenoxyphenoxy)benzoic acid into PEEK in trifluoromethanesulphonic acid (Scheme 2).

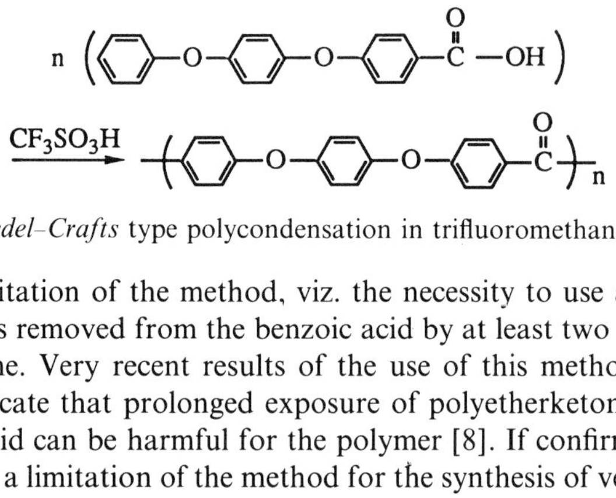

Scheme 2 *Friedel–Crafts* type polycondensation in trifluoromethanesulphonic acid

The known limitation of the method, viz. the necessity to use a monomer where the phenoxy end is removed from the benzoic acid by at least two phenylene groups, was thus overcome. Very recent results of the use of this method at a laboratory scale seem to indicate that prolonged exposure of polyetherketones to trifluoromethanesulphonic acid can be harmful for the polymer [8]. If confirmed, this degradation can represent a limitation of the method for the synthesis of very high molecular mass PEEK.

Other methods based on the acylation way seem worth being mentioned. They make use of a mixture of phophorus pentoxide/methanesulphonic acid as catalyst and solvent [9, 10]. The polycondensation of 4(4'-phenoxyphenoxy)-benzoic acid following the reaction of scheme 2 as well as of dicarboxylic acid containing phenyl ether structure with diphenoxybenzene are reported to produce PEEK.

2.1.2 Nucleophilic Substitution

Probably the most widespread synthesis procedure for PEEK is the nucleophilic substitution of a dihalogenobenzophenone by a bisphenate deriving from hydroquinone.

For kinetic reasons, the reaction using 4,4'-dichlorobenzophenone does not proceed far enough to lead to polymers with high molecular mass. However, successful uses of Ni [11] or $Cu^{II}/SiOH$ [12] catalysts were recently reported to allow polyetherketone synthesis starting from dichloro aromatics. Nevertheless, the expensive 4,4'-difluorobenzophenone is generally used, at least at an industrial scale.

Bisphenates from hydroquinone exhibit an extreme sensitivity to air oxidation. Therefore, the preferred synthetic way for PEEK is the use of hydroquinone and potassium or sodium carbonate which generate in situ bisphenate. The latter reacts then rapidly with difluoroketone (Scheme 3).

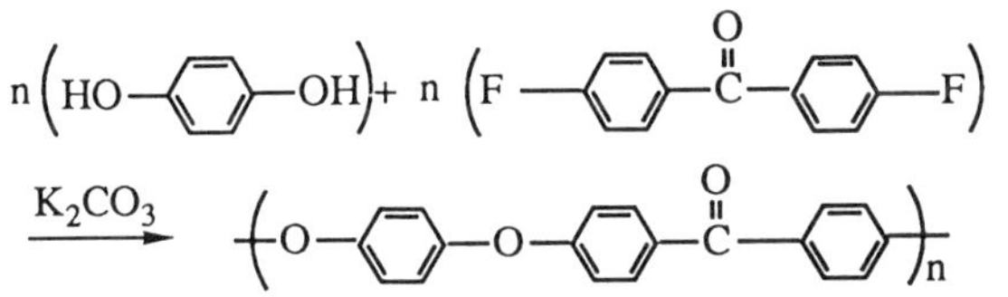

Scheme 3 PEEK synthesis by nucleophilic substitution

Owing to the poor solubility of PEEK, and to the need for a solubilization (at least a partial one) of the carbonate, the selection of the synthesis solvent is quite crucial. DMSO and sulpholane were proved to be unsuccessful. Benzophenone, but mainly diphenylsylphone which is the industrial reaction medium, and very high temperatures ($>300\,°C$) are required in order to reach high molecular masses. A slight excess of difluorobenzophenone is generally used to control the final molecular mass, leading to fluorine terminated chains.

Some consecutive hydrolysis of fluorine chain end or the use of other means to control the molecular mass (phenolic compounds for instance) may lead to different chain ends (see part 3.2).

Mullins [1] also reported a modification of the system using a large excess of carbonates and bicarbonates as reaction medium in place of diphenylsulfone. The mixture is brought to higher temperatures ($>350\,°C$) but remains a free-flowing powder during the full course of the reaction .

A convenient alternative to the use of alkaline bisphenoxides is provided by *Kricheldorf* [13] who used difluoroketones together with well soluble silylated bisphenols. The nucleophilic substitution is induced at high temperature by catalytic amounts of cesium fluoride.

Alternative reaction paths, using t-butyl substituted hydroquinone [14] or aniline-modified difluorobenzophenone (ketimine) [15] were recently reported to produce PEEK precursors soluble in more common solvents. Molecular mass fractionation is even reported [16]. However, a final synthesis step is needed to obtain PEEK polymer.

Ether interchange reactions are known to occur when polyethersulphones (PES) and polyetherketones (PEK) are exposed to nucleophiles at high temperatures. This problem leads for example to recommend the use of sodium carbonate for PEK synthesis in place of the potassium equivalent used for PEEK. Potassium salts seem to be more active for transetherification [17]. Addition of lithium chloride to the reaction medium is also reported [18] in order to exchange its cation with the by-product potassium fluoride, another proved ether exchange catalyst [19]. The problem of ether interchange is however of minor importance in the case of PEEK synthesis compared with the PES or PEK ones.

2.2 PEEK Solubility

PEEK solvents are reported in this paragraph. While most of the systems encountered in the literature are mentioned, this list is probably not fully exhaustive. Nevertheless, this review is aimed to present solvents and solvent mixtures actually used for spectroscopy (UV-visible;NMR), light scattering and chromatography (SEC) characterizations.

Most of the solvents can be classified into three groups:

- concentrated acids, the only products allowing the dissolution of PEEK at room temperature,
- organic media which dissolve the polymer at temperature above 300°C,
- mixtures of solvents which necessitate a dissolution procedure at temperatures between 160 and 220°C to obtain PEEK solutions stable at temperatures below 120°C.

At room temperature, PEEK is insoluble in classical organic solvents and only dissolves in the following concentrated acids: chlorosulphonic acid (HSO_3Cl) [20], sulphuric acid (H_2SO_4) [20], methanesulphonic acid (CH_3SO_3H) [16] and a mixture of methanesulphonic acid and methylene chloride (4/1 in vol.) [4], trifluoromethanesulphonic acid (CF_3SO_3H) [21], hydrofluoric acid [20]...

Protonation of carbonyls contributes to the solubility of PEEK in these acids, but in some cases, chemical modifications occur like sulphonation and cross-linking reactions [20, 22].

Some of these room temperature solvents of PEEK are of interest in the field of UV-visible spectroscopy [16, 20], ^{13}C NMR [21, 23], ^{19}F NMR [22], viscometry [16, 20, 22, 24] and light scattering studies [16, 20, 22, 24].

Some pure organic solvents dissolve PEEK at temperatures around its melting point (~345 °C). This is the case of high boiling point esters [7], diphenylsulphone [7], benzophenone [7, 22–27] and 1-chloronaphtalene [27].

Since the first industrial production of PEEK, several studies were undertaken in the hope of discovering useful solvents compatible with the commercially available SEC instrumentation. For this purpose researchers have found different organic solvent mixtures in which PEEK remains dissolved at temperatures below 120°C after a preliminary dissolution at higher temperatures. *Devaux* [22] described the dissolution procedure at the boiling point (viz. 183°C) of a phenol and 1-2-4-trichlorobenzene mixture (50/50 in weight) and proposed the use of these solvents as mobile phase for SEC analysis at 115°C. *Hay* [28] selected a mixture of 4-chlorophenol and 1,2-dichlorobenzene as PEEK SEC eluent at 40°C. *Roovers* [15] measured PEEK intrinsic viscosities in a mixture of the same solvents (60/40 in weight) at 35°C. To achieve a complete dissolution, the polymer was first swollen overnight in the solvent mixture and then heated for ~5 minutes at 160°C. *Day* [29] and *Roovers* [16] recorded UV-visible absorption spectra of PEEK dissolved in this mixture. Boiling phenol has been used to prepare PEEK films for infrared spectroscopic characterization (see part 3.2.3).

Finally, dichloroacetic acid-chloroform mixed solvent is used by *Ishiguro* [30] for SEC analysis of PEEK at room temperature. In these conditions, the sample is heated in dichloroacetic acid at 220°C for 30 minutes and after cooling, the solution is diluted with chloroform.

2.3 PEEK Sulphonation upon Dissolution in Concentrated Sulphuric Acid

With a very limited choice of room temperature solvents, some researchers, like *Bishop* [20] and *Devaux* [22], investigated the molecular mass determination and the conformation in dilute solutions using sulphonating strong acids as solvents. *Jin* [31] and *Bailly* [32] described controlled sulphonation and neutralization procedures which give rise to partially or totally sulphonated PEEK. In these papers, the influence of the chemical modification on thermal stability, thermal transition behaviour, water absorption, solubility and crystallinity was more particularly examined. Among other results, it was found that the degree of crystallinity significantly decreases with increasing sulphonation levels, thereby enhancing solubility. For example, fully sulphonated PEEK dissolves in several strong polar organic solvents at room temperature. On this basis, as the limited solubility of PEEK in common solvents is a major obstacle to perform routine SEC analyses, *Daoust* [33, 34] proposed to use sulphonation as a sample preparation method preliminary to room temperature SEC.

Therefore, it is essential to summarize the available scientific information concerning:

- the nature of the most suitable sulphonating acid,
- the chemical modification and the change in molecular mass induced by the sulphonation reaction,
- the sulphonation kinetics,
- the operating conditions followed to produce a fully sulphonated PEEK,
- the solubility of sulphonated PEEK.

2.3.1 Selection of a Sulphonating Agent

Sulphonation is a powerful method to modify aromatic polymers. For instance, polystyrene [35-39], poly(2,6-dimethyl-1,4-phenylene oxide) (PPO) [40, 41], and polysulphone [42-44] have already been modified. Various sulphonation methods are reported in the literature, including sulphonation by concentrated sulphuric acid [31, 45], by chlorosulphonic acid [46, 47], by pure or complexed sulphur trioxide [42, 44, 47, 48] and by acetylsulphate [49].

Sulphonation is an electrophilic reaction and the severity of the experimental conditions depends on the degree to which the ring is activated or deactivated. Electron-donating substituents will favour the sulphonation process whereas electron-withdrawing groups will not.

Bishop [20], *Jin* [31] and *Devaux* [22] have reported the sulphonation of PEEK upon dissolution in chlorosulphonic and sulphuric acids. In addition to sulphonation, *Bishop* and *Jin* revealed that chlorosulphonic acid induces crosslinking of PEEK via the condensation of $-SO_3H$ groups and formation of sulphone links. They also suggested that the same reactions might occur with 100% sulphuric acid. However, these authors asserted that sulphone formation and chemical degradation of PEEK dissolved in aqueous H_2SO_4 have never been detected even upon standing for long periods. From this viewpoint, concentrated sulphuric acid, as far as the concentration of the acid is kept below 100% to prevent crosslinking, can be considered as a suitable reactive giving rise to linear sulphonated PEEK without changing the molecular distribution of the original polymer. For this reason, it has been proceeded using such conditions in this study.

2.3.2 Sulphonation Mechanism

Sulphonation of PEEK at room temperature in concentrated sulphuric acid can be represented as follows:

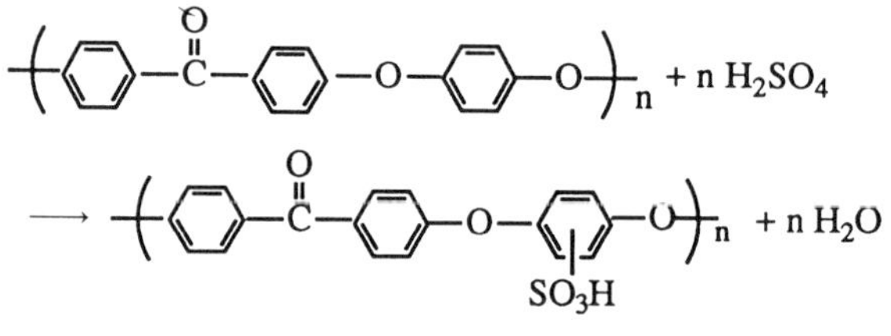

Scheme 4 PEEK sulphonation from concentrated sulphuric acid

The sulphonation process is limited to one sulphonate group per repeat unit and is located at one of the four chemically equivalent positions of the phenyl ring surrounded by two ether linkages. The other phenyl rings are deactivated by the neighbouring ketone [31, 45].

This latter statement is supported by the observation that poly(ether ketone) molecules (PEK) which contain rings only substituted by one ether and one carbonyl functions remain unsulphonated under the same conditions (Scheme 5). Indeed the ^1H NMR spectrum of PEK shows no evidence of sulphonation having occured in sulphuric acid [50]. Very similar conclusions were obtained by *Noshay* in the case of poly(arylsulphones) [42].

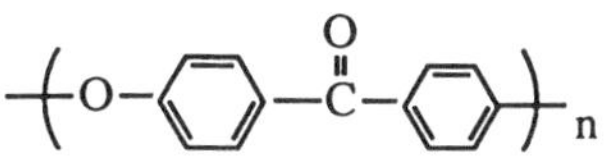

$$\text{-}\!\!\left[\!\text{O}\!-\!\!\bigcirc\!\!-\!\!\overset{\overset{\displaystyle O}{\|}}{C}\!-\!\!\bigcirc\!\!\right]_n$$

Scheme 5 Chemical structure of poly (ether ketone) (PEK)

If evidence of this chemical modification emerged from the infrared and ^{13}C nuclear magnetic resonance spectroscopic studies, one of the first confirmations of the maximum attainable level of PEEK sulphonation in concentrated sulphuric acid was given by *Devaux* [21]. A kinetic analysis of sulphonation of PEEK PKFF-2[1] in 99.5% concentrated sulphuric acid was undertaken at room temperature. The sulphur concentration determined by elemental analysis and the acid content measured by titration in the recovered derivative polymer are reported in Table 1. These results showed an equilibrium sulphonation level corresponding to one $-SO_3H$ per structural unit quickly achieved in the experimental conditions. Afterwards, the stability of the sulphonation level and of the molecular mass is supported by ^{13}C NMR and SEC characterization of polymers dissolved for more than four months in H_2SO_4 [34].

Table 1.1 PEEK PKFF-2 Sulphonation Kinetics in 99.5% H_2SO_4 [22] (Polymer Concentration: 5 g/dl)

Elapsed time* (days)	$-SO_3H$ equivalents per repeat unit	
	From elemental analysis	From acid titration
0.6	1.00	1.12
1	1.18	1.02
2	0.78	0.89
3	1.07	0.97

* Elapsed times are measured from dissolution to precipitation.

Due to sulphonation, the molecular mass and thereby the concentration of a PEEK sample increase as soon as a macromolecule is dissolved in H_2SO_4. This change in concentration must be taken into account when H_2SO_4-PEEK solutions are used for UV-visible (absorption coefficient determination), light scattering (weight average molecular mass, refractive index increment, second virial coefficient measurements) or viscometry (reduced and inherent viscosities calculations) characterizations. Therefore, it is of the greatest importance to ensure that the sulphonation process has reached its completion before any experiment.

---

[1] In appendix, Table 18 reports the main characteristics of the experimental and commercial PEEK grades used throughout this chapter.

A correction of the concentration due to the sulphonation can easily be applied. Indeed, the concentration in sulphonated species (C_s) is related to the initial concentration in unsulphonated species (C_u) by

$$C_s = k\, C_u \tag{1}$$

where k is a constant equal to 1.278 obtained from the ratio between the sulphonated repeating unit molecular mass (368) and the unsulphonated one (288).

2.3.3 Sulphonation Kinetics

Although the reaction time, the temperature, the acid and polymer concentrations are important variables in a kinetic study, these were so far not systematically explored in the literature.

Bailly [32] reported the sulphonation rate of PEEK in solution in 96.4% H_2SO_4 and in various mixtures of methanesulphonic acid and 96.4% H_2SO_4. These media were chosen to develop sulphonation methods that produce random copolymers over the entire available sulphonation range of zero to one sulphonate group per repeat unit. The sulphonation levels were determined from the sulphur to carbon ratios by elemental analyses.

In a paper to be published, *Daoust* [51] proposed a more detailed study of the sulphonation kinetics of PEEK PKFF-2 and of model compounds of this polymer carrying fluoroarylketone chain ends. The influence of different parameters such as the reaction time, the acid strength and the nature of the chain end are examined using especially ^{13}C and ^{19}F NMR spectroscopic measurements. When using similar concentrated H_2SO_4 solutions, it is more particularly observed:

- a lowering of the sulphonation rate at levels above roughly 50%. This effect can be explained by the electron-withdrawing nature of the $-SO_3H$ group giving rise to a decrease of the electronic density of the two adjacent units.
- above 50%, a constant rate following a first order equation. This invariability of the reaction rate is unambiguously proved by the kinetic results obtained from the sulphonation in the same conditions of a model compound representative of the repeat unit of the polymer,
- a supplementary retarding effect on the sulphonation velocity of the chain end repeating unit due to the fluorine atom in the protonating solvent.

From the use of H_2SO_4 of different concentrations, the results emphasize especially the fact that one of the principal factors governing the sulphonation kinetics is the concentration of the electrophile.

Fig. 1 represents the kinetic curves obtained from sulphonation experiments performed in 99.5, 98.7, 96.8 and 95.9% H_2SO_4 solutions at room temperature. The polymer PKFF-2 concentration is 2.8 g/dl. The origin of the time scale is taken one hour after the initial addition of the polymer powder into the acid solvent and

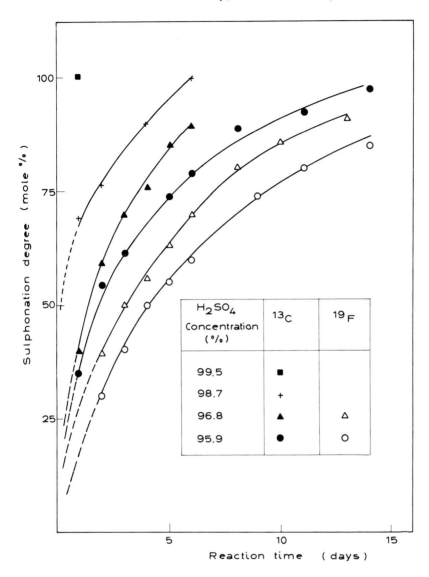

Fig. 1.1 PEEK PKFF-2 sulphonation kinetics in concentrarted sulphuric acid form ^{13}C and ^{19}F NMR experiments. (Polymer concentration: 2.8 g/dl.).

corresponds to the approximate time for complete dissolution. These curves clearly demonstrate the effects of H_2SO_4 concentration: a higher H_2SO_4 concentration leads to a faster sulphonation rate and decreases the reaction time required for reaching the 100% level. This conclusion is essential for selecting effective operating conditions allowing the preparation of a fully sulphonated PEEK.

2.3.4 Selected Procedure for Fully Sulphonated PEEK Preparation

For this purpose, PEEK (around 3 g/dl) is dissolved in 99.5% concentrated sulphuric acid and reacts for periods ranging from 3 to 4 days.

PEEK dissolution is not instantaneous and sulphonation only occurs after dissolution. Consequently heterogeneous mixtures can be obtained if care is not taken. At least, the dissolution process must be as short as possible (less than one hour) relative to the total sulphonation time. PEEK powder is thereby used and the medium is vigorously stirred during the gradual addition of sulphuric acid and during the whole reaction time. The importance of an efficient agitation is well confirmed by *Groggins* [52].

The obtained sulphonated polymer is precipitated with at least a five-fold volume of a stirred mixture of deionized water and hydrochloric acid (50/50 in vol.) at 0°C. The polymer is then shredded to obtain a fine powder, filtered, thoroughly washed with water/HCl to eliminate H_2SO_4 residue and dried at room temperature during around 2 days. The so-prepared sulphonated PEEK (H-SPEEK) can be used in this free acid form for room temperature SEC analyses.

If H-SPEEK has to be stocked for a long time or has to undergo thermal effects in the solid state, it is preferable to perform a neutralization of the polymer. For example, to obtain a Li-SPEEK sample, a stirred H-SPEEK/water mixture (5 g/dl) is neutralized with LiOH (0.1 M). This Li-SPEEK mixture is then filtrated and washed with cold water. Afterwards, the polymer salt can be dried overnight at 100 to 120°C in vacuum.

The sulphonation level can be checked by ^{13}C NMR experiments.

2.3.5 Sulphonated PEEK Solubility [31, 32, 34]

Sulphonated PEEK is soluble:

- Above 30% sulphonation, in several hot strong polar organic solvents such as N-methyl-2-pyrrolidone (NMP), dimethylformamide (DMF), dimethylacetamide (DMAc), dimethylsulphoxide (DMSO) and pyridine.
- Above 40% sulphonation, in the same solvents but at room temperature.
- Above 70% sulphonation, in methanol.
- At 100% sulphonation, in hot water.

Jin [31] has also found that the solubility of the sulphonated polymer does not depend on the neutralization level.

Among these solvents, DMF [31] and NMP were found to be good solvents for the preparation of solution cast films for IR experiments (see part 3.1.3). DMSO [31,33] is used as solvent for ^{13}C NMR characterization (see part 3.1.2) and NMP for room temperature SEC analyses (see part 4.1.3.2).

3 Spectroscopic Characterization of PEEK and Sulphonated PEEK

In a first part, the reader will be informed about the molecular structure characterization of the main chain by UV-visible, ^{13}C nuclear magnetic resonance (NMR) and infrared (IR) spectroscopies of neat and sulphonated PEEK samples of experimental and commercial grades. The interpretation of the different spectra is principally based on data reported in published papers.

The second part is devoted to the identification of the various types of chain ends that experimental or commercial PEEK may contain. Hydroxyaryl and fluoroarylketone chain ends can result respectively from the use of hydroquinone and 4-4′ difluorobenzophenone in the synthesis. Phenoxy end groups can result from the addition of phenol during the synthesis to control the molecular mass. Phenate chain ends can result from a reaction of hydroxyaryl chain ends and the sodium or potassium carbonate additive used in the synthesis.

The assignment of the different peaks detected was favoured by the use of model compounds with well defined chain ends.

Lastly, an endeavour of a quantitative evaluation of diverse chain ends is made.

3.1 Chain Characterization

3.1.1 UV-visible Spectroscopy

UV-visible spectroscopy is in many respects an interesting technique in the field of the PEEK characterization but gives unusual spectra due to the use of protonating strong acids (methanesulphonic and concentrated sulphuric acids).

It is well-known that the exact features of UV-visible spectra depend by complicated ways on acid strength, method of solution preparation, residence time after mixing (solution stability), concentration and nature of the sample. But the main object of this section is to draw the attention of the reader to the influence of the nature of the solvent, the acid strength and the chemical modification reactions (viz. sulphonation and crosslinking) on the UV-visible spectroscopic characteristics. Furthermore, UV-visible studies are also indispensable preliminaries to light scattering experiments since absorbance affects measured intensities. Finally, as UV spectrometers are SEC detectors, this method allows the determination of the most appropriate wavelength.

a) PEEK Characterization When PEEK is dissolved in a protonic acid like methanesulphonic acid (MSA), the solution appears yellowish-orange due to strong absorption in the blue and green regions of the visible spectrum. Fig. 2a shows a typical spectrum of PEEK PKFF-2 in MSA. The colour of the solution is correlated with the large absorption peak in the 350–450 nm resulting from electron orbital modification linked to protonation. The maximum wavelength (λ_{max}) is found at 413.8 nm with an absorption coefficient ε_{max} equal to 15270 l/g m or 44000 l/cm.mole of repeat units. This value is comparable to that of carbonium ions surrounded by phenyl groups suggesting that each ketone function is protonated [16, 53].

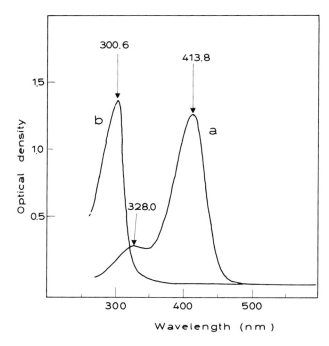

Fig. 1.2 UV-visible spectra. a. PEEK PKFF-2 in methanesulphonic acid. b. totally sulphonated PEEK PKFF-2 in N-methyl-2-pyrrolidone-LiBr (0.1 M).

Roovers [16] has used PEEK/MSA solutions for the determination of weight average molecular masses by light scattering. Since PEEK/MSA solutions have an absorption band which extends into the red, this author suggested a correction for absorption of observed scattered light intensities. On the other hand, he also recorded UV-visible spectra of PEEK dissolved in a mixture of p-chlorophenol and o-dichlorobenzene (60/40 in weight). In this mixed solvent, PEEK reveals a weak structureless spectrum which seems to exclude the presence of specific protonation.

Day [29] studied thermal degradation of PEEK. The extent of the degradation has been notably monitored by UV-visible spectroscopy in MSA and in a mixture of p-chlorophenol and o-dichlorobenzene (60/40 in weight).

b) Partially and Totally Sulphonated PEEK Characterization As the sulphonation level increases, the spectra of solutions of sulphonated PEEK in MSA exhibit a decrease of the λ_{max} values from 413.8 nm for an unsulphonated sample to 405.1 nm for a 100% sulphonated polymer (Table 2).

Solutions of completely sulphonated PEEK in concentrated sulphuric acid present UV-visible spectra very similar to the ones observed for PEEK in MSA. Furthermore, in aqueous H_2SO_4 solutions, as acid strength and associated protonation change with the amount of water, totally sulphonated PEEK spectra reveal a slight blue shift of λ_{max} with decreased absorbance. For example, a completely sulphonated PEEK in a 99.5% H_2SO_4 solution has a λ_{max} of 411.2 nm with an ε_{max} of 12970 l/g.m and in a 95.9% H_2SO_4 solution, a λ_{max} of 410.0 nm with an ε_{max} of 12370 l/g.m.

Table 1.2 UV-visible Spectroscopic Characteristics of Solutions of PEEK PKFF–2 Samples with Various Sulphonation Levels in Methanesulphonic Acid

Sulphonation level (mole%)	λ_{max} (nm)
0	413.8
60	408.8
80	406.4
100	405.1

Totally, sulphonated PEEK in solution in N-methyl-2-pyrrolidone + LiBr (0.1 M), which is the solvent for SEC characterization of the derivative polymer, shows a UV-visible spectrum (Fig. 2b) very similar to the one obtained by *Roovers* [16] for PEEK in a p-chlorophenol/o-dichlorobenzene mixture. Figure 2b also displays that 300 nm can be fixed as an adequate wavelength for UV-visible SEC detection.

Bishop [20] has also investigated the UV-visible behaviour of PEEK solutions in concentrated sulphuric acid (H_2SO_4) and chlorosulphonic acid (HSO_3Cl). H_2SO_4 solutions were characterized in order to apply eventual corrections to light scattering measurements. The experiments performed in HSO_3Cl, in which chemical reactions give rise to "aggregation", lead to the conclusion that the increasing absorption at 514.5 nm with time is probably correlated with the increase in sulphone crosslinks or with the increase in chlorosulphonation.

3.1.2 ^{13}C Nuclear Magnetic Resonance (NMR) Spectroscopy

This section is devoted to the interpretation of ^{13}C NMR spectra of a neat PEEK and of a 80% and a 100% sulphonated PEEK.

For sulphonated products, deuterated dimethylsulphoxide (DMSO-d_6) was used as solvent. This solvent allows to reach concentrations ranging from 5 to 7 g/dl at room temperature. Methanesulphonic acid, a non-sulphonating PEEK solvent, was used for the NMR study of neat PEEK. In this case, the concentration of the solution was around 4 g/dl. Solubility problems prevent to increase the concentration of PEEK in this solvent, which could be interesting to reduce the total acquisition time when observing small features like chain ends peaks[1].

Other non-sulphonating room temperature PEEK solvents were discarded for the following reasons: trifluoromethanesulphonic acid, a much stronger acid than CH_3SO_3H and also a much better PEEK solvent, has four strong resonances peaks

---

[1] Note that it has been claimed by *Colquhoun* [4] that the mixture CH_3SO_3H/CD_2Cl_2 (4/1, in vol.) is also a convenient spectroscopic PEEK solvent, and that a much higher concentration of polymer can be obtained in this mixture than in pure methanesulphonic acid.

Table 1.3 Signal Assignments for the Carbons of the Repeating Units of PEEK and Sulphonated PEEK

Sample	Structural unit	Solvent	^{13}C peak assignment (ppm)										
			C1	C2	C3	C4	C5	C6	C7	C8	C9	C10	C11
Neat PEEK	(structure)	CH$_3$SO$_3$H	199.64	123.96	139.14	118.07	168.35	151.23	123.01	123.01	123.01	123.01	151.23
Totally sulphonated PEEK	(structure)	DMSO-d$_6$	193.35 193.40 193.47	132.15 132.28 132.28 132.49	131.09 131.31 131.78 131.88	117.27 117.46	160.79 160.91 162.06 162.18	151.07	120.07	141.69	123.98	122.49	148.21
80% sulphonated PEEK	(structure)	DMSO-d$_6$	193.31 193.37 193.44	132.13 132.20 132.26 132.35 132.46	131.07 131.29 131.76 131.81 131.86 131.90	117.10 117.24 117.43	160.77 160.90 160.97 161.09 162.05 162.17 162.19	151.08	120.08	141.71	123.96	122.46	148.19
	(structure)						151.70	122.06	122.06	122.06	122.06	151.70	

in a zone in which some PEEK aromatic carbons resonate (between 114 and 122 ppm); hydrofluoric acid is not very attractive due to its toxicity!

a) ^{13}C NMR Spectrum of Neat PEEK A typical ^{13}C NMR spectrum of commercial grade PK150P is given in Fig. 3. Small peaks emerging from the noise are due to chain ends (see section 2.2.1).

Table 3 gives the assignments for the carbon signals of the chain repeating unit in methanesulphonic acid. These assignments are identical to results published by others [4].

b) ^{13}C NMR Spectra of Sulphonated PEEK Partially resolved ^{13}C spectra of sulphonated PEEK have first been assigned by *Jin* [31]. His study confirmed among others the maximum attainable level of PEEK sulphonation in concentrated sulphuric acid. Somewhat later, *Bunn* [54] and *Abraham* [21, 23, 55–57] have published

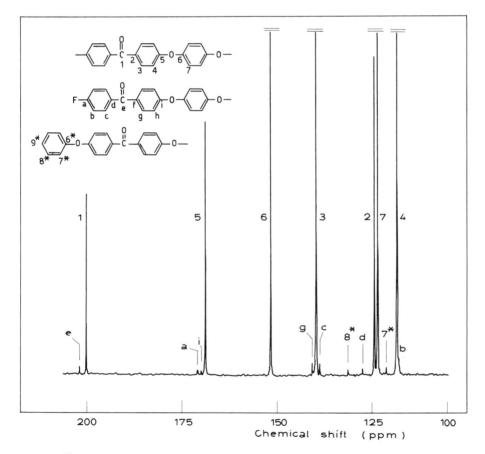

Fig. 1.3 ^{13}C NMR spectrum of commercial PEEK PK150P in methanesulphonic acid. (Instrument: BRUKER AM500; reference TMS; broadband mode).

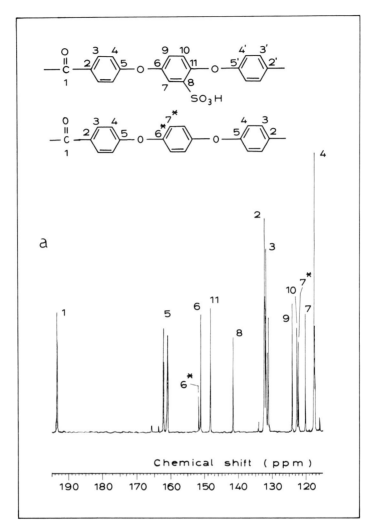

Fig. 1.4 ^{13}C NMR spectra of sulphonated PEEK PKFF-2 in dimethylsulphoxide-d$_6$ (DMSO-d$_6$). a. 80% sulphonated sample. b. 100% sulphonated sample. (Instrument: BRUKER AM500; chemical shifts refer to central resonance of DMSO-d$_6$ multiplet, taken as 39.6 ppm; inverse gate mode).

more detailed assignments on which most of the informations reported below refer.

^{13}C NMR spectra of the 80% and the 100% sulphonated PEEK PKFF-2 dissolved in DMSO-d$_6$ are presented in Fig. 4a and 4b. The assignments for the carbon signals of the repeating units are shown on the spectra and in Table 3.

For the fully sulphonated polymer, the resonances at 120.07, 122.49 and 123.98 ppm are only consistent with substitution on the oxy-1-4-phenylene rings. The resonances due to the carbons of the oxy-1-4-phenylene-carbonyl rings give rise to four

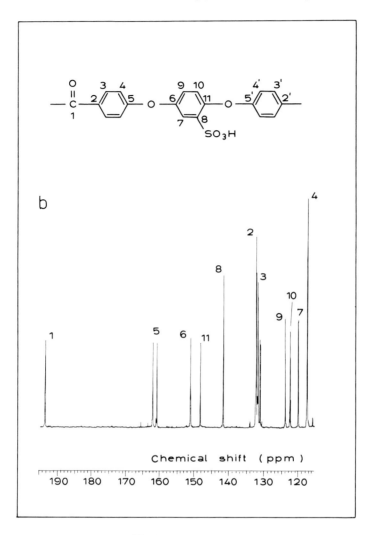

Fig. 1.4 (*Continued*)

(for C2, C3 and C5) and two (for C4) peaks. This observation can be attributed to transmission of electronic effects between aryl rings via bridging groups (—O—; —CO—) and to the subsequent effects that a substituent on one ring (—SO_3H) may have on the chemical shifts of the carbons of the neighbouring rings. Indeed it must be taken into account that if the four sulphonating positions in the di-oxy substituted ring in PEEK are clearly chemically equivalent, there are however two non-equivalent sulphonation sites with respect to the chemical shift of C6 for example. These sites can be called ortho and meta, depending on their orientation with respect to C6. As explained by *Abraham* [57], the transmission medium for these electronic effects is the π-electron system of the sulphonated chain.

For the partially sulphonated polymer (see Table 3), the resonances at 122.06 and 151.70 ppm reveal respectively the carbons C7, C8, C9, C10 and C6, C11 of the unsubstituted phenyl rings surrounded by two ether linkages. It is obvious that the measurements of the peak areas due to the C6 and C11 atoms of the sulphonated and unsulphonated rings allow the determination of the sulphonation level.

An important difference of chemical shifts between the C2, C3 and C5 carbons of the unsulphonated repeating units of PEEK and of partially sulphonated PEEK can also be observed. Following *Abraham* [57], such a behaviour is due to the different natures of the π systems of the chains in the two different media (protonation of the carbonyl function occurs in methanesulphonic acid).

3.1.3 IR Spectroscopy

a) Infrared Spectrum of Neat PEEK   Such spectra were often reported as reference for chemical assessment of synthesis products or as blanks for chemical modification studies. Also, crystallization measurements make wide use of PEEK infrared spectra.

It is the purpose of this section to provide a tentative assignment for the most important infrared bands because some of them will be used hereafter in crystallinity and derivatization measurements. Main references for infrared assignments are taken from *Bellamy* [58] and, as PEEK is a fully aromatic polymer, from *Varsanyi* [59]. Numeric references will be sometimes given, referring to the classification of the latter author.

An infrared spectrum of a self-supporting PEEK film, cast from a phenol solution, is reported in Fig. 5. Due to the technique used, the film should be amorphous, but, as an annealing can induce spectral modifications, some crystalline features can possibly be observed. The reader can refer to the paper of *Nguyen* [60] or to chapter 3 of this book to obtain more informations about crystallization-induced infrared modifications.

For the assignment of the different C—H and C—C phenyl vibrations of PEEK, it has to be taken into account that two species of phenyl rings exist in the molecule: those of the first kind are unsymmetrically para-disubstituted, and will be referred to as K (for ketone) phenyls while the symmetrically substituted ones will be referred to as E (for ether) phenyls.

In the spectral region above 1800 cm^{-1}, very few bands are observable. Three weak ones, between 3100 and 3000 cm^{-1} arise from C—H in-plane vibrations (20a and 20b). Other ones are very weak and have to be assigned to harmonics.

C=O stretching vibration of neat PEEK occurs at 1649 cm^{-1}. According to *Nguyen* [60], the very strong absorption band at 1229 cm^{-1} is assigned to the asymmetric stretching mode of the diphenyl ether group.

Aromatic C—C in-plane vibrations (aromatic triplet bands) can be found at 1600 cm^{-1} (8a), 1493 cm^{-1} (19a) and 1413 cm^{-1} (19b). The latter can be assigned to K phenyls while its sister-band, for E phenyls, could give rise to the shoulder observed around 1450 cm^{-1}.

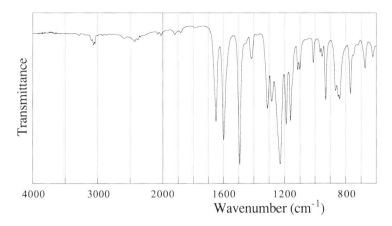

Fig. 1.5 Infrared spectrum of PEEK PKFF-2. (Instrument: PERKIN ELMER FT-IR 1760 X; self-supporting film cast from phenol).

Bands at 1311 cm^{-1} and 1284 cm^{-1} are in-plane C—C vibrations, possibly (3) and (13) or (7a) in the classification of *Varsanyi*, for both K and E phenyls. Peaks at 1190 cm^{-1} and 1162 cm^{-1} can be tentatively assigned to in-plane (9a) vibrations of K and E phenyls respectively. The C—C in-plane vibration 18b seems to split into two peaks with maxima at 1115 cm^{-1} and 1101 cm^{-1} which can possibly be attributed to K and E phenyls respectively. The assignment of the band at 1012 cm^{-1} to the (18a) in-plane phenyl vibration seems straightforward while discrimination between the doublet at 966–953 cm^{-1} and the more intense band at 930 cm^{-1} is more difficult. They are certainly due to C—H out-of-plane vibrations (17a) and (5). It can be proposed to assign the splitting of the doublet to some kind of conjugation effect, which is often observed for diphenyl ether-type molecules [61]. Therefore, this doublet should be attributed to E phenyls. However, influence of crystallization induces more complexity to this region of the spectrum (see chapter 3). The complex band around 850 cm-1 is assigned to the (17b) and possibly (10a) out-of-plane C—H vibrations. Following *Varsanyi*, the former one occurs above 850 cm^{-1} only when there is an electron-withdrawing substituent. Therefore, the 866 cm^{-1} peak has to be assigned to the (17b) out-of-plane C—H vibration of the K phenyls. For chemical reasons which will be seen in the paragraph on derivatization herebelow, the assignment of the doublet 848-838 cm^{-1} to the (17b) C—H out-of-plane vibration of E phenyls seems the most probable, with a splitting due to the above-mentioned conjugation phenomenon. It is finally proposed to assign the peaks at 769 cm^{-1} and 624 cm^{-1} respectively to the in-plane (1) and (6b) vibrations of the K phenyls, while the 674 cm^{-1} peak should arise from the (4) vibration of the same K phenyls.

b) Modifications of the Infrared Spectrum of PEEK upon Sulphonation   It was established from ^{13}C NMR study that sulphonation of PEEK results from the electrophilic substitution of a H atom of the phenyl flanked by two oxygen atoms (E phenyls-see hereabove) by one $-SO_3H$ group.

Therefore, as already pointed out by *Jin* [31], absorptions related to the presence of these sulphonic acid groups have to be observed. *Colthup* [62] reports that these bands should be observed in three regions: 1260–1150 cm^{-1}, 1080–1010 cm^{-1} and 700–600 cm^{-1}. *Bellamy* [58], while citing *Colthup*, reports also comparable ranges from other authors. No significant differences in assignments are reported for acid and salt forms. Spectra of 80% and 100% sulphonated PEEK samples cast from NMP solutions onto KBr pellets are reported in Fig. 6(B and C). In these spectra, new bands are indeed observed at 1254 cm^{-1}, 1077 cm^{-1} and 708 cm^{-1} which were tentatively assigned by *Jin* [31] respectively to O=S=O asymmetric and symmetric stretchings and to S—O stretching. Also a broad band around 3450 cm^{-1} can be assigned to O—H vibrations. The splitting of the carbonyl stretch vibration, with the arising of a shoulder at 1652 cm^{-1} is a consequence of the sulphonation, possibly due to hydrogen-bonding effects but more probably to differences in electron-withdrawing effects acting on the C=O bond strength. The latter explanation seems to be supported by the observation of two shoulders in that region for the partially sulphonated sample.

The modification of the 1493 cm^{-1} and 1413 cm^{-1} in-plane phenyl vibrations which were observed in the neat PEEK spectrum can be tentatively explained. In the PEEK spectrum, (19a) and (19b) vibrations occurred respectively around 1500 cm^{-1} and 1410 cm^{-1} for K phenyls and around 1490 cm^{-1} and 1450 cm^{-1} for E phenyls. Both (19a) peaks superimpose to each other. After sulphonation, only the bands of E phenyls should be significantly affected. (19b) shifts to 1473 cm^{-1} while (19a) shifts to about 1420 cm^{-1} where it superimposes to the (19b) peak of the K phenyl. This behaviour seems supported by the observation of a peak remaining at 1473 cm^{-1} on the spectrum of a partially sulphonated PEEK.

Correlation between the absorbance ratio A1500/A930 with sulphonation levels measured by ^{13}C NMR provides an alternative way to determine the sulphonation content [51].

3.2 Chain End Characterization

3.2.1 Qualitative Identification by ^{13}C NMR

The identification and assignment of chain ends peaks were performed in two ways. First, by comparing the NMR spectra of monodisperse PEEK oligomers of increasing chain length, having welldefined chain ends (fluoroarylketone chain ends, phenoxy chain ends or arylketone chain ends) [2]. Second, by recording the NMR spectra of experimental grades of PEEK, whose main chain ends are known by synthesis, and comparing these spectra with those of model compounds of low molecular mass (see Fig. 3). Experimental grades having only fluoroarylketone chain ends (PKFF-2) or hydroxyaryl chain ends (PKOH) were obtained presumably by using an excess of 4,4'-difluorobenzophenone or of hydroquinone for the polymer synthesis. From the fluoroarylketone-ended PEEK PKFF-2, phenoxy-ended grades were obtained by reacting the polymer with an excess of sodium–phenate or sodium–phenoxyphenate in diphenylsulphone, from analogy with the usual PEEK synthesis [63].

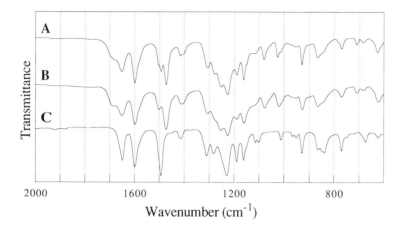

Fig. 1.6 Infrared spectra of neat and sulphonated PEEK PKFF-2. A. IR spectrum of PEEK. B. IR spectrum of a 80% sulphonated sample. C. IR spectrum of a 100% sulphonated sample. (Instrument: PERKIN ELMER FTIR 1760 X; A, self-supporting film cast from phenol; B and C; films cast on KBr from NMP solution).

Table 4 gives the results of the ^{13}C NMR chain ends characterization. Some of the chain ends peaks observed in the oligomers spectra are not detectable in the spectrum of the polymer having the same chain ends. This is due to the fact that a carbon of the repeating unit is resonating in close proximity to the chemical shift of the chain end carbon, or to a lower signal-to-noise ratio in the broadband mode. Fluorine-carbon coupling results sometimes in peak splittings. As noted elsewhere [2], the strong electronegativity of fluorine results in long range effects on the carbon shifts. This is the reason for the high number of chain end peaks observed on the F-ended polymer compare to the other polymers.

Chain end peaks of the commercial PEEK PK150P are found to be due to fluoroarylketone chain ends [FΦCO-, ppm: 202.0 (C1), 171.0 (C5), 170.0 (C9), 140.8 (C7), 138.5 (C3), 127.5 (C2) and 118.0 (C4)] and phenoxyarylketone chain ends (ΦOΦCO-, ppm: 131.1 (C8), 121.2 (C7)]. The absence of hydroxyaryl chain ends in the spectrum is striking: the synthesis must be performed with an excess of 4,4'-difluorobenzophenone. The reason of the suppression of hydroxyaryl chain ends will become obvious in chapter 2 when the degradation of PEEK will be discussed in relation to the nature of chain ends. The presence of phenoxyarylketone chain ends indicates that some phenol is added during the synthesis, probably to control molecular mass, and perhaps also to limit the final content of fluorine in the polymer (toxic fumes).

^{13}C NMR allows essentially a qualitative evaluation of the PEEK chain end content. No quantitative estimation of the chain end content can be performed by this technique, because of differences in relaxation times. Attempts to characterize the polymer chain ends by quantitative proton NMR were unsuccessful, due to the low resolution of the ^{1}H NMR spectra. IR and ^{19}F NMR spectroscopies will be shown

Table 1.4 ^{13}C NMR Shifts of Typical PEEK Chain Ends in CH$_3$SO$_3$H (Reference: TMS). The Assignments were Performed Either by an Analysis of the NMR Spectra of Oligomers as a Function of Chain Length, or from an Analysis of the NMR Spectra of Polymers Having Well-known Chain Ends

Chain end type	Peak alignment from	Chemical shift (ppm)												
		C1	C2	C3	C4	C5	C6	C7	C8	C9	C10	C11	C12	C13
F–⌬–C(=O)–⌬–··· (4,3,2,1,5,6,9,7,8)	Oligomers	202.0	127.5	138.5	118.0 118.2	168.8 171.0	124.5	140.6	118.8	170.0				
	Polymer	202.0	127.5	138.5	~118.0	171.0	*	140.8	*	170.0				
HO–⌬–O–⌬–C(=O)–··· (8,7,6,5,2,1,4,3)	Polymer	*	*	*	*	*	*	122.9	118.0	*				
⌬–C(=O)–··· (5,4,3,2,1)	Oligomers	203.7	141.2	130.3 and 130.8	134.8									
⌬–C(=O)–⌬–··· (9,8,7,6,5,2,1,4,3)	Oligomers	199.9	124.1	139.6	118.5	169.1	153.9	121.1	131.1	126.6				
	Polymer	*	*	*	*	*	154.0	121.2	131.1	126.4...127.2				
HO–⌬–O–⌬–C(=O)–⌬–··· (13,12,11,10,9,8,7,6,5,2,1,4,3)	Oligomers	199.5	124.4	139.6	118.2	169.4	149.3	122.9	121.1	155.4	157.3	119.3	130.6	124.0
	Polymer	*	*	*	*	*	*	*	121.1	*	*	119.2	130.6	*

* Undetected.

below to be adequate analytical tools to characterize quantitatively at least some of the usual PEEK chain ends.

3.2.2 Estimation of the Hydroxyaryl Chain End Contents by IR

From a study of the IR spectra of monodisperse oligomers [2], it was found that the fluoroarylketone chain ends give rise to a shoulder around 820 cm^{-1} in the IR spectrum of the oligomers, while the phenoxy chain ends give rise to various monosubstituted phenyl IR bands, whose the most distinct ones are located at 1073 and 693 cm^{-1}. However, none of these bands can be observed in the polymer IR spectrum, because they are masked by broad and stronger absorptions of chemical groups of the repeating units. Thus, polymer chain ends normally cannot be characterized by IR spectroscopy.

The situation is much different for hydroxyaryl chain ends. Indeed, the strong O-H stretching absorption is situated somewhere between 3300 and 3700 cm^{-1}, in a region where the polymer repeating unit has only few weak absorption bands. Moreover, it could be possible to exchange the terminal proton by deuterium, in a way similar to what has been done on hydroxyl and carboxylic chain ends of poly(ethylene terephthalate) (PETP) [64] and of poly(butylene terephthalate) (PBTP) [65]. The procedure used for PETP by *Patterson* [64] was the following: thin amorphous sheets of polymer were pressed, then carefully dried. The IR spectra of the polymer was then recorded, and the polymer was immersed into deuterated water during many days. It was finally dried, and the spectrum of the deuterated polymer was recorded. Upon deuteration of these polymers, the absorption frequency of the O-H group was suppressed, and a new O-D stretching absorption appeared around 2600 cm^{-1}. By performing a subtraction between the spectra before and after deuteration, it can be possible to detect even very small amounts of hydroxyl chain ends.

Various PEEK samples were pressed into films having a thickness of ~ 150 μm, and directly quenched into cold water from the molten state, to obtain fully amorphous samples. Since the requirement to have an absolutely dry polymer film is stringent, the drying of the PEEK films was examined carefully. It was performed by heating the films in a vacuum oven at 100°C, while continuously pumping on the oven with a rotary pump, during one night. The samples were directly transferred from the oven to a Perkin–Elmer model 580B IR spectrometer, equipped with a dry air circulator. The scans were taken between 4000 and 2000 cm^{-1}, with a resolution of 2 cm^{-1}. Fig. 7 gives the IR spectrum of the fully F-ended PEEK PKFF-2 before and after drying.

Water in PEEK is observed to result in the appearance of two peaks, the first one situated at 3660 cm^{-1}, the second one at 3530 cm^{-1}. Unbonded water is known [58] to absorb at 3760 cm^{-1}. The presence of two peaks is thus due to hydrogen bonding of water onto various chemical groups of PEEK, perhaps onto the ether and ketone bridges (the resonance stabilization afforded by the bonding to the ketone is expected to decrease the absorption frequency more than the bonding to the ether linkage). The same drying procedure has been applied to the fully hydroxyaryl-ended PEEK PKOH. The resulting IR spectrum is shown in fig. 8a. A broad and strong new peak

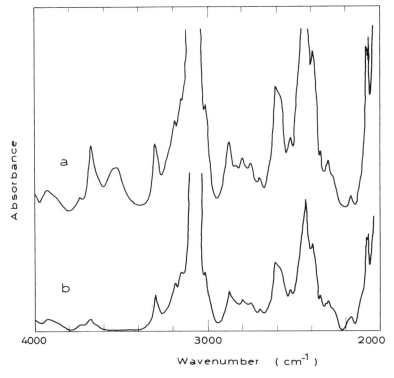

Fig. 1.7 Infrared spectra of an amorphous film of the fully fluoroarylketone ended PEEK PKFF-2. a. undried sample. b. dried at 100° C under primary vacuum during one night. (Instrument: PERKIN ELMER IR 580 B spectrophotometer).

is observed around 3380 cm^{-1}, not present in the fully fluorine-ended PEEK. The broadness and location of this peak is indicative of polymeric O—H groups association [58].

The OH-ended sample was immersed during a week in deuterated water (99.8% D), then dried as usual, reimmersed in fresh D$_2$O during one week, dried again, and its IR spectrum was taken in dry air (Fig. 8b). An important decrease of the IR absorption is observed between 3600 and 3200 cm^{-1}, together with an increase between 2600 and 2400 cm^{-1}. These changes are more easily detected by subtracting the IR spectrum after deuteration from the initial one (Fig. 8c). A strong positive peak is obtained, centered around 3380 cm^{-1}. Corresponding to this positive peak, a negative peak is observed, centered around 2510 cm^{-1}. This is in the region where O—D stretching vibrations are known to absorb. It has been shown elsewhere [64] that the theoretical value for the ratio between the O—H and the O—D stretching frequency is 1.37. The ratio between 3380 and 2510 cm^{-1} is 1.34, in good agreement with the attribution of the 2510 cm^{-1} to the O—D stretch. Thus, upon immersion in deuterated water, hydroxyl chain ends are actually converted into deuteroxyl chain ends.

The deuteration procedure has been applied to other PEEK grades. Although no 3380 cm^{-1} peak could be observed in the original IR spectra of these grades, the

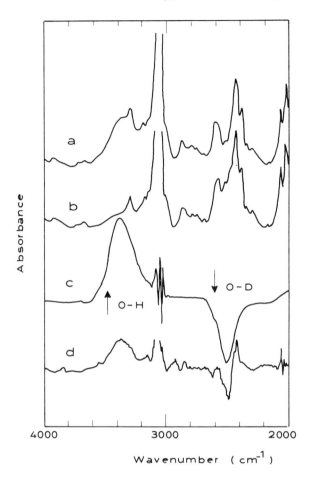

Fig 1.8 Infrared spectra of amorphous films of PEEK. a. Fully hydroxyaryl ended PEEK PKOH, after drying and before deuteration. b. Same sample after 1 day deuteration. c. Difference between both spectra. The disappearance of the O—H stretching absorption and the appearance of the O—D stretching absorption are evident. d. Difference between the IR spectra of the dry undeuterated commercial PEEK PK150P film and the same sample after deuteration. A small O—H absorption is detected by the difference procedure. (Instrument: PERKIN ELMER IR 580 B spectrophotometer).

subtraction of the spectra obtained before and after deuteration revealed in each cases (except one) a positive difference peak due to the O—H stretching, and a negative difference peak due to the O—D stretching. An example of such a difference spectrum, obtained for the commercial polymer PK150P, is shown in Fig. 8d. In such samples, the amount of hydroxyl chain ends is very low, as expected from the absence of characteristic peaks in the ^{13}C NMR spectra; the difference spectra are thus noisy. Increasing the sample thickness reduces the noise, but the 3065 cm^{-1} peak saturates, and the difference spectrum cannot be zeroed correctly in this region (see Fig. 8d).

The above procedure is adequate to detect qualitatively the presence of even very small amounts of hydroxyl chain ends in PEEK. A quantitative estimation of the amount of such chain ends is possible, provided one knows the calibration factor between the chain ends concentration and the area of the 3380 cm^{-1} peak. Since the hydroxyl chain ends content is low, the expected accuracy will be also low due to noise and saturation problems. Only a rough figure of the hydroxyl chain ends content can thus be extracted from the IR analysis.

It was assumed that the absorbance area ratio between the O—H 3380 cm^{-1} peak, and a reference peak, is linearly dependent on the hydroxyl content of the polymer. The small 3919 cm^{-1} absorbance peak was chosen as a reference, because it is well defined and separated from other absorption peaks. The 3380 cm^{-1} area was computed after subtraction of the deuterated dry polymer IR spectrum from the undeuterated one.

The fully hydroxyl-ended sample seemed to be not completely deuterated after 15 days immersion in heavy water, presumably because of the high hydroxyl chain ends content. For this sample, the 3380 cm^{-1} area was thus computed after subtraction of the scaled spectrum of the fully fluorine-ended PEEK from the spectrum of the fully hydroxyl-ended sample. Evidently, only one positive peak located at 3380 cm^{-1} was obtained after this subtraction. From the knowledge of the number average molecular mass ($\bar{M}_n$) of the fully hydroxyl-ended PEEK, determined by size-exclusion chromatography as described in detail in part 4.1.3.2, the proportionality constant between the area ratio and the chain ends content could be computed. It was found that:

$$[-OH] \cong 5.74 \; 10^{-6} \frac{A_{3380} \; cm^{-1}}{A_{3919} \; cm^{-1}} \; (mol \; g^{-1}) \tag{2}$$

where [—OH] is the hydroxyaryl chain ends concentration expressed in mol per gram of PEEK. With this value, the hydroxyl chain ends content of the commercial grade PK150P was found to be of the order of 2% of the total chain ends content. This is very low, as expected. A full characterization of the chain ends of this polymer grade, along with some experimental grades, will be presented in chapter 3 of this book.

From the above IR study, it is concluded that IR spectroscopy is a valuable tool to determine at least roughly the amount of hydroxyl chain ends in PEEK. Although the measured values are usually very low, it will be shown (chapter 3) that such small amounts can be of importance in the determination of the PEEK thermal stability. In this respect, the IR determination of the OH content of PEEK will probably gain some importance in the future.

3.2.3 Quantitative Determination of Fluoroarylketone Chain Ends by ^{19}F NMR

Due to the synthesis process, a PEEK sample contains a percentage of fluoroarylketone chain ends ranging from 0 to 100%.

Both previous methods of end group characterization, IR and ^{13}C NMR spectroscopies, have clearly demonstrated their limits in order to quantify the amount of fluorine atom chain ends.

Even in the case of a fully fluoroarylketone ended PEEK, these spectroscopic techniques have not enough sensitivity and resolution to allow the accurate C—F bond analysis at the very low concentration levels in medium or high molecular masses samples.

Unfortunately, as fluorides can be by-products of synthesis, elemental analysis is not an appropriate tool for the determination of the total fluorine content since residues will interfere and give erroneous results.

However, quantitative ^{19}F NMR spectroscopy has the advantage of high sensitivity and absolute selectivity in fluorine atoms leading to enhanced accuracy. From this point of view, ^{19}F NMR appears to be a powerful tool for the characterization of PEEK fluorine chain ends.

At the present time, only one paper describing a work of *Devaux* [66] was published in this area. Thereby, this paragraph summarizes the main results reported in this publication dealing first with the interpretation of ^{19}F NMR spectra, and secondly with a quantitative evaluation. Some complementary data to the work of *Devaux* are also given.

a) ^{19}F NMR Spectra of PEEK and Sulphonated PEEK in Strong Protonating Media Like Methanesulphonic (MSA) and Sulphuric (H_2SO_4) Acids Fig. 9 illustrates the ^{19}F NMR spectra of respectively the neat PEEK, the 50% and the 100% sulphonated PEEK PKFF-2 dissolved in MSA, solvent known to be unable to sulphonate PEEK. The internal reference selected was 4,4'-difluorodiphenylsulphone (FΦSO$_2$ΦF). Three different peaks noted 1, 2 and 3 can be distinguished. Their assignments were performed thanks to the use of the organic product FΦCOΦOΦOΦCOΦF representative of the PEEK repeat unit.

For unsulphonated PEEK (Fig. 9a), only one peak (peak 1) is observed in the 9.43 ppm region. It is attributed to the fluorine atoms belonging to the unsulphonated chain end repeating unit (Scheme 6).

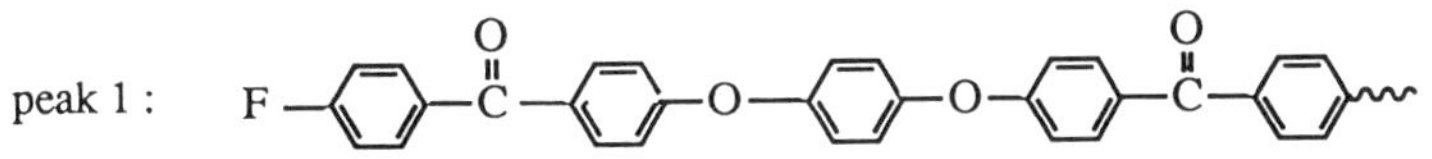

peak 1: F—⌬—C(=O)—⌬—O—⌬—O—⌬—C(=O)—⌬~~

Scheme 6 Chain end unsulphonated repeating unit of PEEK

During sulphonation, peak 1 decreases and two new peaks (2 and 3) arise around 9.88 ppm and 10.03 ppm. These can be correlated with the two non-equivalent sulphonated sequences shown in scheme 7.

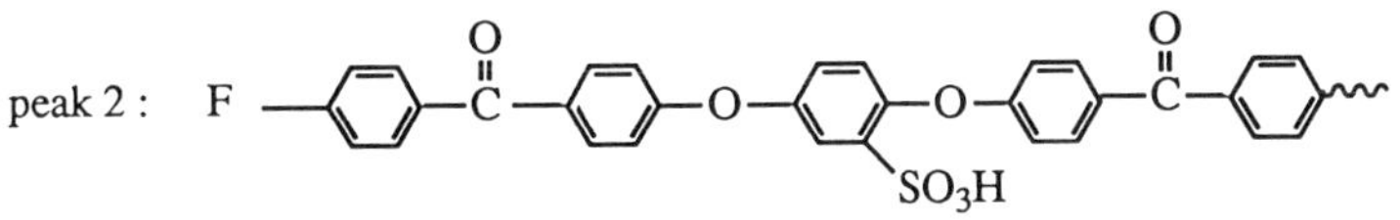

peak 2: F—⌬—C(=O)—⌬—O—⌬(SO$_3$H)—O—⌬—C(=O)—⌬~~

Scheme 7 Chain end sulphonated repeating units of PEEK

peak 3: [Chemical structure: F–Ph–C(=O)–Ph(SO₃H)–O–Ph–O–Ph–C(=O)–Ph~]

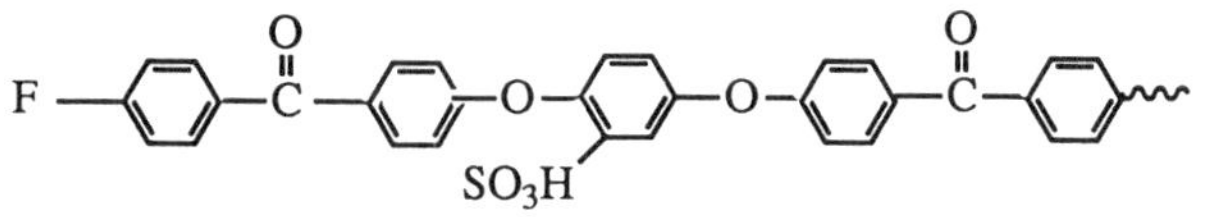

(**Scheme 7** *Continued*)

Indeed, in a PEEK macromolecule, there are two possible non-equivalent sulphonation positions, with respect to each F endgroup. The difference in size of peaks 2 and 3 demonstrates an electronic long range action of the end group on the sulphonation reaction itself. In fact, the inductive nature of the fluorine atom in a solvent which protonates the carbonyl oxygen lowers the reactivity of the neighbouring phenyl ring to electrophilic substitution. Scheme 8 shows this effect.

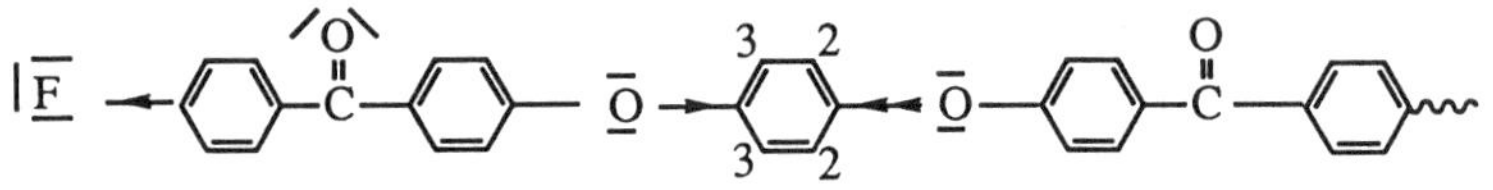

[Chemical structure: $\overline{|F|}$ –Ph–C(=O with arrow)–Ph– $\overline{O}$ – Ph(positions 3,2 / 3,2)– $\overline{O}$ –Ph–C(=O)–Ph~]

Scheme 8 Illustration of the electronic long range effect of the chain end F atom (protonation effect is not indicated)

Due to this inductive effect of the fluorine atom, the free electron pairs of the ether function are less available at the left side than at the right side of the aromatic ring. Consequently, the positions 3 are less reactive than the positions 2 to the electrophilic substitution by SO_3 or $SO_3H^\oplus$. For this reason, the more intense peak 2 in the ^{19}F NMR spectra is attributed to species sulphonated in position 2 and the less intense peak 3 to the sequences sulphonated in position 3 (see Scheme 7).

The fact that each of these sulphonated PEEK chain ends is characterized by a different chemical shift is a consequence of a long range shielding effect on fluorine atoms which has been extensively studied [67–72]. It is known to be due to electronic transmission from one aromatic ring to the following through the π-electron framework. This mechanism is markedly enhanced for compounds like benzophenone by the formation of adducts with Lewis acids [72]. In the PEEK macromolecule, methanesulphonic acid protonates the carbonyl and this protonation increases the π-bonding at the carbonyl-phenyl links.

When PEEK is dissolved in concentrated sulphuric acid, sulphonation appears. The same peaks are obviously detected by ^{19}F NMR experiments. Their chemical shifts are nevertheless different and depend on the water concentration of the aqueous sulphuric acid solution (see Table 5).

b) Quantitative Fluorine Analysis As NaF or KF can be a by-product of the synthesis, it must be noticed that the fluoride ion (or HF in CH_3SO_3H or in concentrated H_2SO_4) gives also rise to a shift, but it is detected outside the range considered in this work and therefore never interferes with the measurements.

Before any experiment on polymer samples, the method was in a first step optimized with model compounds like 4-fluorobenzophenone FΦCOΦ and 4,4'-difluorobenzophenone FΦCOΦF. 4,4'-difluorodiphenylsulphone (FΦSO₂ΦF) was chosen as internal standard.

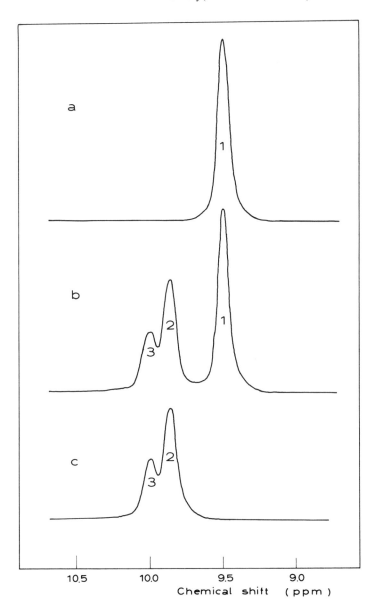

Fig. 1.9 ^{19}F NMR spectra of PEEK PKFF-2 in methanesulphonic acid as a function of the sulphonation degree. a. 0%. b. 50%. c. 100%. (The peak areas are represented in arbitrary units; instrument: BRUKER VM 250; the chemical shifts refer to the internal reference FΦSO$_2$ΦF).

Table 1.5 ^{19}F NMR Analyses of PEEK and Sulphonated PEEK PKFF-2 in Various Solvents (Chemical Shifts Refer to FΦSO$_2$ΦF)

Solvent	Sulphonation	Chemical shift (ppm)		
	No	—	—	9.43
CH$_3$SO$_3$H	Partial	10.03	9.88	9.43
	Complete	10.03	9.88	—
H$_2$SO$_4$ (98.6%)	Partial	12.10	10.87	9.42
	Complete	12.10	10.87	
H$_2$SO$_4$ (97.7%)	Partial	12.01	10.81	9.39
H$_2$SO$_4$ (96.8%)	Partial	11.87	10.70	9.33
H$_2$SO$_4$ (95.9%)	Partial	11.80	10.66	9.30

For quantification of the fluorine content, the total peak area corresponding to the analyzed sample is integrated and compared with the peak area of the internal standard. It has been found that the best accuracy ($\pm 5\%$) is obtained when the internal standard concentration is adjusted in order to produce a peak area of the same order of magnitude as the area due to the sample chain ends.

In a second step, the results obtained with unsulphonated and variously sulphonated samples of the same polymer grade were compared (Table 6).

This study showed that the accuracy is of the same order of magnitude whatever the sulphonation degree. Thus, as expected, the number of ^{19}F NMR peaks has no influence on the total integrated area.

On this basis, the ^{19}F NMR technique appears as a very reliable method for quantitative fluoroarylketone chain end determination and it was thus used to measure the F end groups concentration in the different PEEK samples which are used throughout this chapter (Table 7).

This characterization by ^{19}F NMR presents other advantages for poly(aryl ether ether ketone) samples containing some fluor atom chain ends. It is a good way to control the synthesis of the polymer and to follow end groups modification, for instance in the field of chemical nucleation [73].

Furthermore, the number average molecular mass ($\overline{M}_n$) can be calculated if the polymer is a completely fluoroarylketone ended sample (see section 4.2.2).

And last but not least, it must be specified that the spectrum of a partially sulphonated PEEK sample dissolved in methanesulphonic acid allows the determination of the sulphonation level of the chain end repeating units by ratioing the areas of the peaks 2 and 3 corresponding to the sulphonated species to the total areas of the three detected peaks.

Table 1.6 Quantitative F End Group Analysis of PEEK PKFF-2 in Various Conditions (Increasing Sulphonation Level)

Solvent	Sulphonation	F end groups concentration (ppm)
CH$_3$SO$_3$H	No	3820
H$_2$SO$_4$	Partial	3710
H$_2$SO$_4$	Complete	3850

Table 1.7 Fluorine End Group Concentration of Various PEEK Samples

Sample	F atom concentration (ppm)
PKFX1	720
PKFX2	810
PKFF2	3800
PKFX7	4620
PKFF4	5940
PKFX3	1700
PKFX4	1900
PKFX6	2300
PKFF1	2900
PKFF3	4520
PKFF5	12670

4 Molecular Mass and Distribution Determination

Relations between molecular mass characteristics and physical properties, reaction mechanisms and polymerization conditions are more and more explored. Therefore, it is no more sufficient to obtain one of the average molecular masses (MM), viz. weight, viscosity or number, but the whole molecular mass distribution (MMD) is required. Furthermore, only absolute results can be treated quantitatively or compared when they are supplied by different research or quality-control laboratories.

For all these reasons, size exclusion chromatography (SEC) has become a widespread technique for assessing molecular masses and their distributions, and its next future is full of promise with the development of on-line viscometer, light scattering and osmometer detectors which transform SEC into an absolute method.

Indeed, one of the great disadvantage of SEC is that, as far as a concentration detector is coupled to the instrument, SEC is a secondary method and requires a preliminary calibration. In practice, the calibration procedure often presents considerable difficulties, especially for the characterization of new polymers. In this case, when a suitable solvent of the polymer is found, some successive steps are required to develop an optimized experimental method. For instance, it needs first the synthesis of calibration agents covering the widest range of molecular masses, secondly the characterization of the standards by light scattering, and finally the establishment of the specific and universal calibration curves. This latter calibration necessitates in particular the determination of Mark–Houwink–Sakurada (MHS) relationships from viscometric experiments.

In a wellplanned SEC development, the determination of the universal calibration curve is essential in order to standardize a new SEC method. Indeed, that is the only one procedure which can be used to determine absolute molecular masses when specific standards are not available from commercial sources.

The last section of this chapter is mainly devoted to a description of the size exclusion chromatography of PEEK. First PEEK light scattering and viscometric analyses performed in view to obtain absolute molecular masses from SEC measurements will be described. Then two SEC methods adapted for PEEK characterization are presented. The first one, proposed by *Devaux* [22], concerns SEC at high temperature in a phenol/1,2,4-trichlorobenzene mixture. The second one, reported by *Daoust* [33, 34], is based on PEEK functionnalization with concentrated sulphuric acid and is performed at room temperature with completely sulphonated PEEK dissolved in a N-methyl-2 pyrrolidone salt medium. For high temperature SEC (HTSEC) as well as for room temperature SEC (RTSEC), the specific and universal calibration curves are established.

In order to assess the validity of the PEEK derivatization procedure, molecular characteristics of various PEEK samples determined from both SEC methods are compared.

Afterwards, number average molecular masses ($\bar{M}_n$) of fluoroarylketone ended PEEK samples were determined by ^{19}F NMR experiments and compared with $\bar{M}_n$ values given by SEC. So the reliability of the different techniques developed for the molecular mass characterization of PEEK will be demonstrated.

Finally, an experimental PEEK-carbon fibre composite is analysed. The fibre content is evaluated and the derivatization method is applied to obtain the average molecular masses by RTSEC.

4.1 Light Scattering, Viscometry and Size Exclusion Chromatography Characterization

4.1.1 Light Scattering

Five PEEK samples were synthetized following a nucleophilic substitution procedure in order to obtain SEC standards covering a range of number average molecular mass ($\bar{M}_n$) from ~5000 to ~30000. The samples were not fractionated and were thereby qualified as broad molecular mass distribution calibration agents.

The light scattering (LS) experiments were performed with 99.5% H_2SO_4 PEEK solutions. The solutions were analysed when a complete level of sulphonation was achieved.

The different weight average molecular masses of the totally sulphonated samples $(\bar{M}_w)_s$ were deduced from the Zimm plots and corrected for optical anisotropy. Fig. 10 illustrates the Zimm plot for sample PKFX-1 and shows a quite surprising dissymetry for rather low molecular masses. This fact is probably related to the semiflexibility of the sulphonated macromolecules in a protonating solvent [74]. The measurements of scattered light intensities gave the $(\bar{M}_w)_s$ values reported in Table 8.

The molecular mass of the unsulphonated PEEK $(\bar{M}_w)_u$ is related to the molecular mass of the totally sulphonated sample H-SPEEK by

$$1.278 \, (\bar{M}_w)_u = (\bar{M}_w)_s \qquad (3)$$

1.278 represents the ratio between the structural unit molecular mass of the sulphonated and unsulphonated species (see part 2.3.2). Using relationship (3), the true

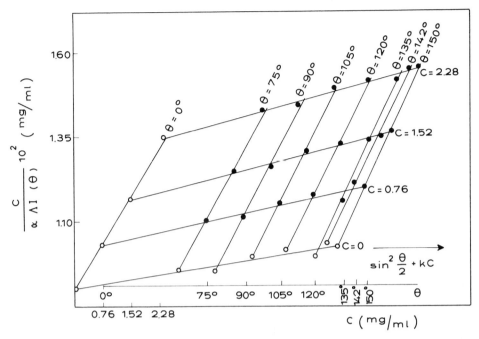

Fig. 1.10 Illustration of a Zimm plot of PEEK PKFX-1 in concentrated sulphuric acid. (Instrument: FICA 42000 Photo Gonio Diffusiometer fitted with a He–Ne laser source of 4 mW at 632 nm wavelength, vertically polarized, $\alpha = \sin\theta$). Points at $C = 0$ and $\Theta = 0$ were calculated from experimental measurements at various concentrations and angles.

masses of the unsulphonated PEEK standards can be determined. The obtained values are gathered in Table 8.

The reader who requires information about other works undertaken in the field of the light scattering characterization of PEEK is referred to papers previously published by *Roovers* [16] and *Bishop* [20].

4.1.2 Viscometry

Viscometry, the most familiar method for polymer molecular mass characterization, is widely used in research in polymer science as well as in industrial fields. This technique is classified as an indirect determination method but has various advantages over direct methods, for instance, rapidity and easiness.

The Mark–Houwink–Sakurada (MHS) relationship between the intrinsic viscosity [η] and the molecular mass offers a convenient mean for determining the molecular mass of a polymer:

$$[\eta] = KM^a \qquad (4)$$

where K and a are constants for a polymer solvent system at a given temperature.

Table 1.8 Characteristic Weight Average Molecular Masses of PEEK Standards from Light Scattering Experiments (solvent: 99.5% H_2SO_4)

Sample	$(\bar{M}_w)_s{}^1$	$(\bar{M}_w)_u{}^2$
PKFX-1	72200	56500
PKFX-2	47500	37200
PKFX-5	36200	28300
PKFX-7	28900	22600
PKFX-4	18300	14300

^{1} and ^{2}: weight average molecular masses of respectively fully sulphonated (H-SPEEK) and unsulphonated PEEK.

a) Mark–Houwink–Sakurada Relationships for PEEK The MHS parameters have been determined for PEEK in various solvents from diluted solution viscometry.

For example, Fig. 11 illustrates the double logarithmic plot of the intrinsic viscosity in methanesulphonic acid at 25°C against weight average molecular mass $(\bar{M}_w)_u$ for four of the five PEEK standards analysed by light scattering [24]. This figure also includes the data of *Roovers* [16] obtained with narrow molecular mass distribution samples of PEEK which have been characterized by light scattering in methanesulphonic acid. The experimental values of *Roovers* agree remarkably well with the results presented in the present work. The similitude of both viscosity laws in the same solvent and at almost the same temperature confirms that reliable and accurate

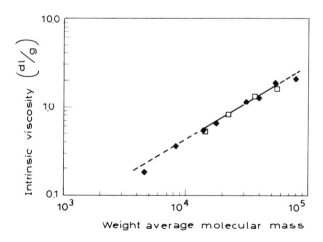

Fig. 1.11 Logarithmic plot of the intrinsic viscosity against the weight average molecular mass in methanesulphonic acid. □: PEEK standards PKFX-1, PKFX-2, PKFX-7 and PKFF-4 at 25°C [24]. ♦: Fractions of PEEK at 30°C [16].

weight average molecular masses of unsulphonated PEEK can be obtained from light scattering experiments performed with PEEK/H$_2$SO$_4$ solutions.

Table 9 summarizes the different viscosity laws available for PEEK in the scientific literature.

b) Mark–Houwink–Sakurada Relationships for Totally Sulphonated PEEK Totally sulphonated PEEK (H-SPEEK) is a polyelectrolyte which carries many ionizable groups. In dilute NMP solution, an organic polar aprotic solvent, ionization transforms the macromolecule into a polyion bearing many charges. This polyion is accompanied by an equivalent number of small counterions, forming an "ionic environment".

In this state, the number of electrical charges within the polymer coil exceeds that in the bulk solvent. Consequently osmotic forces drive solvent into the coil and cause the counterions to diffuse out away from the backbone chain into the bulk portions of the solvent. This process leaves a net residue of negative charge on the polymer chain. These charged groups remaining on the SPEEK macromolecule are responsible for large intramolecular electrostatic repulsive forces and lead to a polyelectrolyte being more expanded than the equivalent non ionic polymer [75–77].

The viscometric characterization of the sulphonated PEEK PKFF-2 in solution clearly shows this particular polyelectrolyte behaviour (see Fig. 12).

The viscometric properties of a polymer notably depend on the coil size. At high concentrations of H-SPEEK in a solvent without added salt (around 1%), the polyelectrolyte molecules overlap one another and no possibility is offered for the counterions to leave the domain of a given macromolecule. The molecules are not appreciably expanded and the specific viscosity is varying as usual.

When the solution is diluted to concentrations in the range of those commonly used for SEC experiments, the molecules are in the above described state. Regions appear which are not occupied by polymer molecules and mobile counterions will diffuse from the molecular domain into the bulk of the solvent. The development of

Table 1.9 Viscosity Laws of PEEK and Sulphonated PEEK

Polymer	Solvent	Temp. (°C)	a	10^5 K*	$\bar{M}_w$ range	Reference
PEEK	Methanesulphonic acid	25	0.842	17.31	56500–14300	24
	Methanesulphonic acid	30	0.818	22.60	79500–4600	16
	parachlorophenol/ o-dichlorobenzene (60/40 in weight)	35	0.702	58.60	79500–4600	16
	Phenol/1,2,4-trichlorobenzene (50/50 in weight)	115	0.670	75.88	56500–14300	22
Totally sulpho- nated PEEK	N-methyl-2-pyrrolidone + LiBr (0.1 M)	25	0.938	4.87	73400–18600	24, 34
	H$_2$SO$_4$ (99.5%)	25	0.939	3.77	72200–18300	22, 24

* the intrinsic viscosity is expressed in dl/g.

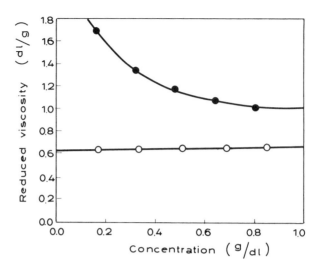

Fig. 1.12 Variation of reduced viscosity versus concentration for H-SPEEK PKFF-2 in N-methyl-2-pyrrolidone (●) and Li-SPEEK PKFF-2 in N-methyl-2-pyrrolidone + LiBr (0.1 M) (○).

this net negative charge due to loss of counterions is responsible for the molecular expansion. Therefore, for the salt-free SPEEK-NMP system (Fig. 12), the reduced viscosity increases markedly with decreasing concentration showing an unusual but typical polyelectrolyte upward curvature.

The addition of an excess of a simple salt (LiBr) produces a deep change of behaviour. The thickness of the "ionic atmosphere" around the chain backbone decreases. The charges on the polymer are to some extent shielded from each other and the degree of expansion decreases. In these conditions the plot of the reduced viscosity against the concentration becomes linear (Fig. 12) and correct intrinsic viscosities [η] can be obtained [75].

However, the determination of [η] also requires the knowledge of an accurate weight concentration of SPEEK. Indeed, SPEEK is hygroscopic [32] and, in addition, due to the sulphonation method followed, may contain some residual impurities. Both moisture and impurities problems were overcome by determining absolute weight concentrations by spectroscopic measurements. This work is described in full detail elsewhere [34].

On another hand, a question should be asked regarding the definition of the correct weight concentration considering polyion alone or with the cationic counter ion Li^+. In this study, all the results obtained have been calculated considering the second case.

Consequently, the weight average molecular mass of the totally sulphonated PEEK $(\bar{M}_w)_{\text{H-SPEEK}}$ has been corrected to obtain the weight average molecular mass of the neutralized sample $(\bar{M}_w)_{\text{Li-SPEEK}}$ taking into account the ratio between the structural units (374/368):

$$(\bar{M}_w)_{\text{Li-SPEEK}} = \tfrac{374}{368}(\bar{M}_w)_{\text{H-SPEEK}} \qquad (5)$$

The molecular masses given by relationship (5) and the intrinsic viscosities of the Li-SPEEK standards dissolved in the NMP-LiBr (0.1 M) solvent are given in Table 10.

These values allow the determination of the molecular mass dependence of [η] reported in Table 9.

Finally, it has to be noticed that *Devaux* [22] (see Table 9) and *Bishop* [20] have characterized PEEK samples in concentrated sulphuric acid by viscometry.

Also, *Karcha* [78] has determined viscosities of a sulphonated PEEK sample ($\bar{M}_w$ = 38600 and $\bar{M}_n$ = 14100) in dimethylacetamide solutions in a work treating of miscible blends of sulphonated PEEK and aromatic polyimides.

4.1.3 Size Exclusion Chromatography

In addition to the PEEK size exclusion chromatography (SEC) methods herebelow developed, papers of *Ishiguro* [30] and *Hay* [28] reported a brief description of two other procedures.

Ishiguro proposed the use of an unusual mobile phase for room temperature SEC, a mixture of dichloroacetic acid and chloroform with a gel of styrene and divinylbenzene as stationary phase. However, in such an acid medium, the chromatography of a PEEK sample exhibits a classical distribution curve eluting between the permeation and exclusion volumes.

In a work devoted to the influence of the molecular mass on the crystallization behaviour of PEEK, *Hay* characterized PEEK samples by SEC in a mixture of para-chlorophenol and ortho-dichlorobenzene at 40 °C and expressed the results in "polystyrene equivalent" molecular masses. In this case, it was only intended to use SEC to compare the molecular mass distributions of samples. Here it is important to state that there could be considerable differences between the "polystyrene equivalent" and the absolute molecular masses.

4.1.3.1 High Temperature SEC (HTSEC) of PEEK   Fig. 13 shows a typical high temperature (T = 115°C) SEC chromatogram and the mass distribution curve of PKFF-4 sample. In addition to the molecular distribution of the polymer, the differ-

Table 1.10 Weight Average Molecular Masses and Intrinsic Viscosities of Li-SPEEK Standards in N-methyl-2-pyrrolidone-LiBr (0.1 M) at 25°C

Sample	$(\bar{M}_w)_{\text{Li-SPEEK}}$	[η] (dl/g)
Li-SPKFX-1	73400	1.68
Li-SPKFX-2	48300	1.28
Li-SPKFX-7	29400	0.78
Li-SPKFF-4	18600	0.47

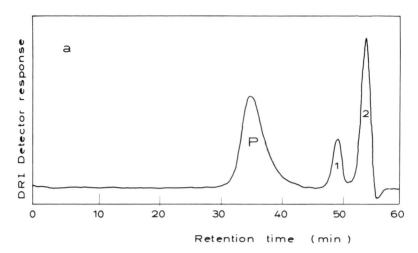

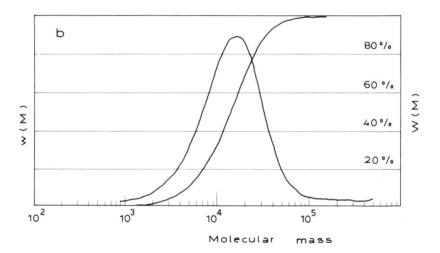

Fig. 1.13 High temperature size exclusion chromatography of PEEK PKFF-4 ($\bar{M}_w = 14300$). a. Chromatogram (P: molecular distribution of the polymer; 1 and 2: solvent peaks). b. Differential and cumulative molecular mass distibutions. (Instrument: 150 C of MILLIPORE-WATERS; solvent: phenol/1,2,4-trichlorobenzene at 115°C; packing: gel of styrene and divinylbenzene; detector: differential refractometer).

ential refractometer detector (DRI) displays also two peaks (1 and 2) due to the mixture of solvents used as eluent. These peaks lie near the permeation volume of the columns set and never interfere with the molecular distribution of the polymer.

As SEC is a relative technique, the system has to be calibrated for molecular mass against retention time (RT). Two usual procedures were considered. The first one is the establishment of the specific PEEK calibration from the five broad MMD standards well characterized by light scattering experiments. The second one is the application of the universal calibration principle.

This last concept assumes that for any given elution time, the hydrodynamic volumes of the considered polymer and of the reference polymer are the same [79]. Using polystyrene (PS) as primary standards, the *Fox* and *Flory* [75] equation can be written:

$$\log [\eta]_{PEEK} M_{PEEK} = \log [\eta]_{PS} M_{PS} \qquad (6)$$

$[\eta]_{PEEK}$ and $[\eta]_{PS}$ being the intrinsic viscosities of PEEK and PS.

Such a relationship can be used if no retention effects of PS and PEEK are occuring during the elution of the macromolecules through the columns.

This method can thereby be applied when the MHS constants are known for the PEEK samples and for the narrow MMD polystyrene standards in the same SEC solvent and at the same temperature.

a) Specific Calibration A classical mathematical iterative process allowed the establishment of the PEEK specific calibration. This calibration was achieved using twelve narrow PS standards with molecular masses ranging from 580 to $3.8 \; 10^6$, five broad MMD PEEK standards (Table 8) and nine PEEK model compounds chosen to extend the calibration range (Table 11).

First the usual PS calibration curve coming from PS retention times (RT) has been realized.

$$\log M_{PS} = f_1 (RT) \qquad (7)$$

Analysing PEEK calibration agents and using the PS calibration relationship (7), $(M_p)_1$ and $(\bar{M}_w)_1$ values are obtained. They are, in fact, "polystyrene equivalent" PEEK molecular masses. These values have to be corrected to take into account the weight average molecular masses $(\bar{M}_w)_u$ given by light scattering characterization.
A first iteration is

$$(M_p)_2 = \frac{(\bar{M}_w)_u (M_p)_1}{(\bar{M}_w)_1} \qquad (8)$$

Reporting the $(M_p)_2$ values together with the model compounds molecular masses as a function of the corresponding retention times, a new calibration curve is obtained

Table 1.11 Model Compounds Used in HTSEC for the Establishment of the Specific Calibration Curve

Chemical structure	Name	Molecular mass
⌬-CO-⌬	Benzophenone	182
⌬-CO-⌬-F	4-Fluorobenzo-phenone	200
F-⌬-CO-⌬-F	4,4'-Difluorobenzo-phenone	218
⌬-CO-⌬-O-⌬	4-Phenoxybenzo-phenone	274
⌬-O-⌬-CO-⌬-O-⌬	4,4'-Diphenoxy-benzophenone	366
F−[⌬-CO-⌬-O-⌬-O]$_n$-⌬-CO-⌬-F		
$n=1$		506
$n=2$		794
$n=3$		1082
$n=4$		1370

which, in turn, enables to calculate $(M_p)_{2'}$ and $(\bar{M}_w)_{2'}$ values used in a following iteration.

Such an iterative process has to be performed until the weight average molecular masses become equal to the values given by light scattering analysis. Generally, four steps are enough to reach an accuracy of around 5%.

The main problem associated with the use of the broad MMD PEEK calibration standards is the restricted range of molecular masses covered. The lack of PEEK standards with $(\bar{M}_w)_u$ above 60000 imposes thereby a strict parallelism between the two successive curves for high molecular masses.

The average molecular masses deduced from the specific calibration are shown in Table 12.

b) Universal Calibration   Universal calibration procedure is very easy to apply when MHS relationships are known for polystyrene and PEEK.

The viscosity law for PEEK is given in Table 8 whereas the MHS equation for PS in the same solvent (phenol/1,2,4-trichlorobenzene) at the same temperature (115°C) is

$$[\eta] = 13.03 \ 10^{-5} \ (\bar{M}_w)^{0.69} \ (dl/g) \ [22] \tag{9}$$

The universal calibration curve was then constructed from PS solutions retention times which lead to the expression

Table 1.12 Comparison Between Molecular Masses Obtained by Light Scattering (LS) and by SEC from Universal and Specific Calibrations

Sample	LS $(\bar{M}_w)_u$	SEC					
		Universal calibration			Specific calibration		
		$\bar{M}_w$	$\bar{M}_n$	H	$\bar{M}_w$	$\bar{M}_n$	H
PKFX-1	56500	67400	25100	2.7	57100	26900	2.1
PKFX-2	37200	44500	16300	2.7	39000	16800	2.3
PKFX-5	28300	33500	11700	2.9	29900	12300	2.4
PKFX-7	22600	24200	10500	2.3	22300	11100	2.0
PKFF-4	14300	13200	6200	2.1	12800	6800	1.9

$$\log [\eta] M = f_2 (RT) \qquad (10)$$

Relationship (10) allows the determination of the molecular masses given in Table 12.

The evaluation of the results presented in Table 12 shows that a more accurate weight average molecular mass measurement is obtained from the specific calibration than from the universal calibration. In fact, the agreement between the $\bar{M}_w$ values obtained by SEC with the universal calibration (solvent: phenol/trichlorobenzene) and the $\bar{M}_w$ values obtained by light scattering (solvent: concentrated sulphuric acid) is fairly good for the low and medium molecular mass samples (<8%) but the difference goes up to 20% for the highest one.

This comparison does not call in question the universal calibration method. It can strictly be exploited for weight average molecular masses below ~30000. This conclusion however draws once again the attention to the problems connected with the use of broad MMD standards of limited molecular mass in the field of viscometric determination of the MHS constants. In this case, no experimental proof is obtained demonstrating the invariability of the MHS parameters against the molecular mass.

4.1.3.2 Room Temperature SEC (RTSEC) of Totally Sulphonated PEEK In the absence of salt, the chromatogram of the SPEEK PKFF-2 dissolved in N-methyl-2-pyrrolidone (NMP) is distorted (Fig. 14). The macromolecules elute predominantly near the exclusion volume. As the sizes of macromolecules of a polyelectrolyte in solution are governed by molecular mass, by the number of attached ionic groups, by the type of counterion (its charge and mobility), by the polarity and electrical screening properties of the solvent [80–85], this unusual behaviour shows that, in this case, there is little or no separation on the basis of molecular mass. Not only chain expansion can give rise to larger hydrodynamic volumes, but also, as reported in the literature even for gel of styrene-divinylbenzene copolymer, repulsion between the polyion and some charged groups on the packing can lead to exclusion of the polyion from the pores. This effect, called "ion exclusion", is a well-known phenomenon in which the diffusion of the ionic species into the porosity of the gel is restricted by electrostatic repulsion [86–89].

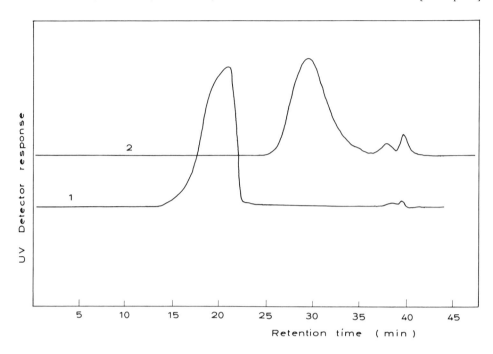

Fig. 1.14 Chromatograms of the totally sulphonated PEEK PKFF-2, ($\bar{M}_w = 24000$) in N-methyl-2-pyrrolidone (NMP) without (curve 1) and with (curve 2) 0.1 M LiBr. (Temperature 25 °C; packing: gel of styrene and divinylbenzene; detector: UV spectrometer).

The addition of a strong electrolyte such as LiBr to the solvent overcomes "ion exclusion" and suppresses the loss of counterions from the charged sites on the polymer. This allows a return of the polymer to normal physical and thermodynamical solution properties. Classical elution and peak shape (see Fig. 14) are then observed.

The SEC experimental conditions of SPEEK being optimized and the viscosity law of SPEEK being determined (see part 4.1.2), the next step was the establishment of the universal and specific calibration curves.

There has been a number of attempts to apply the "universal calibration" concept to polyelectrolyte when an organic polar solvent containing a salt is used as mobile phase and while a successful size exclusion separation can be made for the polyelectrolyte [90–93]. The difficulty generally stays in the search of an appropriate polymer standard which will not show unwanted partition with the column packing selected. In this case, the first approach was the calibration of the columns set with narrow polystyrene. This work has revealed affinity of this polymer with the styrene divinylbenzene copolymer gel attributed to "solvophobic" or "reverse phase" interaction.

As the use of polystyrene as standard polymer didn't provide pure size exclusion, a search for other macromolecules for calibration was performed. It was found that

Li-SPEEK and PMMA obeyed the same relationship between [η] M and the elution volume. The viscosity law of PMMA was then determined in N-methyl-2-pyrrolidone-LiBr (0.1 M) at 25°C. The MHS relationship is

$$[\eta] = 5.33 \ 10^{-5} \ M^{0.790} \ (dl/g) \tag{11}$$

After the establishment of the universal calibration with the help of the MHS parameters of sulphonated PEEK (see table 9), the four Li-SPEEK standards were characterized and their average molecular masses calculated. The results are presented in Table 13.

The specific calibration was obtained following the iterative procedure described in part 4.1.3.1 using the PMMA calibration and the five standards shown in Table 14. It allowed the calculation of the results reported in Table 13.

In order to assess the reliability of the determination of absolute molecular mass of totally sulphonated PEEK measured by the SEC universal calibration method, six other experimental samples were analysed. Their molecular data are also presented in Table 13.

Table 1.13 Comparison Between Molecular Masses Obtained by SEC from Specific and Universal Calibrations

Sample	Specific calibration			Universal calibration		
	$\bar{M}_w$	$\bar{M}_n$	H	$\bar{M}_w$	$\bar{M}_n$	H
Li-SPKFX-1	73000	35200	2.1	71900	31300	2.3
Li-SPKFX-2	49700	23200	2.1	50900	23100	2.2
Li-SPKFX-7	30500	17400	1.8	31600	16300	1.9
Li-SPKFF-4	18100	9800	1.9	18800	10000	1.9
Li-SPKFX-3	48900	20700	2.4	47500	21000	2.3
Li-SPKFX-6	35200	17600	2.0	34300	17400	2.0
Li-SPKFX-4	31300	15000	2.0	30200	15600	1.9
Li-SPKFF-3	24000	12700	1.9	23200	12200	1.7
Li-SPKFF-2	23900	12900	1.9	24000	12300	2.0
Li-SPKFF-5	8500	4400	1.9	8200	5100	1.6

Table 1.14 Characteristic Molecular Values of Li-SPEEK and Model Compounds Used as Standards for RTSEC Specific Calibration

Sample	$\bar{M}_w$*	M
Li-SPKFX-1	73400	
Li-SPKFX-2	48300	
Li-SPKFX-7	29400	
Li-SPKFF-4	18500	
LiOΦCOΦOΦSO₃Li		382

* From light scattering experiments.

4.2 Correlation Between the Various Molecular Characteristics

4.2.1 On the Reliability of the Derivatization Procedure

The reliability of the derivatization procedure will be confirmed if RTSEC supplies identical MM averages and MMD curves as HTSEC. For this purpose, seven experimental PEEK samples were analysed by RTSEC and HTSEC.

From the $M_{Li\text{-}SPEEK}$ values obtained by RTSEC, absolute PEEK molecular masses (M_{PEEK}) could be calculated if the correction for sulphonation and neutralization was taken into account. As the sulphonation level of each sample is one -SO_3Li pendant group attached to the polymer chain backbone per repeat unit, the increase in molecular mass or the ratio between $M_{Li\text{-}SPEEK}$ and M_{PEEK} is

$$k = \frac{M_{Li\text{-}SPEEK \ unit}}{M_{PEEK \ unit}} = \frac{374}{288} = 1.299 \qquad (12)$$

Therefore $M_{Li\text{-}SPEEK}$ results can be converted in M_{PEEK} values by applying the correction factor of 1.299.

The average molecular masses given by both SEC techniques are quoted in Table 15.

Taking into account the accuracy of SEC analysis, the agreement between RTSEC and HTSEC absolute molecular masses can be regarded as very satisfactory. Indeed, the difference between both average molecular masses in number and in weight remains in the limit of the experimental accuracy (between 5 and 10%). Furthermore, it was also demonstrated that both methods give the same molecular mass distribution curves [94].

This overall agreement confirms the validity of the assumption that functionnalization of PEEK upon dissolution in concentrated sulphuric acid, as far as the H_2SO_4 concentration is kept below 100%, is an appropriate way to obtain soluble sample useful for RTSEC characterization.

4.2.2 Comparison Between Number Average Molecular Masses Obtained by SEC and ^{19}F NMR

From a quantitative point of view, ^{19}F NMR is a very reliable method for the determination of the F end group concentration in PEEK samples (see part 3.3.3).

When all the PEEK chain ends are aromatic fluorine, their quantitative analysis provides a way to determine the number of molecules and thereby the number average molecular mass ($\bar{M}_n$). This procedure was followed for the molecular characterization of five fully fluoroarylketone ended PEEK samples. The solvent used was methanesulphonic acid and the internal standard was $F\Phi SO_2\Phi F$. The results are given in Table 16 together with the $\bar{M}_n$ values determined by SEC.

It can be seen that the experimental values agree very well keeping in mind the

Table 1.15 Molecular Masses Obtained by High Temperature SEC (HTSEC) and Room Temperature SEC (RTSEC)

Sample	HTSEC*			RTSEC*		
	$\bar{M}_w$	$\bar{M}_n$	H	$\bar{M}_w$	$\bar{M}_n$	H
PKFX-3	41400	17300	2.5	37600	15600	2.4
PKFF-1	34500	15700	2.3	31900	15500	2.0
PKFX-6	29000	15000	1.9	27100	13600	2.0
PKFX-4	25100	12300	2.2	24100	11600	2.0
PKFF-2	19500	10000	2.0	18400	9900	1.9
PKFF-3	19000	9300	2.0	18500	9800	1.9
PKFF-5	6700	4000	1.7	6500	3400	1.9

* From the specific calibration.

Table 1.16 Number Average Molecular Masses ($\bar{M}_n$) from ^{19}F NMR and SEC Experiments

Sample	^{19}F NMR	SEC*
PKFF-1	13100	15500
PKFF-2	10000	9900
PKFF-3	8400	9800
PKFF-4	6400	7500
PKFF-5	3000	3400

* From RTSEC of sulphonated PEEK.

accuracy of both techniques in the field of the number average molecular mass determination. This observation corroborates the validity of the SEC methods developed for the molecular mass characterization of PEEK.

4.3 PEEK-Carbon Fibre Composite Characterization

This paragraph is devoted to an application of the PEEK derivatization procedure in the field of the characterization of PEEK APC composites.

Aromatic Polymer Composites (APC) are a new class of advanced structural material based on continuous carbon fibres embedded in a PEEK matrix [95].

The present analysis is dealing with a prototype of the particular commercial grade APC-2, where the matrix is "Victrex" PEEK[1]. Two samples were characterized: an APC-2 sample "as received" and an APC-2 sample treated one hour at 400 °C in air. The molecular mass parameters and the fibre content were determined following a H_2SO_4 extraction procedure described in Fig. 15.

---

[1] Imperial Chemical Industries (ICI) trade mark.

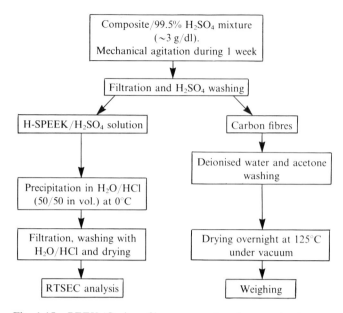

Fig. 1.15 PEEK/Carbon fibre composites characterization procedure.

The RTSEC analyses are illustrated with chromatograms of Fig. 16 and the molecular mass results are shown in Table 17.

To check the validity of the H_2SO_4 extraction method for the fibre content determination, an "acid digestion" procedure described elsewhere [34] was also used. The agreement between the results obtained is quite satisfactory: 66.0 and 65.5 weight percent of fibre content respectively for the acid extraction and "digestion" methods.

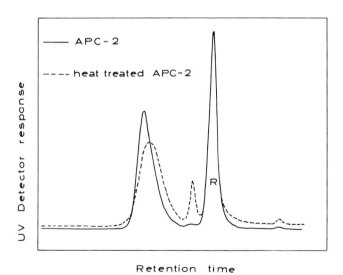

Fig. 1.16 Room temperature size exclusion chromatography of APC-2 and APC-2 treated one hour at 400°C in air. (R: internal reference peak).

Table 1.17 Comparison Between Molecular Masses of APC-2 and Heat Treated APC-2

Sample	$\overline{M}_w$	$\overline{M}_n$	$\overline{M}_z$	M_p*	H
APC-2	25500	11500	40700	26600	2.3
Heat treated APC-2**	22400	7200	68700	18100	3.1

* Molecular mass at the top of the peak.
** 1 hour at 400 °C in air.

Conclusion

This work was devoted to the characterization of PEEK chemical structure and molecular mass distribution. The development of analytical methods to achieve this aim was difficult due to the high solvent resistance of PEEK.

Confronted to the problem of the limited number of suitable solvents, the scientists either have brought themselves to the use of corrosive strong protic media, or have succeeded in finding very original mixtures of solvents or have decided in favour of a PEEK derivatization procedure.

This latter choice was feasible owing to the sulphonation of the polymer upon dissolution in concentrated sulphuric acid at room temperature. The mechanism of this reaction is an electrophilic substitution of an hydrogen of the phenyl flanked by two oxygen atoms and the totally sulphonated PEEK is an ionomer which possesses monosulphonated repeating units. This derivatization method presents many advantages:

- the complete sulphonation level is easily achieved,
- the behaviour of PEEK in concentrated sulphuric acid is free of degradation as far as a small amount of water is present to prevent crosslinking via sulphone formation,
- SPEEK is soluble in a variety of polar aprotic solvents and classical SEC techniques become available.

Due to the interest of the derivative polymer in the field of PEEK characterization, neat PEEK and sulphonated PEEK samples have been analysed by UV-visible, ^{13}C NMR and IR spectroscopies. It appears that the interpretation of the different spectra is actually relatively well mastered.

The main chain characterization study notably emphasizes that:

- ^{13}C NMR and IR spectroscopies can be used after the recovering of the sulphonated polymer for the quantitative evaluation of the sulphonation level. But ^{13}C NMR is the only direct method,
- recent results seem to show that UV spectroscopy can be considered as a potential method for following the sulphonation kinetics of the polymer dissolved in concentrated sulphuric acid.

Table 1.18 Main Characteristics of PEEK Samples Used in Chapter 1

Sample	Melt viscosity[3] (KN s m^{-2})	Intrinsic viscosity[4] (dl/g)	Chain ends[7] (%) FΦ(CO)—[5]	HOΦ—[6]	$\bar{M}_n$ ^{19}F NMR	$\bar{M}_n$ SEC	$\bar{M}_w$ LS	$\bar{M}_w$ SEC
PK150P	0.150		58	2	—	1030		26800
PKOH				~100	—	9850		20400
PKFF-1			~100		13100	12000		24100
PKFF-2			98		10000	9900		18100
PKFF-3			~100		8400	9800		18500
PKFF-4	0.015	0.53	~100		6400	7000	14300	13800
PKFF-5			~100		3000	3400		6500
PKFX-1	1.120	1.64			—	26700	56500	56200
PKFX-2	0.450	1.32	33	4.7	—	17000	37200	38200
PKFX-3					—	15600		37600
PKFX-4					—	15500		31900
PKFX-5	0.200				—	12300	28300	29900
PKFX-6					—	13600		27100
PKFX-7	0.090				—	12700	22600	23400
APC-2[2]					—	11500		25500

[1] Only PK150P is a commercial grade, the other samples are experimental ones. Both letters after PK mean the nature of the two chain ends: OH = hydroxyaryl-; F = fluoroarylketone-; X = unknown.
[2] APC-2 (PEEK-carbon fibre) prototype sample.
[3] Measured at 400°C at a shear rate of 1000 s^{-1}.
[4] In CH$_3$SO$_3$H at 25°C.
[5] From ^{19}F NMR analyses.
[6] From IR experiments.
[7] For other chain ends see chapter 3.

The chain end spectroscopic identification elucidates the different possible natures of chain ends created during the polymerization process of experimental and commercial PEEK grades analysed in this chapter, which is a nucleophilic substitution procedure. Fluoroarylketone, hydroxyaryl, and phenoxy end groups have been detected but only fluoroarylketone chain ends have been quantified according to the sensitivity of the ^{19}F NMR spectroscopy.

Several methods usable for the determination of molecular mass (MM) and molecular mass distribution (MMD) were developed.

Until recently, the molecular mass distribution of PEEK by SEC was feasible only under drastic conditions involving the use of a phenol/trichlorobenzene mixed solvent at high temperature. This high temperature process also involves dissolution and filtration of the solution carried out at a temperature above 150°C. Furthermore, the only apparatus commercially available to perform the analysis is equipped with a differential refractometer (DRI) detector having a poor baseline stability. However accurate absolute MM of unmodified PEEK samples can be obtained using the universal calibration till $\bar{M}_w$ around 30000.

This work draws the attention to the development of a new reliable SEC for PEEK based on a derivatization procedure described in full details. This method requires dissolution of the sample in concentrated sulphuric acid, complete sulphonation and recovery of the sulphonated polymer. Several solvents of sulphonated PEEK are compatible with the packings of the columns used in SEC (NMP, DMF, DMSO, DMAc,...). The dissolution process is easily accomplished in N-methyl-2-pyrrolidone–LiBr. Owing to the UV-visible properties of the sulphonated PEEK-NMP solution, the macromolecules can be detected by a UV spectrometer which has a higher baseline stability and sensitivity than the DRI one. Furthermore, if this method seems to solve more satisfactorily the problem of accurate determination of absolute MM by using the universal calibration method, improvements of the SEC analysis could be obtained either by the use of PEEK calibration agents of higher molecular masses, in order to cover the entire MMD, either by the use of viscometer or light scattering detectors connected on-line with the SEC instrument.

Beside the SEC methods, absolute weight average MM measurements by light scattering of a series of unfractionated PEEK samples have been performed. Beyond their interest as standards for SEC, they have allowed the standardization of a viscometric MM determination in methanesulphonic acid at room temperature.

Number average MM ($\bar{M}_n$) of fluoroarylketone ended PEEK have also been determined with the aid of ^{19}F NMR. The agreement between these results and $\bar{M}_n$ values from SEC confirms the validity of the different methods developed in this chapter for the MM and MMD characterization of PEEK.

And, last but not least, preliminary results show that the derivatization procedure can be extended to PEEK-carbon fibre composite characterization.

The presently available PEEK characterization methods, even if further improvements are needed, already provide to the scientists the necessary analytical tool for any fundamental or applied research concerning the neat material or the matrix of composites.

Acknowledgments

The authors are greatly indebted to Dr. C. Bailly, Dr. E. Nield, Dr. P.T. McGrail, Dr. A. Bunn and Dr. Ph. Staniland for their constant interest in this work and encouraging discussions. They would like to express their thanks to the "Fonds National de la Recherche Scientifique (Belgium)" and to the "Service de Programmation de la Politique Scientifique (PAI, Belgium)" for the financial support of this research. They wish also to express their gratitude to Dr. C. Strazielle for his contribution in the critical evaluation of molecular mass by light scattering. They thank Dr. J.M. Dereppe and Mrs. C. Dereppe for their help in obtaining meaningful NMR spectra. The authors acknowledge the contribution of Miss C. Fagoo for her skilled and careful UV-visible spectroscopic, viscometric and chromatographic works. They would like to thank Imperial Chemical Industries for supplying the polymers used in this study. Finally, they would like to express their thanks to Mrs. A. Becquevort for her patient manuscript typing and to Mr. F. Gerardis for drawing most of the figures.

References

1. Mullins, M.J. and Woo, E.P.: JMS-REV. Macromol. Chem. Phys., **C27**(2), 313–341, 1987.
2. Jonas, A., Legras, R., Devaux, J.: "Synthesis and characterization of monodisperse PEEK oligomers" presented at the IUPAC Int. Symp. on New Polymers, Kyoto (Japan) Nov.30–Dec. 1, 1991.
3. Goodman, I., Mc.Intire, J.E. and Russel, W., British Patent 971, 227, 1964.
4. Colquhoun, H.M., Lewis, D.F.: "Synthesis of aromatic polyetherketones in trifluoromethanesulfonic acid", Polymer, **29**, 1902–1908, 1988.
5. Marks, B.M., US Patent 3 441 538, 1969.
6. Jansons, V., US Patent 4 361 693, 1982.
7. Rose, J.B. and Staniland, P.A., US Patent 4 320 224, 1982.
8. Charlier, Y. private communication.
9. Ueda, M. and Oda, M., Polymer Journal, **21**(9), 673–679, 1989.
10. Ueda, M. and Sato, M., Macromolecules, **20**, 2675–2678, 1987.
11. Ueda, M. and Ichikawa, F., Macromolecules, **23**, 926–930, 1990.
12. Tanabe, T., Fukawa, I., Donozo, T., Proceedings IUPAC, Montreal, 1990.
13. Kicheldorf, H.R., New Polymer Mater., **2**, 127, 1988.
14. Sogah, D.Y. and Risse, W., presented at the "2nd Pacific Polymer Conference", Otsu (Japan), nov. 26–29, 1991.
15. Mohanty, D.K., Sachdeva, Y., Hedrick, J.L., Wolfe J.F. and McGrath, J.E., Polym. Prep., Am. Chem. Soc., **25**, 19, 1984.
16. Roovers, J., David Cooney, J. and Toporowski, P.M., Macromolecules, **23**, 1611–1618, 1990.
17. Daoust, D., Devaux, J., Legras, R, Mercier, J.P. and Nield, E., GB 8401411.
18. Kelsey, D.R. E.P., 0211 693 A1. Union Carbide.
19. Carlier, V., Devaux, J., Jambe, B., Legras, R. and McGrail, P.T., Submitted to Macromolecules.
20. Bishop, M.T., Karasz, F.E., Russo, P.S. and Langley, K.H., Macrmolecules, **18**, 86, 1985.
21. Abraham, R.J., Bunn, A, Haworth, I.S. and Hearmon, R.A., Polymer, **31**, 728–735, 1990.

22. Devaux, J., Delimoy, D., Daoust, D., Legras, R., Mercier, J.P., Strazielle, C. and Nield, E., Polymer, **26**, 1994–2000, 1985.
23. Abraham, R.J., Bunn, A., Haworth, I.S. and Hearmon, R.A., Polymer **30**, 126–129, 1990.
24. Daoust, D., Fagoo, C., Godard, P. and Strazielle, C., to be published.
25. Nguyen,H.X. and Ishida, H., Die Makromol. Chem., Macromol. Symp., **5**, 135, 1986.
26. Nguyen, H.X. and Ishida, H., Polymer Composites, **8**(2), 57–73, 1987.
27. Lovinger, A.J. and Davis, D.D., Polymer Communic., **26**, 322, 1985.
28. Hay, J.N. and Kemmish, D.J., Plastics and Rubber Processing and Applications, **11**, 29–35, 1989.
29. Day, M., Sally, D. and Wiles, D.M., Journal of Applied Polymer Science, **40**, 1615–1625, 1990.
30. Ishiguro, S., Yonemoto, S. and Moriguchi, S., International GPC Symposium 1989, Newton, USA, October 1–4, 1989.
31. Jin, X., Bishop, M.T., Ellis, T.S. and Karasz, F.E., Br. Polym. J., **17**, 4, 1985.
32. Bailly, C., Williams, D.J., Karasz, F.E. and MacKnight, W.J., Polymer, **28**, 1009–1016, 1987.
33. Daoust, D., Spectra 2000, **122**, 40–41, Annexe II, 1987.
34. Daoust, D., Devaux, J., Fagoo, C., Godard, P. and Legras, R., to be published.
35. Lundberg, R.D., Am. Chem. Soc., Div. Polym. Chem. Polym. Prepr., **19**(1), 455, 1978.
36. Rigdahl, M. and Eisenberg, A., Am. Chem. Soc. Div. Polym. Chem. Polym. Prepr., **20**(2), 269, 1979.
37. Yarusso, D.J., Cooper, S.L., Knopp, G.S. and Georgopulos, P., J. Polym. Sci., Polym. Lett. Edn., **18**, 557, 1980.
38. Weiss, R.A., Turner, S.R. and Lundberg, R.D., J. Polym. Sci., Polym. Chem. Ed., **23**, 525, 1985.
39. Forsman, W.C., MacKnight, W.J. and Higgins, J.S., Macromolecules, **17**, 490, 1984.
40. Chalk, A.J. and Hay, A.S., J. Polym. Sci.(A-1), **7**, 691, 1968.
41 Xie, S., MacKnight, W.J. and Karasz, F.E., J. Appl. Polym. Sci., **29**, 2679, 1984.
42. Noshay, A. and Robeson, L.M., J. Appl. Polym. Sci., **20**, 1885, 1976.
43. Sivashinsky, N. and Tanny, G.B., J. Appl. Polym. Sci., **28**, 3235, 1983.
44. Johnson, B.C., Ylgor, I., Igbal, M., Wrightman, J.P., Lloyd, D.R. and McGrath, J.E., J. Polym. Sci., Polym. Chem. Edn., **22**, 721, 1984.
45. Rose, J.B., Eur. Pat. Appl. 8894 and 8895, 1980.
46. Lee, J. and Marvel, C. S., J. Polym. Sci., Polym. Chem. Edn., **22**, 295, 1984.
47. Litter, M.I. and Marvel, C.S., J. Polym. Sci., Polym. Chem. Edn., **23**, 2205, 1985.
48. Ogawa, T. and Marvel, C.S., J. Polym. Sci., Polym. Chem. Edn., **23**, 1231, 1985.
49. Makowsky, H.S., Lundberg, R.D. and Singahl, G.M., US Pat. 3 870 841, 1975.
50. Iwakura, Y., Uno, K. and Takiguchi, T., Polym. Sci., A-1, **6**, 3345, 1968.
51. Daoust, D., Devaux, J. and Godard, P.,to be published.
52. Groggins, P.H., "Unit processes in organic synthesis", McGraw Hill, New York and London, 3rd Edn., 260–337, 1947.
53. RAO, C.N.R., "Ultra-violet and visible spectroscopy" Third Edition Butterworth and Co. Ltd (1975) London.
54. Bunn, A., British Polymer Journal, **20**(4), 307–316, 1988.
55. Abraham, R.J., Haworth, I.S., Bunn, A. and Hearmon, R.A., Polymer, **29**, 1110–1117, 1988.
56. Abraham, R.J., Bunn, A., Haworth, I.S. and Hearmon, R.A., Polymer, **30**, 1969–1972, 1989.
57. Abraham, R.J., Bunn, A., Haworth, I.S. and Hearmon, R.A., Polymer, **32**, 1414–1419, 1991.
58. Bellamy, L.J., "The infrared spectra of complex molecules", Vol. 1, 3d edition, Chapman and Hall, London, 1975.
59. Varsanyi, G., "Assignments for vibrational spectra of seven hundred benzene derivatives", Vol. 1 and 2, Adam Hilger, London 1974.

60. Nguyen, H.X. and Ishida, H., Journal of Polym. Sci., part B. Polym. Physics, **24**, 1079–1091, 1986.
61. Green, J.H., Spectrochim. Acta., **24a**, 1627–1637, 1968.
62. Colthup, N.B., Daly, L.H. and Wiberley, S.E., "Introduction to infrared and raman spectroscopy", Academic Press, New York and London, 1964.
63. Attwood, T.E., Dawson, P.C., Freeman, J.L., Hoy, L.R.J., Rose, J.B., Staniland, P.A., "Synthesis and Properties of Polyaryletherketones", Polymer, **22**, 1096–1103, 1981.
64. Patterson, D., Ward, I.M., "The assignment of the carboxyl and hydroxyl absorptions in the infrared spectrum of polyethylene terephthalate", Trans. Faraday Soc., **53**, 291–294, 1957.
65. Kosky, P.G., McDonald, R.S., Guggenheim, E.A.: "Determination of end-group concentrations and molecular weight of poly(butylene terephthalate) by solid-state Fourier transform infrared spectroscopy", Polym. Eng. Sci., **25**, 389–394, 1985.
66. Devaux, J., Daoust, D., Legras, R., Dereppe, J. M. and Nield, E., Polymer, **30**, 161–164, 1989.
67. Pews, R. G., Tsuno, Y. and Taft, R.W., J. Am. Chem. Soc., **89**, 2391, 1967.
68. Giam, C.S. and Taft, R.W., J. Am. Chem. Soc., **89**, 2397, 1967.
69. Ager, I.R., Phillips, L. and Roberts, S.J., J. Chem. Sco., Perkin II, 1972, 1988.
70. Dayal, S.K., Ehrenson, S. and Taft, R.W., J. Am. Chem. Soc., **94**, 9113, 1972.
71. Mitchell, J.P., Phillips, L., Roberts, S.J. and Wray, V., Org. Mag. Res., **6**, 126, 1974.
72. Fukunaga, J. and Taft, R.W., J. Am. Chem. Soc., **97**, 1612, 1975.
73. Legras, R., Leblanc, D., Daoust, D., Devaux, J. and Nield, E., Polymer, **31**, 1429–1434, 1990.
74. Chu, B., Ying, Q., Wu, C., Ford, J.R., Dhadwal, H., Qian, R., Bao, J., Zhang, J. and Xu, C. Polymer Communic., **25**, 211, 1984.
75. Flory, P.J. "Polymer chemistry", Cornell University, Flory Press, 1978.
76. Mandel, M., "Encyclopedia of Polymer Science and Engineering", **11**, 739–829, J. Wiley and Sons, New York, 1988.
77. Champetier, G. and Monnerie, L., "Introduction à la chimie macromoléculaire", 340–342, Masson, Paris, 1969.
78. Karcha, R.J. and Porter, R.S., J. of Polymer Sci., Part. B., Polymer Physics, **27**, 2153–2155, 1989.
79. Benoit, H., Rempp, P. and Grusibic, Z., J. Polym. Sci., **5**, 753, 1967.
80. M.G. Styring and A. Hamielec, "Determination of molecular weight", J. Wiley and Sons, New York, 290-291, 1989.
81. Kato, Y., "Size exclusion chromatography", Blackie and Son, 170–188, 1989.
82. Yau, W.W., Kirkland, J.J. and Bly, D.D., "Modern size exclusion chromatography", John Wiley and Sons, 397–399, 1979.
83. Desbrières, J., Mazet, J. and Rinaudo, M., Eur. Pol. J., **18**, 269–272, 1982.
84. Rochas, C., Domard, A. and Rinaudo, M., Eur. Pol. J., **16**, 135–140, 1980.
85. Pasch, P., Dairanieh, I.S. and Khan, Z.H., Journal of Polym. Sci., Part. A, Polym. Chem., **28**, 2063–2074, 1990.
86. Cha, C.Y., J. Polym. Sci., Part. B, **7**, 343, 1969.
87. Stenlund, B., Adv. Chromatog., **125**, 231; **14**, 37, 1976.
88. Stenlund, B. and Forss, K.G., J. Polym. Sci. Symp., **42**, 951, 1973.
89. Scheuing, D.R., Journal of Applied Polymer Science, **29**, 2819–2828, 1984.
90. Ludlam, P.R. and King, J.G., Journal of Applied Polymer Science, **29**, 3863–'3872, 1984.
91. Coppola, G., Fabbri, P. and Pallesi, B., Journal of Applied Polymer Science, **16**, 2829–2834, 1972.
92. Stickler, M. and Eisenbeiss, F., Eur. Pol. J., **20**(9), 849–853, 1984.
93. Siebourg, W., Lundberg, R.D. and Lenz, R.W., Macromolecules, **13**, 1013–1016, 1980.
94. Daoust, D. and Godard, P., to be published.
95. Blundell, D.J. and Wilmouth, F.M., Sampe Quaterly, **17**(2), 50–57, January 1986.

Chapter 2
PEEK Degradation and Its Influence on Crystallization Kinetics

A. Jonas and R. Legras

Abstract

The PEEK molecular mass distribution has been measured after various melt holding conditions in air and in an inert atmosphere. A rapid molecular mass increase is observed in air, leading to the production of insolubles. From experimental evidences, the PEEK degradation in air is suggested to be a random chain scission process, followed by reaction of one of the produced fragments on a near chain, leading to branchings. The polymer crystallization kinetics are shown to be rapidly slowed down by the reticulation process. A similar mechanism is observed in nitrogen, though at a much slower rate. A significant crystallization rate decrease is also observed as a consequence of the reticulation. A purification procedure in order to remove additives in commercial PEEK is also described. The purification is shown to decrease dramatically the polymer thermal stability in nitrogen. A thermal stability equivalent to the commercial polymer can be recovered by addition of small amounts of a neutral buffer to the purified PEEK. Finally, hydroxyaryl and phenate chain ends are shown to have a detrimental influence on PEEK thermal stability. One of the roles of the neutral buffer is to deactivate such chain ends.

1 Introduction

Processing of PEEK requires to melt the polymer around 400°C [1, 2]. The use of such high temperatures when processing an organic material demands much caution since most covalent bonds become unstable or at least very reactive at these temperatures: the occurrence of degradation mechanisms is to be expected. The molecular modifications brought by the degradation will in turn modify the crystallization rate of the polymer, thereby changing the final morphology of the polymer sample.

Alain Jonas, Research Assistant of the Belgian National Fund for Scientific Research, Laboratoire de Physique et de Chimie des Hauts Polymères, Université Catholique de Louvain, Place Croix du Sud 1, B-1348 Louvain-la-Neuve, Belgium; Roger Legras, Laboratoire de Physique et de Chimie des Hauts Polymères, Université Catholique de Louvain, Place Croix du Sud 1, B-1348 Louvain-la-Neuve, Belgium.

Ultimately, the structural properties of the sample could be different from the ones expected without degradation. A clear technological importance is thus attached to the knowledge of the PEEK thermal stability in the molten state.

Moreover, in any study of the crystallization kinetics of a polymer, material scientists have to select adequate melting conditions before recording the crystallization of the polymer under study. It is indeed necessary to ensure that no significant degradation occurs during the prior melt holding time, and that a true homogeneous liquid polymer without any trace of local crystalline order is obtained. The study of molten PEEK thermal stability is thus a necessary step to undertake before any crystallization rate study.

The main purpose of this paper is to describe the molecular mass distribution evolution of the PEEK held in the molten state in different atmospheres, and to relate it to the crystallization kinetics of the polymer. Part of the results of this study have been published previously [3]; since this publication, complementary results obtained at Louvain-la-Neuve have supported the main conclusions of this previous study. Also, recent results from another group [4] have contributed to forge a better picture of the PEEK degradation. The present work will offer a phenomenological description of the degradation proces(es) in PEEK and some general features of the degradation mechanisms will be outlined (random chain scission, followed by branch formation and crosslinking). However no detailed study of the chemical reaction paths followed during the degradation will be given here. To our knowledge, such a chemical study has not been undertaken so far.

Most of the results presented here were obtained on the commercial powder as received from I.C.I. However, some measurements have been performed after removing additives or impurities present in the commercial PEEK. It will be shown that the purification of commercial PEEK leads to a decrease of its thermal stability in nitrogen. The addition of a simple neutral buffer will be demonstrated to delay strongly the degradation of the purified polymer. Finally, a study of the degradation behaviour of various purified PEEK grades as a function of the nature of their chain ends will be presented.

2 Thermal Stability of Molten PEEK in an Oxidative Atmosphere

In the following section, the molecular mass evolution of PEEK held in the molten state in air will be assessed. A very rapid reticulation mechanism will be evidenced, and its consequences on the crystallization kinetics examined.

2.1 Molecular Mass Evolution with Melt Holding Conditions

The evolution of the PEEK molecular mass distribution has been studied as a function of the melt holding conditions, as reported elsewhere [3]. Dried commercial

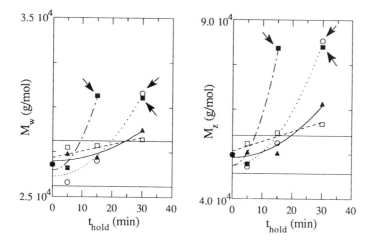

Fig. 2.1 Evolution of M_w and M_z with time (t_{hold}) for PEEK held in the molten state in air at 385°C (□, - - - -), 400°C (▲, cont. line), 420°C (○, ...), 440°C (■, - . - . -). The horizontal lines represent the estimated SEC maximum uncertainty on molecular mass. Arrows indicate samples with insolubles.

PEEK samples (grade 150P)[1] were held in the molten state in air, at temperatures T_{hold}s ranging from 385 to 440°C and for times between 5 and 30 minutes. These conditions were selected since they are similar to usual melt processing conditions. The PEEK powder was introduced into 5 mm inner diameter glass tubes. After thermal exposition, the samples were sulphonated, and their molecular mass distributions were measured using size exclusion chromatography (SEC) as detailed elsewhere [6]. At least two measurements were performed, and the average value was taken.

Results are shown in Fig. 1, where the second moment (M_w) and the third moment (M_z) of the molecular mass distribution are presented versus the melt holding time t_{hold}. The first moment of the distribution (the number molecular mass, M_n) was found to be constant within the precision of the technique (ca. 10%). A dramatic increase of M_w and M_z is observed as t_{hold} evolves. Faster increase rates are observed for higher T_{hold}s, as expected. For the strongest conditions, insolubles were collected after sample dissolution in concentrated sulfuric acid (up to 5%). These samples are indicated by an arrow in Fig. 1.

The results demonstrate unambiguously that a branching process is operative when PEEK is held above 385°C in an oxidative atmosphere. This process leads rapidly to an increase of M_w and M_z, and results ultimately in the apparition of insoluble fractions in the polymer. Experiments performed with samples having larger surface

---

[1] This polymer grade was selected because of the close proximity of its molecular mass distribution to the molecular mass distribution of PEEK used in the carbon fibre prepregs produced by I.C.I. (APC-2) [5].

to volume ratios show that the degradation rate is much increased [3]. The oxygen diffusion into the molten polymer is thus one of the degradation rate-controlling parameter. The same conclusion was also drawn by others on the basis of non-isothermal thermogravimetric experiments [7].

The existence of crosslinking reactions for PEEK held at high temperatures in air has been recognized by others [4, 8–10]. Spectroscopic and viscosimetric studies have been performed by *Day et al.* [4] on air-treated samples. A striking feature of the results of *Day et al.* is the decrease of the intrinsic viscosity of the soluble fraction of their PEEK samples, with t_{hold} at 400°C. This seems contradictory with our results. Note that their measurements are essentially related to longer t_{hold}s, for which important amounts of insolubles were measured (more than 25%).

This apparent discrepancy can be resolved under the hypothesis of a random chain scission process, producing two chain segments, one of which reacts with a near molecule to produce a branching. Although a smaller molecule is produced by this mechanism (the unreacted chain segment), there is no change in the total number of molecules, and M_n is kept constant during degradation. However, the process results in the formation of some heavier branched molecules, which are predominant in the determination of M_z and M_w; hence, these two parameters increase with t_{hold}. From the branched molecules, insolubles finally appear; the soluble fraction consists at that time of all the unreacted shorter chain segments produced by the degradation, plus branched molecules not yet making part of the insolubles. These branched molecules progressively join the insoluble fraction; consequently, the soluble fraction is more and more enriched in small molecules and its intrinsic viscosity ultimately decreases with t_{hold}, in agreement with the results of *Day et al.* [4].

The hypothesis of a random chain scission process has been put forward previously [3, 4, 11], but there is so far little experimental evidence about the chemical mechanism involved. *Hay and Kemmish* [11] found conflicting evidence about which of the links, the ether or the carbonyl, is broken during the chain scission. *Day et al.* [4] suggested the carbonyl to be the more plausible one on the basis of their spectroscopic results. There is at the present time some need to perform more elaborated studies on model compounds to elucidate the exact chemical mechanism of degradation. Nevertheless, the consequences of the oxidative degradation process on the polymer molecular mass distribution are already clearly understood.

2.2 Volatiles Emission

Isothermal thermogravimetric (TGA) experiments have been performed on molten PEEK in air, for T_{hold}s near 400°C. The weight loss is found to be strong and to increase linearly with t_{hold} [3]. Experiments performed using various initial surface to volume ratios confirm the rate-controlling character of the oxygen diffusion into the molten sample: the total absolute amount of volatiles is almost independent of this ratio, but the relative weight loss (expressed as a percentage of the initial sample weight) increases strongly with the sample surface to volume ratio (Fig. 2). This reflects the fact that volatiles emerge from a zone of roughly the same size, whatever

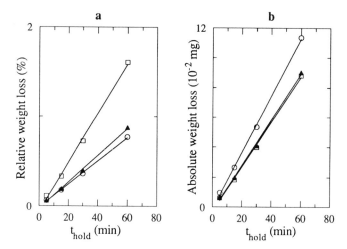

Fig. 2.2 Volatile emission of PEEK held at 420°C in air. Various surface to volume ratios were used, by changing the initial sample weights in the TGA pan (largest surface to volume ratio: □, lowest: ○, intermediate: ▲). Fig. a represents the relative weight loss, and fig. b the absolute weight loss of the samples.

the sample surface to volume ratio. This result exemplifies the role of oxygen diffusion.

Activation energies (E_a) for the degradation process in air are reported to be 105 ± 25 kJ/mol [3], 105 kJ/mol [12] and 100–150 kJ/mol [7]; all these values were calculated from TGA results, using various determination procedures. Actually, as pointed out by *Jonas and Legras* [3], since the recorded volatile emission is a complex function of the oxygen diffusion rate into the molten PEEK, of the oxidative degradation kinetics, and of the back-diffusion rate of the created volatiles to the free sample surface, the calculated E_as are only apparent global activation energies. No simple interpretation can thus be given to them.

Volatiles were identified by *Prime and Seferis* [12] by mass spectrometry. Benzoquinone, benzophenone, phenol and diphenyl ether were the main observed volatiles in air. These are chain fragments, produced during the random chain scission step of the degradation, probably when the scission occurs near a chain end. Actually, thermogravimetric studies do not bring much insight into the degradation process of PEEK. The main result is the confirmation of the importance of the oxygen diffusion on the degradation rate.

2.3 Consequences on Crystallization Kinetics

The crystallization kinetics of PEEK after various melt holding conditions in air were studied using differential scanning calorimetry (DSC). The samples were molten in air at T_{hold} during t_{hold}, and then cooled in the DSC at −10 K/min in nitrogen from T_{hold} to ambient temperature. A crystallization exotherm is detected during the cooling scan. Its temperature location is characterized by the peak temperature $T_{cc,peak}$ (corresponding to the maximal heat production) and the onset temperature $T_{cc,ons}$ (corresponding approximately to the beginning of the crystallization exotherm). The higher T_{cc}, the faster the crystallization kinetics. It has indeed been shown previously [13] that T_{cc} correlates with the crystallization half-times measured during isothermal crystallization experiments. The end-of-run crystallinity is estimated by the area under the crystallization exotherm, ΔH_{cc}. Results are obtained after averaging the values of at least five measurements.

Results are given as a function of t_{hold} in Fig. 3, for various T_{hold}s. The crystallization rate is dramatically depreciated by the holding conditions in air. A similar decrease of the crystallization rate for PEEK held in the molten state in air has been also observed by *Day et al.* [10]. These results are the logical consequence of the branching mechanism detected by SEC. The crystal growth rate is dependent upon a thermodynamic driving force resulting from the free energy difference between the liquid polymer and the crystalline phase ($\Delta G_{l-c}(T)$), and upon a transport factor representing the ease with which the molecules can move in the liquid phase to reach the growing crystal front (the so-called 'molecular mobility') [14]. A linear molecular mass increase is known to decrease the molecular mobility through an increase of the entanglement density [15]; a branching process reduces also strongly the molecular mobility. Moreover, a branching process decreases the configurational entropy of the melt, since it reduces the number of available conformations for some molecules; as a consequence, ΔG_{l-c} decreases. Globally, the two effects result in a crystallization rate decrease[1].

The end-of-run crystallinity is also strongly reduced by the air melting conditions. The branches that appear along the chains are structural defects that cannot be matched into the crystallographic lattice of the polymer; they are rejected outside the crystalline lamellae. For a sufficiently high branch concentration, this mechanism reduces the maximal attainable degree of crystallinity. Moreover, since the crystallization rate is decreased by the presence of the branches, the amount of polymer which is able to crystallize during the time length of a cooling scan is decreased. These two effects result in an end-of-run crystallinity lower for the degraded polymer than for the initial one. The DSC results are thus fully in agreement with the previous observations.

2.4 Conclusions

The thermal stability of PEEK held in air in the molten state has been assessed by SEC, TGA and DSC, for conditions similar to processing conditions. A fast and strong branching process is observed, leading to crosslinking, and reducing the PEEK

---

[1] The predominant cause of the crystallization rate reduction is the molecular mobility decrease.

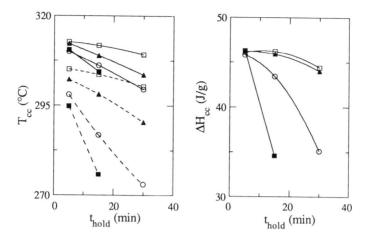

Fig. 2.3 Evolution of the PEEK crystallization temperature on cooling (T_{cc}) and crystallization enthalpy (ΔH_{cc}) with t_{hold} in air, for $T_{hold} = 385°C$ (□), 400°C (▲), 420°C (○) and 440°C (■). Dashed line: $T_{cc,peak}$; cont. line: $T_{cc,ons}$.

crystallization rate and end-of-run crystallinity. An important volatiles emission is recorded. Both TGA and SEC show that the oxygen diffusion is one of the rate-controlling parameters of the polymer degradation. From these results, it is evident that processing PEEK in air must be absolutely avoided, especially for samples of low surface to volume ratio (prepregs,...). Otherwise, the physical properties of PEEK will be much different as the expected ones, owing to the progressive loss of its thermoplastic character, and to a badly-controlled crystalline morphology.

It has been suggested that the degradation mechanism is a random chain scission promoted by oxygen, followed by reaction of one of the produced chain segment on a near chain to produce a branching point. This mechanism explains the initial independence of M_n on melt holding conditions and the increase of M_w and M_z. The intrinsic viscosity decrease of the soluble PEEK fraction when the insoluble fraction reaches high values, can also be explained in the frame of this model. Model compound studies are now required in order to gain a more detailed picture of the reaction path.

3 Thermal Stability of Molten PEEK in Inert Atmosphere

In this section, the molecular mass evolution of commercial PEEK when held in inert atmosphere (vacuum or nitrogen) will be examined. The evolution of the crystallization kinetics with melt holding conditions will then be discussed. Finally, the influence of some additives, impurities, cyclic oligomers and chain ends on thermal stability will be assessed.

3.1 Molecular Mass Evolution with Melt Holding Conditions

The same procedure as described in the previous section has been used to study the evolution of the PEEK molecular mass when the polymer is held in the molten state in an inert atmosphere. Evacuated sealed glass tubes were used, except for two samples (400°C/180 min and 400°C/420 min) which were molten in nitrogen. The results are given in Table 1, as the ratios of M_n, M_w and M_z to the corresponding molecular masses of the as-received polymer[1].

For melt holding times shorter than 30 minutes, no significant molecular mass variation can be detected, except at 440°C where a small but significant increase of M_w and M_z is observed. For the longest t_{hold}s at 400°C, a significant decrease of M_w (and M_z) is observed, together with the apparition of a large amount of insolubles.

No doubt a reticulation process is also operative when PEEK is held in the molten state in an inert atmosphere, although at a much slower rate than in air. This is supported by the small increase of the molecular mass with holding time at 440°C, and by the large amount of insolubles collected after long t_{hold}s at 400°C. The lower M_z and M_w of the soluble fraction for these last conditions are to be interpreted similarly to what has been done in the previous section.

For the short t_{hold}s of the present study, SEC does not detect any molecular mass variation at T_{hold}s lower than 420°C. But the relatively low sensitivity of SEC ($\sim 10\%$) precludes to gain any information on low branching amounts by this technique. Isothermal TGA experiments have been thus performed to have some insight of the degradation at short times in nitrogen. No significant weight losses were observed at $T_{hold} \leq 420°C$, for t_{hold}s up to one hour [3]. A small volatile emission was recorded at 440°C, with a constant emission rate of $\sim 0.1\%$/hour. The degradation does not lead to an appreciable volatile emission for short t_{hold}s.

The PEEK degradation in an inert atmosphere bears some phenomenological similarity to the degradation occurring in air. For instance, M_w and M_z of the PEEK soluble fraction decrease after long t_{hold}s at 400°C in both atmospheres, suggesting that the branching mechanism in vacuum or nitrogen is also preceded by a random chain scission process. Some experiments were performed on monodisperse linear PEEK oligomers to examine further the random chain scission hypothesis. The synthesis and properties of these monodisperse oligomers have been described elsewhere [16, 17]. Non-isothermal TGA was performed in nitrogen on these oligomers by heating them at 10K/min from room temperature up to 900°C. A typical weight loss curve is presented in Fig. 4, together with the three parameters extracted from these experiments: T_{first}, T_{ons}, and ΔW_{900}.

T_{first} is the temperature at which the first deviation from the zero weight loss baseline is observed. Actually, it characterizes the temperature at which the volatile emission rate is sufficiently large to be detected at 10 K/min heating rate. T_{ons} is the onset temperature of the volatiles emission, i.e. the intersection of the zero weight loss baseline with the steeper tangent to the weight loss curve. It represents the

---

[1] Averaged SEC results performed on six 150P-grade PEEK samples give $M_n = 10300$ g/mol, $M_w = 26800$ g/mol, and $M_z = 52200$ g/mol. The maximum deviation is of the order of 10%.

Table 2.1 Evolution of the PEEK Molecular Mass with Melt Holding Conditions in Nitrogen

$T_{hold}(°C)$	t_{hold} (min)	M_n/M_{n0}	M_w/M_{w0}	M_z/M_{z0}	Insolubles (weight %)
385	15	0.96	0.99	0.99	0
385	30	1.01	1.01	1.02	0
400	5	0.98	0.97	0.95	0
400	15	0.98	1.00	1.02	0
400	30	1.09	1.06	1.04	0
400	180	—	0.81	0.91	11
400	420	—	0.88	0.99	17
420	5	1.12	1.02	0.96	0
420	15	0.91	1.03	1.07	0
420	30	1.00	1.03	1.05	0
440	5	1.05	1.00	0.97	0
440	15	0.99	1.01	1.04	0
440	30	1.09	1.06	1.10	0

temperature from which an important volatiles emission is recorded, and it is thus related to the activation energy of the degradation process. $\Delta W900$ is the total weight loss recorded between room temperature and 900°C. The three parameters are plotted versus molecular mass in Fig. 5. Data obtained on the commercial polymer have also been added to this figure[1].

For the two lightest oligomers, a complete volatilization occurs during the heating scan; T_{ons} is strongly lower than for the other oligomers or the polymer. Actually, these two oligomers evaporate during the heating scan before reaching a sufficiently high temperature to experience a significant degradation. The T_{ons} values of the other oligomers are near the value measured for the polymer ($\sim 600°C$); they are all comprised between 550 and 600°C, and most of them are identical to the polymer value. This is due to the expected fact that the degradation activation energies are very similar for the polymer and the oligomers. However, T_{first} values are significantly lower for all the oligomers than for the polymer. This is evidence that the volatiles are produced after a chain scission. Indeed, in this hypothesis, a volatile is emitted by the polymer only after a scission near a chain end, whereas the probability to produce volatiles from oligomers is much higher. The volatile emission rate will thus be large enough to be detected at a lower temperature for the oligomers than for the polymer.

All the observations thus converge to indicate that a random chain scission process indeed occurs in nitrogen in molten PEEK. Reticulation then proceeds. The whole process is thus similar to what has been observed in air, although the chemical reactions involved are certainly different, and the reaction rate slower. In the next section, the consequences of this degradative process on the PEEK crystallization kinetics are examined.

---

[1] M_w was used to plot this point in Fig. 5.

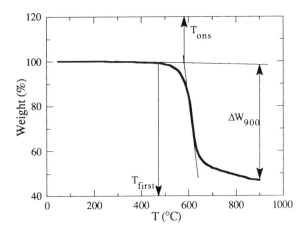

Fig. 2.4 Typical weight loss curve of a monodisperse PEEK oligomer in nitrogen. Three parameters are defined from this curve: T_{first}, T_{ons} and ΔW_{900} (see text).

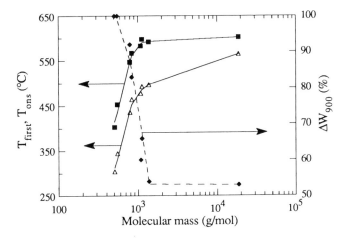

Fig. 2.5 Evolution of T_{first} ($\triangle$), T_{ons} ($\blacksquare$) and ΔW_{900} ($\blacklozenge$) with molecular mass. TGA measurements were performed at 10K/min in nitrogen, on monodisperse oligomers. The last point was obtained on a polydisperse PEEK sample.

3.2 Crystallization Kinetics Evolution with Melt Holding Conditions

Since the crosslinking process is much slower in nitrogen than in air, its consequences on PEEK crystallization rate are accordingly expected to be much less dramatic; consequently, phenomena other than degradation could dominate the crystallization kinetics. Before discussing the evolution of the PEEK crystallization rate with melt holding time in nitrogen, it is thus necessary to review briefly mechanisms able to affect polymer crystallization.

The overall crystallization kinetics are controlled by the nucleation rate and the crystal growth rate [14]. As stated above, the crystal growth rate is determined by the 'molecular mobility' and by a free energy term related to secondary nucleation. The first term is directly affected by variations in the molecular mass distribution, while the second one is less dependent on it. The nucleation rate depends on the presence of nucleants and affects strongly the overall crystallization rate. Actually, the nucleation density depends on many factors, like the crystallization temperature [14], the number of solid impurities present in the melt and the nature of their surface [18], the chain ends nature [13, 19], the melt holding conditions [18],... Changing the melt holding temperature before crystallization could affect the nucleation density for various reasons:

- When T_{hold} is lower than the thermodynamic melting temperature of the polymer (T_m°, which is the melting temperature of the extended infinite chains crystal), although higher than the observed melting temperature T_m, local regions having a crystalline order can survive in the melt [14]. These regions are able to act as nucleating seeds. Since their exact number decreases with t_{hold} and T_{hold}, the overall crystallization rate also decreases with T_{hold} and t_{hold} if such seeds are present. This phenomenon is called 'self-nucleation'.

  Self-nucleation is most conveniently studied with the aid of optical microscopy. It has been observed by this technique [3] that the nucleation density is large for $T_{hold} \leq 385°C$, and lower and roughly constant for $T_{hold} \geq 400°C$. The self-nucleation phenomenon is thus suppressed by melt holding the polymer 5 min. above a temperature comprised between 385 and 400°C. This is in accordance with published PEEK T_m° values (395°C [20], 389°C [21]), as well as with DSC results of *Lee et al.* [22] and infrared spectroscopy results of *Nguyen et al.* [23].

- Above T_m°, all the crystalline remnants are normally destroyed. However, some of them can survive at these temperatures, provided they are retained on a solid heterogeneous surface which stabilizes them against melting [18, 24]. Their progressive disparition with stronger T_{hold} conditions leads to a progressive decrease of the crystallization rate with these conditions. Elemental analysis performed on our PEEK samples reveal that the inorganic impurities level in them is very low (Table 2). This mechanism is thus not expected to be of importance for this PEEK grade.

- In another paper [3], we pointed out that the melting of the polymer synthesis powder could lead to an inhomogeneous melt. Indeed, the initial crystalline pow-

Table 2.2 Elemental Analysis of As-received Commercial PEEK (Grade 150P)

Element	Al	Ca	Fe	K	Na	P
ppm	9	45	10	–	110	34

der is probably inhomogeneous, since segregation of short chains could arise [14] during the polymer cooling in the synthesis solvent. Upon melting, these inhomogeneities would subsist in the melt during a given amount of time, depending on self-diffusion coefficients. Provided the segregated oligomers crystallize faster than the polymer, and provided the resulting seeds are able to act as nucleants for polymer crystals, the polymer nucleation density could depend on (T_{hold}, t_{hold}) conditions.

The presence of an oligomer peak in the PEEK molecular mass distribution has been detected previously [3]. However, we demonstrated later that this peak is due to cyclic oligomers, crystallizing very slowly in a crystalline lattice different from linear PEEK [25]. Furthermore, linear oligomers having an extended chain length up to ~ 7 nm crystallize more slowly than the polymer [17]. Thus, the previous hypothesis of segregated oligomers acting as PEEK nucleating agents can be discarded.

Hence, we can safely conclude that for $T_{hold} \geq 400°C$ the PEEK nucleation density is not dependent on the selected melt holding condition, provided no degradative process occurs in parallel. Any variation of the PEEK overall crystallization rate thus reflects alterations of the PEEK chains during the melt holding conditions. From SEC results, it is thus expected that the PEEK crystallization rate will decrease with t_{hold} for $T_{hold} \geq 400°C$, due to the slow branching mechanism.

DSC experiments were performed to check this prediction. A PEEK sample was repeatedly molten in nitrogen during various t_{hold} values at a constant T_{hold}, and scanned between each melting from T_{hold} to $200°C$ at -10 K/min. The crystallization exotherm was characterized as described in the previous section. Results are given in Fig. 6, for different T_{hold}s values (a new sample was taken for each T_{hold}).

As expected, T_{cc} decreases with t_{hold}, and the decrease rate is higher the higher T_{hold}. Such a crystallization rate decrease with T_{hold} and t_{hold} in nitrogen has been reported by others [10,22]. The end-of-run crystallinity is also decreasing with t_{hold}. The DSC results are fully in agreement with the degradation process evidenced above. Exactly the same comments as made when studying the PEEK crystallization kinetics after melting in air, can be done for the kinetics performed under nitrogen; the major difference consists only in the degradation rate, much higher in air than in nitrogen. It is interesting to observe that the crystallization rate is very sensitive to small amounts of branchings; indeed, T_{cc} decreases long before any significant molecular mass evolution can be detected with SEC. This results from the low sensitivity of SEC, especially to branchings which increase less the hydrodynamic volume of the chains than a linear molecular mass increase.

An important observation can be made from inspection of Fig. 6: there is a small but significant crystallization rate decrease with t_{hold} for $T_{hold} = 385°C$. Although self-

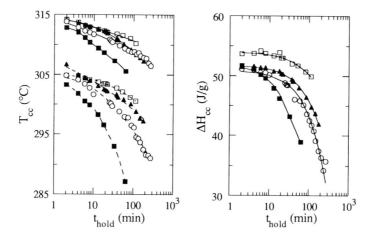

Fig. 2.6 Evolution of T_{cc} and ΔH_{cc} with t_{hold} in nitrogen for the as-received commercial PEEK. $T_{hold} = 385°C$ (□), $400°C$ (▲), $410°C$ (○), $440°C$ (■). Continuous line: $T_{cc,ons}$, dashed line: $T_{cc,peak}$.

nucleation is not suppressed at this temperature, the similarity of the results obtained at this temperature with results obtained at higher T_{hold}s, suggests that the degradation mechanism is already operative at 385°C. Rigorously speaking, this means that it is impossible to have an homogeneous melt without simultaneously changing slightly the structure of the chains. Evidently, the thermodynamic melting temperature of PEEK is so high that some of the constituting chemical bonds reach their stability limit before $T_m^°$. However, these effects are weak, and melting the polymer in nitrogen 5 min at 400°C does not decrease significantly the crystallization kinetics or the end-of-run crystallinity.

The consequences of the reticulation process on the PEEK crystallization kinetics are illustrated in Table 3, where the $T_{cc,peak}$ decrease and the end-of-run degree of crystallinity decrease, computed assuming a melting enthalpy of 130 J/g for the fully

Table 2.3 Crystallization Temperature Decrease and End-of-run Degree of Crystallinity Reduction, for PEEK Held 30 min in the Molten State in Nitrogen and in Air, at Various T_{hold}s

T_{hold} (°C)	In nitrogen		In air	
	$T_{cc,peak}$ decrease (°C)	crystallinity decrease (%)	$T_{cc,peak}$ decrease (°C)	crystallinity decrease (%)
385	1.5	1.3	5.1	1.5
400	3	1.5	13.5	1.5
410	4.5	2.2	–	–
420	–	–	29.8	8.5
440	12	6	> 30	> 8.5

crystalline polymer [20], are indicated for a 30 min–melting in air and in nitrogen, at various T_{hold}s.

Although t_{hold} is small, and of the order of usual processing times, a significant variation of the crystallization kinetics is detected by DSC for all the processing temperatures. However, the branching process is slow at 400°C in nitrogen, and only minor alterations of the PEEK structure are expected for short processing times. These alterations are besides undetectable with SEC; it is only the very high sensitivity of crystallization kinetics to molecular mobility which allows to detect them. This is an important observation, since the crystallization kinetics can now be used as a tool to monitor PEEK degradation, in an easier way as the less sensitive chromatographic technique. In the rest of this paper, this property will be extensively used. However, it has to be kept in mind that this is valid only if the nucleation density is constant.

3.3 Effects of Impurities on Degradation and Crystallization Kinetics

So far, the thermal stability and crystallization of commercial PEEK only has been described. Like every commercial polymer, PEEK contains traces of various impurities. It contains intrinsic impurities, e.g. chain ends, which can be removed or modified only through a chemical reaction. It contains extrinsic impurities, e.g. catalyst residues, synthesis solvent and by-products, and various additives. Physical treatments like refluxes are normally able to remove these impurities.

The purpose of this section is to detect if some of these impurities play a role in the PEEK degradation in nitrogen. A purification process of commercial PEEK will be first described, and the effect of some extrinsic impurities on thermal stability and crystallization discussed. Then, the role of chain ends will be presented.

3.3.1 Extrinsic Impurities Effects on the PEEK Degradation

Removal of extrinsic impurities followed by introduction of selected additives is the simplest way to obtain informations on the role of extrinsic impurities in PEEK crystallization and degradation kinetics. Hence, a purification process of commercial PEEK has to be designed first. This requires some knowledge about the nature of extrinsic impurities in this polymer:

- Elemental analysis performed on commercial PEEK reveals the presence of various cations, mainly sodium, phosphorus and calcium (Table 2). Sodium ions are due to small traces of sodium fluoride which is a synthesis by-product, to sodium carbonate which is the synthesis catalyst, or to phenate chain ends [6]. Phosphorus and calcium are probably due to inorganic additives. Refluxing the polymer into water is a way to remove some of these impurities.
- Infrared analysis performed on commercial PEEK reveals the presence of two small absorption bands located at 2919 and 2845 cm^{-1}. These are two typical C–H stretching frequencies of methylene units [26]. From the known presence of

calcium, one can infer the probable presence in PEEK of calcium stearate, a well-known dye lubricant [27]. The solubility of such a compound, quite low in most usual organic solvents, is very high in hot pyridine.

- Small amounts (0.1% max) of diphenylsulphone (DPS) have been found in PEEK [3]. These are removed by an acetone reflux.
- Cyclic oligomers can be extracted from the polymer by acetone extraction with a Soxhlet, during one week [25].

The selected purification process consists thus in three 2-hour refluxes successively in boiling pyridine, demineralized water and distilled acetone. A final cyclic oligomer extraction is eventually performed during one week in a Soxhlet with acetone. High purity grades solvents are used throughout.

Fig. 7 presents the evolution of T_{cc} and ΔH_{cc} as a function of melt holding time in nitrogen at 400°C for both commercial and purified PEEK. The purification process increases the crystallization kinetics for the shortest t_{hold}s. However, the crystallization temperature decreases much more rapidly with t_{hold} for purified PEEK. A parallel stronger decrease of the crystallization enthalpy is observed.

The T_{cc} increase observed at short t_{hold}s upon purification is related to variations in the nucleation density brought by the purification process. The reason for this phenomenon will be discussed in section 3.3.3. The dramatic decrease of T_{cc} with t_{hold} after purification is due to an acceleration of the crosslinking degradation process. This is evidenced by the large amount of insolubles found in purified PEEK samples, hold for various times in nitrogen at 400°C, as compared to as-received PEEK (Table 4).

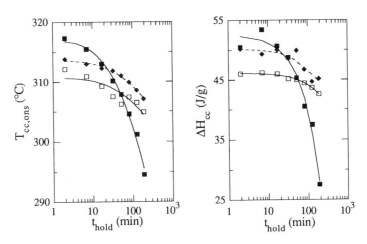

Fig. 2.7 Evolution of $T_{cc,ons}$ and ΔH_{cc} with t_{hold} in nitrogen at $T_{hold}=400°C$. ♦: as-received commercial PEEK, ■: purified PEEK, □: purified PEEK after addition of 0.8 weight % of a neutral phosphate buffer. A dramatic decrease of the PEEK thermal stability is brought by the purification process. Upon phosphate buffer addition, the purified PEEK is thermally stabilized again.

Table 2.4 Insolubles Collected in the As-received and Purified PEEK Powders, after Holding the Polymer in Nitrogen at 400°C during Various t_{hold}s. Samples were Stirred in Concentrated Sulfuric Acid during One Week before Filtration

t_{hold} (min.)	Insolubles (weight %)	
	As-received PEEK	Purified PEEK
180	11	32
420	17	76

A stabilizer has clearly been removed from commercial PEEK by the purification procedure. The thermal stability of the purified PEEK is very bad, and proper processing of this unstabilized PEEK would be very difficult. To identify the PEEK stabilizing agent, various substances have been added to the purified PEEK by a slurry method, and the effects of such an addition on the crystallization kinetics determined.

The introduction of a phosphate buffer (0.8 weight % NaH_2PO_4/Na_2HPO_4, pH = 7, slurry in water) into the purified polymer results in a clear-cut thermal stabilization of the purified PEEK. Fig. 7 illustrates this effect: the evolution with t_{hold} of both crystallization enthalpy and crystallization temperature is positively affected by the neutral buffer; actually, buffer addition brings the purified polymer back to the thermal stability of the as-received polymer. However, the values obtained for T_{cc} after buffer addition are lower than the original values of the as-received PEEK grade. Actually, the buffer has a negative effect on the crystallization kinetics at short t_{hold}s, but a positive effect on the thermal stability at long t_{hold}s. Since the presence of sodium and phosphorus was detected in the commercial PEEK by elemental analysis, it is highly probable that the additive used to stabilize commercial PEEK against thermal branching is the phosphate buffer. The T_{cc} difference at short t_{hold}s between as-received and buffer-stabilized purified PEEK could be due to differences in the amount of added buffer, since it is clear that the buffer addition decreases the PEEK T_{cc} at short t_{hold}s.

The stabilizing role of the neutral buffer suggests that the degradation process in PEEK could be accelerated by the presence of (Lewis-type) acids or bases. Bases are expected to affect molecular mass, since they are able to perform nucleophilic attacks on ether linkages [28], or to react with PEEK fluorine chain ends. But PEEK reticulation promoted by bases at high temperatures has not yet been demonstrated, although it has been shown that benzophenone in presence of sodium hydroxide at 400°C gives rise to small amounts of triphenylcarbinol [28]. Still less knowledge is at the present time available concerning a possible degradation process involving initiation by a Lewis acid. One could think for instance to the formation of a carbocation by acid-base reaction of the Lewis acid with the basic PEEK ketone linkage. This cation would be stabilized by resonance (this is similar to the usual mechanism of PEEK protonation in strong acids). It could then in turn initiate an electrophilic attack of a near chain.

The main PEEK impurities of acidic character are the hydroxyaryl chain ends, whose presence in commercial PEEK has been detected [6]. Phenate chain ends and sodium carbonate residues are most probable basic impurities. It has to be suspected that the two hours water reflux of PEEK during its purification is too short to wash out all the Na_2CO_3, which is intricately mixed into the polymer (contrary to additives like the phosphate buffer, probably added only after synthesis). Chain ends effects will be detailed in section 3.3.2.

Other additives or impurities exert a smaller influence on the PEEK thermal stability. For instance, upon addition of calcium stearate to purified PEEK (0.4 weight %, slurry in methanol), the crystallization temperature is reduced of ~7–10°C for short t_{hold}s at 400°C. Then it decreases with t_{hold} in a fully similar fashion as for the purified polymer. Thermogravimetric experiments reveal that the aliphatic chain of calcium stearate decomposes around 360°C. It is thus not surprising to observe a crystallization kinetics decrease associated to short melt holding times. This is probably due to interaction of the produced aliphatic fragments with the PEEK chains. It seems thus questionable to use such a dye lubricant in this high melting point polymer.

However, no significant change in the crystallization kinetics results from the addition of the previously extracted cyclic oligomers (0.2 weight %, slurry in acetone). As already stated, these oligomers, which crystallize very slowly [25], do not play any role in the nucleation mechanism. They do not seem either to initiate the degradation process. This means that no important cycle strain exists even in the lightest of these PEEK cyclic oligomers.

3.3.2 Chain Ends Effects

3.3.2.1 Samples Chain Ends Characterization Spectroscopic methods to characterize PEEK chain ends have been presented elsewhere [6]. Possible chain ends of PEEK synthesized by the nucleophilic route are:

- fluoroarylketone chain ends (FΦ(CO)—), resulting from the use of 4,4'-difluorobenzophenone in the PEEK synthesis. A quantitative evaluation of these chain ends is conveniently performed by ^{19}F NMR [6].
- hydroxyaryl chain ends (HOΦ—), resulting from the use of hydroquinone in the PEEK synthesis. A roughly quantitative evaluation of HOΦ—chain ends content is possible by infrared spectroscopy (IR) [6]. Only small amounts of such chain ends were found in the commercial polymer by this spectroscopy.
- phenate chain ends (NaOΦ—), resulting from the reaction of hydroxyaryl chain ends with the sodium carbonate catalyst used in the synthesis. The Na content of the polymer is due to traces of sodium carbonate or sodium fluoride, phenate chain ends, and phosphate buffer. An upper bound to the phenate chain ends content can thus be obtained from elemental analysis, after subtraction of the Na content due to the phosphate buffer (computed from the polymer P content).

- phenoxy chain ends (ΦOΦ—), probably resulting from addition of phenol during the synthesis to control molecular mass [6]. A qualitative detection of such chain ends is possible by ^{13}C NMR; however, a quantitative evaluation is not possible by this technique, due to the low proportion of ^{13}C spins as compared to ^{12}C spins, and to the low chain ends content of the polymer.

Chain ends analyses were performed on four different PEEK grades. Results are given in Table 5.

Some comments have to be made on the measurements reported in Table 5:

- The percentage of chain ends of a given type is obtained by ratioing the measured number of chain ends of this type to the total number of chain ends computed from M_n.
- No elemental analysis was performed on the PK–OH grade, because the available amount of sample was too low for such an analysis.
- It has to be stressed again that the indicated phenate chain ends content is an upper bound to the actual phenate chain ends content.
- As stated above, the phenoxy chain ends content of the polymer was qualitatively evaluated by ^{13}C NMR. For PK–OH and PK–F grades, no phenoxy chain ends resonances were observed. The PK–HMW grade did not show any ^{13}C NMR chain ends peak. The chain ends signals are probably too low, as compared to the resolution of the analog to digital converter of the NMR apparatus, due to the higher M_n of this sample. In such a case, no matter how long one accumulates: no chain ends signal will ever be added in the computer memory [29]. A solution could be to increase the polymer concentration in solution, but this was not possible in CH_3SO_3H, the NMR solvent. However, a signal was detected by ^{19}F NMR, because of the stronger sensitivity of this spin. The fluoroarylketone content of this PEEK grade could thus be computed. It represents less than half the chain ends content estimated from M_n. From IR spectroscopy, the hydroxyaryl chain ends content was assessed, and found to be low. It is then assumed that the rest of the chain ends are of the phenoxy-type.

3.3.2.2 Role of Chain Ends in the PEEK Thermal Stability In order to convert possible small amounts of phenate chain ends into hydroxyaryl chain ends, we used a procedure developed elsewhere [13] to acidify PEEK sulfonate chain ends. So, purified PEEK powders were refluxed in a 14/86 vol/vol HCl/acetone mixture; this reflux was followed by two other refluxes in distilled acetone, to remove acid traces[1]. The crystallization kinetics of the acidified sample were subsequently recorded as a function of the melt holding time t_{hold} in nitrogen at 400°C.

Results are given in Fig. 8. The crystallization enthalpies have been normalized to the maximum value of each sample, in order to allow an easier comparison.

For short holding times, the crystallization temperature on cooling decreases with increasing molecular mass, as reported by others [13, 15]. However, this scaling

---

[1] The acid reflux is also expected to wash out sodium carbonate traces, if any.

2. PEEK Degradation 75

Table 2.5 Chain-end Characterization of the Four PEEK Grades Used in This Study. Experimental Grades were Provided by I.C.I.

Grade	M_n (g/mol)	M_w (g/mol)	[FΦ(CO)—] (% chain ends)	[ΦOΦ—]	[HOΦ—] (% chain ends)	[NaOΦ—] (% chain ends)
PK-F (experimental)	9 080	18 600	98	not detected	0	<2
PK-OH (experimental)	9 850	20 400	0	not detected	100	—
PK 150P (commercial)	10 300	26 800	58	important amounts	2	<2
PK-HMW (experimental)	15 500	40 900	33	not detected	5	<2

relation is no more followed after 20 min. melt holding time, due to the very rapid T_{cc} decrease of the fully hydroxyaryl-ended sample (PK-OH). This suggests that the crosslinking mechanism is accelerated by the presence of hydroxyaryl chain ends. The total amount of hydroxyaryl chain ends after polymer acidification is equal to the sum of the initial amounts of phenate and hydroxyaryl chain ends. Since only an upper bound for the phenate chain ends content is known, it is only possible to compute lower and upper bounds for hydroxyaryl chain ends contents in acidified polymers. These bounds are given in Table 6.

It appears from consideration of Table 6 and Fig. 8 that the degradation rate of acidified polymers is correlated to the total amount of hydroxyaryl chain ends. For

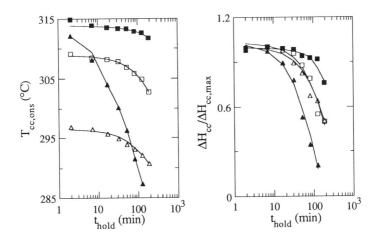

Fig. 2.8 Crystallization temperature on cooling and ratio of crystallization enthalpy to maximum crystallization enthalpy, as a function of melt holding time in nitrogen at 400°C. (Purified then acidified polymers of Table 5).
(■): PK-F; (▲): PK-OH; (□): PK150P; (△): PK-HMW.

Table 2.6 Lower and Upper Bounds for the Hydroxyaryl Chain-ends Content of the Four Polymers of Table 5, after Acidification

Polymer grade	[HOΦ—] in acidified polymer (10^{-6} mol/g)	
	lower bound	upper bound
PK-F	0	4
PK-OH		200
PK150P	4	7
PK-HMW	6	8

a very large hydroxyaryl chain ends content (PK—OH), T_{cc} decreases extremely rapidly. Samples having similar hydroxyaryl chain ends contents, but differing in other respects, show an identical degradation behaviour (samples PK150P and PK-HMW). The sample having the smallest hydroxyaryl chain ends content (PK-F) is the most thermally stable.

Hence, hydroxyaryl chain ends are clearly able to accelerate the PEEK crosslinking degradative process. However, it should be pointed out that the presence of these chain ends in the as-received commercial PEEK is not able to explain alone the poor thermal stability of the purified polymer. This is best seen in Fig. 9, where we report the T_{cc} and ΔH_{cc} evolution with t_{hold} of purified, purified then buffer-stabilized, and purified then acidified samples of this grade. The T_{cc} of the acidified sample does not

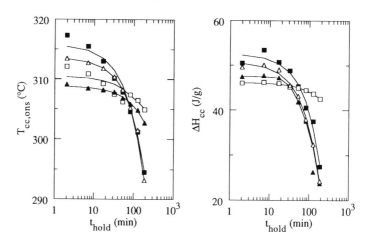

Fig. 2.9 Evolution of the crystallization temperature and enthalpy with melt holding time in nitrogen at 400°C, for
(■): purified PEEK (commercial 150P grade)
(□): purified PEEK, stabilized by 0.8 weight % of phosphate buffer
(▲): purified and acidified PEEK
(△): previous sample, refluxed in a basic solution.

behave identically to the T_{cc} of the purified sample, although the crystallization enthalpies of both samples evolve similarly. Also, the amount of insoluble fractions collected in acidified samples after long t_{hold}s is nearer to the stabilized than to the purified polymer.

Refluxing the acidified polymer into a 0.4 M sodium acetate/distilled methanol solution, followed by two refluxes in pure methanol, brings the acidified polymer nearly back to the behaviour of the purified polymer before acidification (Fig. 9). Such a treatment is expected to convert at least part of the hydroxyaryl chain ends into phenate chain ends. It seems thus probable that phenate chain ends are also able to induce PEEK crosslinking at high temperature. This would explain the differences observed between purified and acidified samples.

In summary, it has been shown that hydroxyaryl chain ends play a negative role in the PEEK thermal stability. Phenate chain ends probably also contribute to such detrimental effects. Phosphate buffer appears to be able to deactivate these chain ends.

3.3.3 Comments on the PEEK Nucleation Density

In the previous sections, the evolution of PEEK crystallization temperature and enthalpy with t_{hold} was used as a tool to investigate the polymer thermal stability in the molten state, after various treatments. The crystallization temperature of the polymer at very short t_{hold}s, where no significant degradation occurs, has been found to be also dependent on these treatments, suggesting parallel variations in the PEEK nucleation density (Fig. 7, Fig. 9). In Fig. 10 we summarize some observations concerning the effect of acid and basic refluxes on the PEEK crystallization kinetics at short t_{hold}s: starting from purified PEEK, the acidification of the chain ends (1) lowers T_{cc}; a subsequent basic reflux (2) increases T_{cc}; a new acid reflux (3) decreases again T_{cc}.

Hydroxyaryl chain ends neutralization (basic reflux) thus increases the PEEK nucleation density, while backconversion of phenate chain ends into hydroxyaryl chain ends (acid reflux) decreases this density. In accordance with results obtained previously on sufonate-ended PEEK [13], this increase could be due to a clustering of the phenate chain ends through electrostatic interactions; nucleation would be initiated by these clusters which are not destroyed upon simple melting. Acidification would reduce the electrostatic interactions of the clusters, thereby allowing the ungrouping of the chain ends upon melting, and causing the observed decrease in the nucleation density.

The existence in the commercial polymer of clusters able to act as crystalline seeds, and not destroyed simply by melting the polymer at 400°C, is confirmed by the following experiment. As-received PEEK was melt-mixed at 400°C in nitrogen in a Brabender mixer (30 rpm), and samples were taken regularly from the melt. The T_{cc} measured on these samples are plotted in Fig. 11 versus mixing time, along with T_{cc} values measured on the initial polymer powder as a function of t_{hold} in nitrogen (without shearing). The T_{cc} of the melt-mixed sample is observed to decrease a little faster with mixing time than the original polymer with t_{hold}. This is probably due to

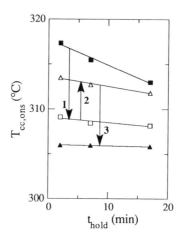

Fig. 2.10 Evolution of $T_{cc,ons}$ with short t_{hold}s in nitrogen at 400°C, for as-received PEEK after various treatments.
(■): purified PEEK; (□): same sample after an acid reflux; (△): previous sample after a basic reflux; (▲): previous sample after an acid reflux.

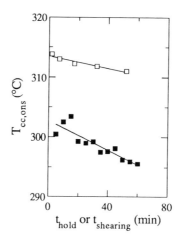

Fig. 2.11 Evolution of $T_{cc,ons}$ with the melt holding time (t_{hold}) in nitrogen at 400°C (□), or with the mixing time ($t_{shearing}$) of the polymer in the Brabender under nitrogen at 400°C (■) (as-received 150P grade PEEK).

the fact that the Brabender mixer was not totally purged of air, a condition difficult to fulfill. However, it is obvious that the large T_{cc} difference observed between the sheared polymer and the unsheared one at short times is not due to a degradation mechanism. Actually, under shear, the PEEK T_{cc} drops immediately of ~10°C. Some kind of nucleating clusters disappear at 400°C in molten PEEK only when the polymer experiences some shear. Since self-nucleation is suppressed at 400°C [3], we suggest that these nucleating clusters are resulting from phenate chain-ends ionic grouping. Further work is required to confirm the possible existence of specific chain ends clusters in molten polymers.

Conclusion

The thermal stability of molten PEEK has been assessed both in nitrogen and in air. From various experimental evidences, a random chain scission mechanism has been suggested to occur first. It is immediately followed by branchings formation and crosslinking. As a consequence, M_w and M_z increase with melt holding time, and important amounts of insolubles are detected for long melt holding times. Phenomenological differences between air and nitrogen (or vacuum) melt degradations consist mainly in the degradation rate, much faster in an oxidative atmosphere.

The crosslinking is responsible for a crystallization kinetics decrease. The very high sensitivity of the PEEK crystallization kinetics to even small amounts of branchings has been pointed out. Actually, the easiest way to monitor the PEEK thermal degradation is to use the DSC technique, provided nucleation density is controlled.

Commercial PEEK has been shown to be stabilized by addition of a neutral buffer. In the absence of such a buffer, the crosslinking reaction is very rapid, and processing of PEEK would be nearly impossible at 400°C, even in nitrogen. The detailed role of this buffer is not clear at the present time, but it could have a relation with the presence in PEEK of chain ends having an acid or basic character. The detrimental effect of hydroxyaryl chain ends on the PEEK thermal stability has been demonstrated. Phenate chain ends probably also contribute to decrease the PEEK thermal stability. However, the grouping of these phenate chain ends into clusters could parallely result in a nucleating effect. The need to precisely control the PEEK chain end nature in order to obtain a processable material is stringent.

The present work has been devoted to phenomenological observations. Further work should concentrate on degradation mechanism, by studying low molecular weight model compounds. In this respect, the possibility to synthesize well-defined monodisperse PEEK oligomers [16] is clearly of much interest. Chain ends effects being amplified by the use of short compounds, confirmation of some observations presented in this paper could be obtained.

Acknowledgments

We are indebted to the Belgian National Fund for Scientific Research for its financial support of this work. The authors wish also to express their gratitude to Dr P.T. Mc Grail and Dr. E. Nield from ICI for their helpful comments, and for providing the PEEK powder. The SEC results of this study were obtained with the aid of D. Daoust and C. Fagoo. A particular thanks has to be given to Dr J. Devaux for enligthning discussions.

References

1. Cattanach, J.B., Cogswell, F.N., "Processing with aromatic composites", in "Developments in reinforced plastics", Pritchard, G., editor, Vol. 5, Elsevier, London, 1986.
2. Blundell, D.J., Crick, R.A., Fife, B., Peacock, J., Keller, A., Waddon, A., "Spherulitic morphology of the matrix of thermoplastic PEEK/carbon fibre aromatic polymer composites", J. Mater. Sci. **24**, 1989, pp 2057–2064.
3. Jonas, A., Legras, R., "Thermal stability and crystallization of poly(aryl ether ether ketone)", Polymer **32**, 1991, pp 2691–2706.
4. Day, M., Sally, D., Wiles, D.M., "Thermal degradation of poly(aryl-ether-ether-ketone): experimental evaluation of crosslinking reactions", J. Appl. Polym. Sci. **40**, 1990, pp 1615–1625.
5. MacGrail, P.T., personal communication.
6. Daoust, D., Devaux, J., Godard, P., Jonas, A., Legras, R., "Poly(ether ether ketone) characterization", this book, chapter 1.
7. Day, M., Cooney, J.D., Wiles, D.M., "The thermal stability of poly(aryl-ether-ether-ketone) as assessed by thermogravimetry", J. Appl. Polym. Sci. **38**, 1989, pp 323–337.
8. Chan, C.M., Venkatraman, S.J., "Crosslinking of poly(aryl ether ketone)s. 1. Rheological behaviour of the melt and mechanical properties of the cured resin", J. Appl. Polym. Sci. **32**, 1986, pp 5933–5943.
9. Cebe, P., "Annealing study of poly(etheretherketone)", J. Mater. Sci. **23**, 1988, pp 3721–3731.
10. Day, M., Suprunchuk, T., Cooney, J.D., Wiles, D.M., "Thermal degradation of poly(aryl-ether-ether-ketone) (PEEK): a differential calorimetry study", J. Appl. Polym. Sci. **36**, 1988, pp 1097–1106.
11. Hay, J.N., Kemmish, D.J., "Thermal decomposition of poly(aryl ether ketones)", Polymer **28**, 1987, pp 2047–2051.
12. Prime, R.B., Seferis, J.C., "Thermo-oxidative decomposition of poly(ether ether ketone)", J. Polym. Sci., Polym. Lett. Edn 24, 1986, pp 641–644.
13. Legras, R., Leblanc, D., Daoust, D., Devaux, J., Nield, E., "Extension of the concept of chemical nucleation to poly(ether ketones)", Polymer **31**, 1990, pp 1429–1434.
14. Wunderlich, B., "Macromolecular Physics, Vol. II. Crystal nucleation, growth, annealing", Academic Press, New York, 1976.
15. Hay, J.N., Kemmish, D.J., "The crystallization of PEEK: molecular weight effects", Plast. Rub. Proces. Appl. **11**, 1989, pp 29–35.
16. Jonas, A., Legras, R., Devaux, J., "Synthesis and characterization of monodisperse PEEK oligomers", presented at the IUPAC Int. Symp. on New Polymers, Kyoto (Japan), 30 Nov.–1 Dec. 1991.
17. Jonas, A., Legras, R., "Structure and thermal behaviour of some monodisperse PEEK oligomers", presented at the 2nd Pacific Polymer Conference, Otsu (Japan), 26–29 Nov. 1991.
18. Binsbergen, F.L., "Heterogeneous nucleation of crystallization" in "Progress in solid state chemistry", McCaldin, J.O., Somorjai, G., editors, Vol. **8**, Pergamon Press, Oxford, 1973.
19. Legras, R., Bailly, C., Daumerie, M., Dekoninck, J.M., Mercier, J.P., Zichy, V., Nield, E., "Chemical nucleation, a new concept applied to the mechanism of action of organic acid salts on the crystallization of polyethylene terephtalate and bisphenol-A carbonate", Polymer **25**, 1984, pp 835–844.
20. Blundell, D.J., Osborn, B.N., "The morphology of poly(aryl-ether-ether-ketone)", Polymer **24**, 1983, pp 953–958.
21. Lee, Y., Porter, R.S., "Double-melting behaviour of poly(ether ether ketone)", Macromolecules **20**, 1987, pp 1336–1341.
22. Lee, Y., Porter, R.S., "Effects of thermal history on crystallization of poly(ether ether ketone) (PEEK)", Macromolecules **21**, 1988, pp 2770–2776.

23. Nguyen, H.X., Ishida, H., "Molecular analysis of the melting behaviour of poly(aryl-ether-ether-ketone)", Polymer **27**, 1986, pp 1400–1405.
24. Turnbull, D.J., "Kinetics of heterogeneous nucleation", J. Chem. Phys. **18**, 1950, pp 198–203.
25. Jonas, A., Legras, R., "PEEK oligomers: a model for the polymer physical behaviour. 3. Nature of oligomers in the PEEK polymer", submitted.
26. Bellamy, L.J., "The infra-red spectra of complex molecules", 2nd edition, Methuen & Co, London, 1958.
27. Seymour, R.B., "Nonfiller additives for plastics", in "Additives for plastics", Seymour, R.B., editor, Academic Press, New York, 1978.
28. Leblanc, D., "Modification par voie chimique de la cristallisation des polyéthercétones aromatiques", Ph.D. thesis, Université Catholique de Louvain, 1988, in french.
29. Derome, A.E., "Modern NMR Techniques for Chemistry Research", Pergamon Press, Oxford, 1987.

Chapter 3

Assessing the Crystallinity of PEEK

A. Jonas and R. Legras

Abstract

The evaluation of PEEK degree of crystallinity is discussed. A comparison of the results given by Differential Scanning Calorimetry (DSC), Infrared (IR) absorbance spectroscopy and gravitometry, is performed. From this, it appears that DSC is not a convenient routine technique to assess the PEEK crystallinity, because of the recrystallization the polymer experiences during heating scans. On the contrary, IR is shown to be a valuable technique. The 947 cm^{-1} absorption band is proved to be sensitive only to the crystalline content of the PEEK samples, while the 965 cm^{-1} band, previously claimed as indicative of crystallinity, is demonstrated to be only a conformational band one should not use to estimate rigorously the degree of crystallinity of PEEK. A discussion of the variation of the crystalline unit cell density with crystallization temperature, is also given in connection with the gravitometric assessment of PEEK crystallinity. Finally, a critical evaluation is done of the two-phase model on which crystallinity determinations rely. It is illustrated that this model is but a gross approximation of the morphology of PEEK. A three-region model is used to interpret specific heat measurements of PEEK. In this respect, the existence of important amounts of a so-called rigid amorphous fraction, sometimes related in the literature to the existence of interfacial crystalline/amorphous regions, is enlightened. The decrease of the configurational entropy of the interlamellar liquid-like regions, compared to the pure amorphous polymer, is also pointed out, on the basis of the observed glass transition variations.

1 Introduction

The physical properties of semi-crystalline polymers are strongly dependent on their morphology. Since this morphology consists in crystalline lamellae embedded in an amorphous background, it is customary to characterize such polymers by a figure

---

Alain Jonas, Research Assistant of the Belgian National Fund for Scientific Research, Laboratoire de Physique et de Chimie des Hauts Polymères, Université Catholique de Louvain, Place Croix du Sud 1, B-1348 Louvain-la-Neuve, Belgium; Roger Legras, Laboratoire de Physique et de Chimie des Hauts Polymères, Université Catholique de Louvain, Place Croix du Sud 1, B-1348 Louvain-la-Neuve, Belgium

giving the volume or weight proportion of the lamellae to the amorphous phase. This figure is currently called the polymer degree of crystallinity or, simply, the polymer crystallinity.

Crystallinity is an important structural parameter to know, since it plays an unnegligible role in many ultimate material properties. It is thus necessary to determine it unambiguously, in order to better understand and control the properties of PEEK and of its composites. This is the reason of this chapter, included after the presentation of the PEEK chain structure, crystallization and thermal stability, and before the discussion of its main mechanical properties.

In this chapter, Gravitometry, Differential Scanning Calorimetry and Infrared Spectroscopy will be each discussed with respect to the determination of the PEEK degree of crystallinity. There will be however in this chapter no specific presentation of the ability of Wide-Angle X-Ray Scattering to evaluate the PEEK crystallinity. The interested reader will find a clear review of the subject in the book of *Balta-Calleja* and *Vonk* [1], and some applications of the technique to PEEK in recent publications (the Ruland's method has been performed by *Kumar et al.* [2], but most of the published works have used simple intensity ratioing procedures [3, 4]). A brief outline of this topic will be given in a next chapter of this book.

The chapter will be ended with some criticism about the simple two-phase model which is the basis of the crystallinity determination. The mutual coupling of crystalline to amorphous zones will be shown to affect the properties of both regions, and the need to resort to more sophisticated analyses to describe the PEEK crystalline/amorphous structure will be pointed out.

2 Density Determination of the Degree of Crystallinity of PEEK

One of the simplest way to estimate the degree of crystallinity of polymers consists in measuring their specific weight ρ. With the aid of a simple two-phase model, in which lamellae of pure crystalline polymer are considered to be embedded in a pure amorphous matrix without any transition layer in-between, the degree of crystallinity by volume, hereafter denoted as ϕ, and the degree of crystallinity by weight (ψ) are easily derived:

$$\rho = \phi\rho_c + (1-\phi)\rho_a \text{ and } \psi = \phi\rho_c/\rho. \tag{1}$$

These simple formulas show that one requires the knowledge of the specific weight of both the pure amorphous phase ρ_a and the pure crystalline phase ρ_c, in order to evaluate the crystallinity ratios.

Since PEEK can be quenched from the molten state to obtain fully amorphous plates, ρ_a is an easily measurable quantity, whose value is reported to be 1.263 g/cm^3 [3, 5] or 1.264 g/cm^3 [4, 6][1]. In this paper, the first value will be used throughout.

---
[1] A somewhat smaller value (1.2604 g/cm^3) has been measured by *Deslandes et al.* [10].

The specific weight of the pure crystalline phase is usually obtained through evaluation of the crystalline unit cell volume and content by using Wide-Angle X-Ray Scattering (WAXS). A description of the PEEK unit cell will be given in chapter 4 of this book. Various values can be found for ρ_c in the literature : they are ranging from 1.341 to 1.415 g/cm^3 [7]. Such important discrepancies cannot be explained by statistical uncertainties only. They reflect a true physical phenomenon. It has indeed be pointed out by various authors that the PEEK lattice parameters are strongly affected by the crystallization conditions, and particularly by the crystallization temperature T_c [8–10]. As this temperature decreases, the packing of the polymer chains into the crystalline lattice becomes less dense (i.e. a and b increases), without much effect on the polymer fiber length (c axis). Especially the a lattice parameter seems to be sensitive to T_c. The overall effect is an increase of ρ_c with the crystallization temperature. This is illustrated in Fig. 1, where crystalline specific weight (derived from unit cell dimension) is plotted versus crystallization temperature. These results are taken from the works of the previously quoted workers.

One clearly observes a quasi-linear increases of ρ_c with T_c, although the absolute values differ somehow from worker to worker. This has probably to be related to various experimental errors more or less taken into account by each research group (transparency error, divergence of primary beam, specimen displacement from the diffractometer axis, 2θ-calibration,...). In this respect, the most careful analysis has been performed by *Hay et al.* [8], who investigated the effect of these errors on the determined PEEK lattice parameters.

Such an increase of ρ_c with crystallization temperature has already been observed for polyethylene by *Davis et al.* [11], although the variations were relatively less important. This increase was attributed mostly to the effect of chain folds at the lamellae surface; actually, a correlation was found between the variation of the long period and the changes in the lattice parameters. However, it was argued by *Hay et*

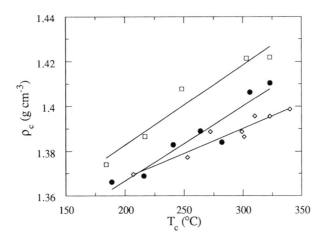

Fig. 3.1 Evolution of the unit cell specific weight with crystallization temperature (□ : from [8], ● : from [9], ◇ : from [10]).

al. [8] that for PEEK the lattice parameters variation could not be due to lamellae surface effects, since PEK, which is very similar to PEEK in many respects, did not show any evidence of such a variation. One must observe however that this last assertion is far from obvious. Indeed, it is based upon measurements on two PEK samples only. And, more importantly, *Hay et al.* consider in this particular case a 2% increase in b_{PEK} (for a 100°C increase of the crystallization temperature) as non significant, although their reported greatest variation of the a axis in PEEK amounts only to 1.9%, and to 0.9% for the b axis. Complementary information is thus required before rejecting the role of the lamellae surface in the PEEK unit cell dimensions variation. Note that such a coupling between lamellae surface and crystalline density would indicate that the used simple two-phase model is only a very rough approximation of reality.

Another hypothesis was put forward by *Abraham et al.* [12]. These authors suggested on the basis of a computer molecular modelling of the PEEK crystal, that the variations in the unit cell dimensions with T_c could be due to improvements in the PEEK chain alignment, i.e. the relative location of ketone and ether bridges of neighbouring chains in the crystal. These bridges are known to be crystallographically equivalent; but it was shown that a more stable crystal is obtained if the chains adopt eclipsed configurations in (2 0 0) planes (each ketone bridge having two other ketone groups on each side along b), while two successive (2 0 0) planes are staggered relatively to each other of one phenyl group in the c-direction (each ketone bridge in the center of the cell being thus surrounded by four ether bridges at the corners of the cell). Starting from a low temperature-crystallized material, having a disordered chain alignement, an increase of the crystallization temperature would result in increased formation of the more stable structure, the stabilization energy manifesting itself in a closer chain packing, and thus in an increase of ρ_c.

However, studies performed on monodisperse PEEK oligomers [13, 14] have revealed that variations of the relative alignement of the chains can at best explain half the maximal variation of a. It is thus very improbable that changes in the chain alignement could alone explain the variation of the unit cell dimensions of the polymer.

Whatever the exact physical reason of this phenomenon, it brings an important practical consequence: the degree of crystallinity of PEEK apparently cannot be determined with precision by densitometry without performing in parallel a WAXS evaluation of ρ_c. This is a somehow heavy procedure to follow. Moreover, the possible presence of carbon fibres renders the task quite more complicated, due to the strong carbon fibre diffraction which has to be subtracted first [15–17]. Finally, if one has nevertheless to make use of WAXS to estimate ρ_c, it seems then more logical to evaluate directly the degree of crystallinity by using an absolute method like the Ruland's one.

However, densitometry is still able to provide an evaluation of the PEEK crystallinity, under certain cautions. One can convince himself of this point by considering the early work of *Blundell and Osborn* [3], who plotted the specific volume of 15 PEEK samples versus a WAXS crystallinity index, calculated by ratioing the area of the crystalline peaks to the area of the amorphous halo between $2\theta = 10°$ and $36°$

(Cu Kα radiation). The crystallinity index varied between 0 and 0.45. A good linear correlation was found between the specific volume and the crystallinity index. This relation extrapolates to a crystalline specific weight ρ_c of 1.401 g/cm³. The authors concluded from the obtained linearity that a simple two-phase model is consistent both with density and WAXS results. This observation shows that selecting a value of 1.401 g/cm³ for ρ_c leads to a correct densitometric evaluation of crystallinity, i.e. giving results consistent with a simple WAXS determination.

This can be explained on the basis of the previously quoted results from *Hay et al.* [8], concerning the variation of ρ_c with T_c. The value 1.401 g/cm³ for ρ_c corresponds to an interpolated crystallization temperature of 250°C. This temperature is situated roughly in the center of the usual PEEK crystallization temperature interval, which grossly runs from 170°C to 320°C. Since the crystalline specific weight evolves almost linearly with crystallization temperature (see Fig. 1), the 1.401 g/cm³ value can thus be thought as a 'kind of mean on a lot of samples crystallized in this temperature interval. With the aid of the Hay's results, it is actually easy to show that by using the value 1.4 g/cm³ instead of the true value of ρ_c, one makes on the degree of crystallinity an error less than 10% (relative) for samples crystallized between 210°C and 290°C. As the crystallization temperature decreases lower than 210°C, the underestimation of the degree of crystallinity increases further, while its overestimation increases for temperatures higher than 290°C.

Thus, for most of the samples dynamically crystallized from the molten state, the densitometric evaluation of the crystallinity must be nearly correct when using this ρ_c value. Indeed, the PEEK crystallization kinetics are not appreciably rapid above ~300°C, and the dynamical conditions will promote a distribution in ρ_c, so that the average 1.4 g/cm³ value is acceptable. For isothermally crystallized samples, the error can be more important. However, for samples crystallized between 200°C and 300°C this error will be less than 3% (absolute) on the degree of crystallinity. This is certainly tolerable, if one reminds that the two-phase model itself is probably but a gross approximation of the reality. For other samples, care must be taken when computing the crystallinity degree from specific weight, as higher errors could be possible.

3 Differential Scanning Calorimetry Determination of the PEEK Degree of Crystallinity

Differential Scanning calorimetry (DSC) is one of the most frequently used techniques to determine polymer crystallinity. In its simplest form, it relies on the assumption that the melting enthalpy (ΔH_m) observed around the melting temperature T_m when heating a sample, is directly connected to the amount of crystalline matter in that sample. If one knows the melting enthalpy of a 100% crystalline sample ($\Delta H_{m,100\%}$) (at the same temperature T_m), the degree of crystallinity by weight is found to be:

$$\psi = \Delta H_m / \Delta H_{m,100\%}. \qquad (2)$$

Very often, one uses the thermodynamic melting enthalpy (ΔH_m^0) of the equilibrium crystal (extended chains infinite crystal) at the equilibrium melting temperature T_m^0 instead of $\Delta H_{m,100\%}$. Since the polymer samples melt at temperatures lower than T_m^0, this introduces an error due to differences between specific heats of the crystalline and liquid phases. Two values for the PEEK thermodynamic melting enthalpy can be found in the litterature, namely 130 J/g [3] and 161 ± 20 J/g [18]. The first value, which is recommended by *Wunderlich et al.* [19], has been obtained from an extrapolation between the melting enthalpy of various PEEK samples and crystallinity (determined by WAXS). As such, it actually represents $\Delta H_{m,100\%}$, at a temperature close to the usual PEEK melting temperature. The second one has been derived from the PEEK equation of state, but seems to be less precise. In this paper, we will use the first value throughout. However, since the precision of this value is unknown, the precision on the crystallinity derived from DSC cannot *a priori* be expected to be higher than ~ 10% (relative).

The main problem of the DSC determination of crystallinity is to evaluate correctly ΔH_m. Polymers indeed usually organize themselves into nonequilibrium structures [19], which are subjected to reorganization during DSC heating scans. Examples of such reorganization are crystallization on heating from the amorphous state, and lamellar thickening or perfection of already crystallized polymers. This results in an ultimate melting enthalpy higher than predicted on the basis of the initial crystallinity, the excess crystallization enthalpy being sometimes spread over all the DSC trace, and being thus mostly undetectable at first sight. This has been shown for instance to be the case for poly(ethylene terephtalate) (PET) [20], which presents on heating in the DSC a double-melting behavior. The first (small) melting endotherm has been attributed to the initial trace of melting of the *thin* lamellae formed during the previous isothermal crystallization. This melting is almost immediately followed by recrystallization of the polymer, since the crystallization kinetics of the polymer is high at this temperature, leading to more stable and thicker lamellae. The corresponding crystallization exotherm masks most of the concurrent melting endotherm; this explains that only a small initial endothermic peak is detected. Then, the subsequent increase in temperature during the scan causes continuous melting and recrystallization of the previously formed lamellae which become unstable at higher temperatures, leading continuously to the formation of thicker lamellae. The net energy balance is small, and no peak can be observed on the DSC trace. This process ends up when the temperature is too high to allow a sufficiently fast recrystallization, and one observes finally the strong melting endotherm of the thicker lamellae resulting from the scan. A similar process has been suggested to occur in PEEK [5, 21], which exhibits also a double-melting behaviour. A more sophisticated view has been offered by other authors [22], in which both reorganization during DSC heating scans and initial existence of different types of lamellae having markedly distinct thicknesses has been postulated. Nevertheless, the existence of PEEK reorganization processes during DSC heating scans seems well established.

To illustrate this point, 12 PEEK samples of (weight) degree of crystallinity varying between 0 and 42.4% have been prepared [23]. The crystallinity was determined gravimetrically, using the above recommended 1.4 g/cm³ value for ρ_c. Some samples were isothermally crystallized from the amorphous glassy state at T_c during t_c; others were dynamically crystallized from the molten state by cooling them as fast as possible to a given crystallization temperature T_c, where they were kept during t_c. All samples were quenched in cold water at the end of the crystallization time. The sample thickness was small, around 200 µm, to obviate thermal conduction problems. The DSC thermograms of these samples are shown in Fig. 2. (*experimental conditions: Perkin-Elmer DSC2, ca. 7 mg sample, scan rate: 10°C/min, nitrogen gas flow, temperature calibration: Zinc and Indium; the energy ordinate was calibrated using the melting enthalpy of Indium; the instrument baseline was compensated for slope and balance (curvature); a baseline was recorded prior to each run with empty pans, and then subtracted from the following sample run performed using the same pans*). Table 1 contains all relevant information about these samples.

The net melting enthalpy of each sample was then simply evaluated by subtracting from the total area of the final melting endotherm the area of the crystallization exotherm, if any. When a small low temperature melting endotherm was observed, it was also added to the total. This is not however considered to be very important, since the area of this endotherm is always very small. The peak integration was performed by tracing a somehow subjective linear baseline under the peak in the way indicated in Fig. 3. With this procedure, account is taken only of the clear-cut phase transitions, but not of continuous reorganization processes.

Results are given in Fig. 4, where the weight degree of crystallinity calculated from this net melting enthalpy (ψ_{DSC}) are plotted versus the weight degree of crystallinity obtained from densitometry.

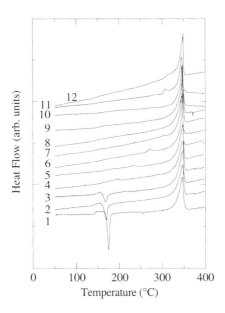

Fig. 3.2 DSC thermograms of the 12 PEEK samples described in Table 1 (10 K/min).

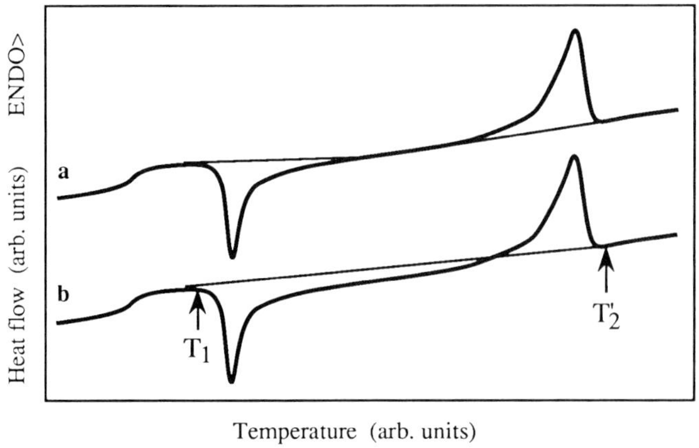

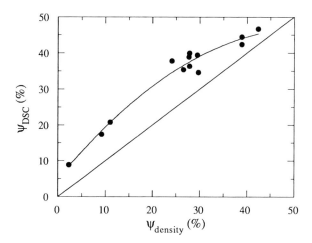

Fig. 3.3 Two tested baseline choices for the assessment of the PEEK crystallinity by DSC. Curve a: simple evaluation of the area of the transition peaks, without taking into account recrystallization during the DSC scans. Curve b: linearly drawn baseline between two temperatures T_1 and T_2'. No universal values of these temperatures could be found in order to match the corresponding DSC-calculated crystallinity with gravitometry results.

Fig. 3.4 PEEK degree of crystallinity, calculated by the DSC technique without taking care of the reorganization processes occurring during the heating scan (ψ_{DSC}) (for the baseline choice: see curve a, Fig. 3), versus the crystallinity determined from gravitometric measurements ($\psi_{density}$). The straight line is the line $\psi_{DSC} = \psi_{density}$.

A clear overestimation of crystallinity by the DSC method is apparent on this figure. The overestimation amounts to $\sim 10\%$ crystallinity, the exact value depending on the initial crystallinity. This results from the reorganization process during the DSC scan; for the most crystalline samples, obtained by crystallization at elevated temperatures, the overestimation is less important, since these samples have more perfect crystalline structures reorganizing less during the scan. It is obvious that this discrepancy cannot be due to a poor evaluation of $\Delta H_{m,100\%}$, nor to errors in ρ_c since the overestimation is by far superior to the $\sim 3\%$ maximal absolute error one expects currently from the use of $\rho_c = 1.4$ g/cm^3 (see the previous section), and since errors in ρ_c would lead both to overestimation and underestimation of the true crystallinity. So a correct DSC evaluation of the PEEK crystallinity must take into account the continuous recrystallization process occurring during the heating scan.

Theoretically, it would be sufficient to integrate all the DSC trace over a correctly drawn baseline, taking into account both heat absorption and production from the initial state to the molten state, to have a correct figure for ΔH_m. This is however hardly possible to perform rigorously, since the baseline has to follow the specific heat of the material at every temperature; this specific heat is in turn a function not only of the crystallinity (changing at every temperature due to reorganization), but also of what has been called 'the rigid amorphous content' of the polymer [24], i.e. amorphous polymer segments more or less immobilized by their proximity to crystallites, and therefore contributing less to the specific heat as the true 'free' amorphous segments. At the present state of affair, prediction of the evolution of the rigid amorphous content of a polymer with temperature is not yet possible. The existence of rigid amorphous fractions in PEEK has been demonstrated on the basis of specific heat measurements[1] [22].

This is the reason why other procedures were developed in the literature [25], based mainly on the computation of sample enthalpy differences between two adequately selected temperatures. Since enthalpy is a state function, it is independent of the path one follows to integrate specific heat. This property can be used to simplify the problem. One of these integrating path choice is particularly interesting since it was shown to give correct results for PET, a polymer reorganizing also during heating scans, as stated above. The procedure, originally described by *Gray* for polyethylene [26], has been discussed by *Blundell et al.* for the PET [27].

Essentially, it consists in drawing a linear baseline between two temperatures T_1 and T_2 *where there is no melting, crystallization or reorganization*. T_1 is chosen after the glass transition temperature, and before the onset of crystallization or reorganization, where the DSC trace corresponds to the specific heat of the initial unreorganized polymer; T_2 is chosen as the temperature where the specific heats of crystalline and amorphous polymers are equal. The whole procedure is thus equivalent to integrate the DSC curve with respect to the baseline one should get without reorganization (constant degree of crystallinity), assuming linear variations of the specific heats with temperature. It was shown theoretically by *Gray* [26] that this procedure gives the

---

[1] This observation is of course not rigorously compatible with the simple two-phase model one used for the evaluation of crystallinity. We shall come later on this point.

correct value for the crystallinity, provided there is no instrumental baseline curvature.

The determination of the PEEK T_2 is possible, since the PEEK amorphous specific heat is experimentally known from 423 to 680K [28], and the crystalline specific heat can be calculated following the addition scheme procedure given by *Cheng et al.* [29]. The total contribution of the group vibrations to specific heat (c_v) was computed from their tabulated single group contribution. The skeletal part of heat capacity (c_v) was derived from the Tarassov equation, using their quoted Tarasov temperatures and values for the Debye functions found in the litterature [30,31]. The computation of c_p from the obtained c_v values was performed through resolution of the Nernst–Lindemann equation, assuming $A_0 = 2.32 \ 10^{-4}$ K mol/J (for one mole of repeating unit) [29] and $T_m^0 = 668$ K. The values of both computed crystalline specific heat and amorphous (liquid) specific heat are presented versus temperature in Fig. 5. The two curves intersect at 599 K. Since this temperature falls for the PEEK in the main melting endotherm, the whole procedure seems to be inadequate for this polymer.

Actually, the true intersection temperature is quite certainly higher, since the Nernst-Lindemann constant A_0 is expected to decrease somehow with temperature for linear polymers [32]. Moreover, the computation of c_v is a semi-empirical procedure, based on experimental measurements performed at temperatures appreciably lower than 600 K; its accuracy could be questioned at high temperatures. One has then attempted to find empirically a temperature T_2' above the PEEK melting peak, for which the procedure would give crystallinities comparable to the crystallinity determined by specific weight. This was done for the samples of Table 1. T_1 was chosen just after the glass transition (which varies following the sample crystallinity and microstructure). An example of such a baseline choice is also given in Fig. 3. (curve b). However, it was impossible to find any universally convenient T_2', so that no further research was carried out in this direction. The failure of such a procedure for PEEK had besides been recognized earlier by *Blundell* [3].

Other rigorous procedures proposed by *Gray* suffer all from the disadvantage that they need a perfectly flat instrumental baseline on a large temperature interval (150–400°C), a condition difficult to fulfill. Attempts to use these procedures on our results were all unsuccessful, probably because of small uncorrected curvature of the instrumental baseline. Also, the fact that the specific heat of the crystalline polymer is not rigorously linear (Fig. 5), is a supplementary source of error.

From the above considerations, one must be convinced that the DSC is not a convenient tool to assess routinely the PEEK degree of crystallinity. Its major drawback lies in the extreme difficulty to trace a correct baseline; a simple baseline cannot be drawn because of reorganization effects during the heating scan. Moreover, the existence of so-called rigid amorphous fractions having unpredicted thermal behaviour prevents to calculate a precise baseline. Methods based on the state function character of enthalpy are not applicable, or requires extreme flatness of the instrument baseline, which renders them not suitable for routine measurements. On the whole, DSC seems not to be an easy crystallinity characterization tool, although it

3. Assessing the Crystallinity of PEEK   93

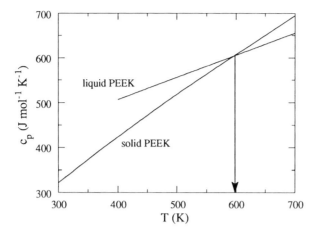

Fig. 3.5 Specific heats versus temperature of crystalline (solid) and amorphous (liquid) PEEK. The crystalline c_p has been computed using a procedure given in reference [29]. The amorphous c_p is taken from [28]. The curves intersect at 599 K.

Table 3.1 Processing Conditions, Specific Weight and Crystallinity, Determined by Specific Weight Measurements or by DSC Using a Simple Peak Area Evaluation (see Text). Samples Isothermally Crystallized from the Amorphous Glassy State at T_c during t_c are Referred to as 'G', While Samples Rapidly Cooled from the Molten State down to T_c, Where They Were Held during t_c, are Referred to as 'M'

Sample number	initial state	T_c (°C)	t_c (min)	ρ (g/cm^3)	$\psi_{density}$ (%)	ψ_{DSC} (%)
1	G	–	–	1.2659	2.3	8.9
2	G	150	30	1.2745	9.2	17.4
3	G	150	30	1.2769	11.1	20.8
4	G	180	30	1.2935	24.1	37.8
5	G	220	30	1.2981	27.7	38.9
6	G	260	30	1.2984	27.8	39.9
7	G	300	30	1.3008	29.7	34.6
8	M	120	15	1.2967	26.6	35.4
9	M	150	30	1.2983	27.8	36.3
10	M	180	30	1.3005	29.5	39.5
11	M	300	15	1.3029	31.3	44.5
12	M	slowly	cooled	1.3177	42.4	46.6

is certainly a sensitive technique if used with great caution. It should be stressed to close this section that some agreement between crystallinities determined by DSC and other methods have sometimes been published in the literature [4, 33]. No baseline choice is specified by the authors; the agreement is thus believed to be coincidental, related to subjective baseline choices.

4 Infrared Absorption Spectroscopy Determination of the Degree of Crystallinity of PEEK

Infrared spectroscopy (IR) is frequently used to determine polymer crystallinity, since it is an easy and widespread technique. It has found two different applications for PEEK. The first one, developed by *Nguyen and Ishida* [34], requires the obtention of the IR absorption spectra of both the purely amorphous polymer, and of the fully crystalline one. This was possible through an interpolation procedure on IR spectra of PEEK films having different degrees of crystallinity. The evaluation of the crystallinity of an unknown sample is then performed by a fitting procedure, assuming the spectrum to be the crystallinity-weighted sum of the two reference spectra. However, the authors state that their procedure overevaluates the true degree of crystallinity, because it is sensitive to all crystalline-like conformers, and not only to true crystalline segments.

The second way to evaluate the crystallinity by IR requires first the identification of at least one absorption band specific to the crystalline phase. In this step, it is essential to find a true crystallinity band, that is an absorption band due to the intermolecular interactions of the polymer chains into the crystalline lattice, or a regularity band, arising from the intramolecular coupling of the oscillations of a regularly aligned chain [35]. Since the interchain interactions are small in polymer crystals, crystallinity bands are ordinarily very weak, and are difficult to detect unambiguously. Regularity bands can be more easily found, since the intrachain coupling is stronger. In this paper, we will call both types of bands 'crystalline bands', since they normally disappear on melting. One should take care however not to confuse these bands with conformational bands, which are connected to the existence of specific polymer conformations, and which are not destroyed upon melting. By ratioing the area of the crystalline band to the area of another band representative from the whole material (a localized group vibration for instance), one get a figure proportional to the weight degree of crystallinity. After performing on well-characterized samples a calibration to obtain the proportionality factor, one is able to compute the crystallinity of any other sample film. This technique is particularly interesting, since, with the advent of IR microscopes, the evolution of crystallinity inside a thin cut slide of a molded specimen could even be followed.

This procedure has been followed for the PEEK by *Chalmers et al.* [36] and *Cebe et al.* [4]. They both used the same reference peak located at 952 cm^{-1}. The first authors found that the height ratio of the 970 cm^{-1} IR peak to the reference peak was a linear function of crystallinity, while the second claimed that the area ratio of the 966 cm^{-1} band to the reference one is a better linear function of crystallinity. Actually, the 966 cm^{-1} peak used by *Cebe* is the same as the 970 cm^{-1} peak used by *Chalmers*, the location difference between them being related to *Chalmers* performing his experiments in reflection instead of transmission.

We recently reexamined these correlations on the 12 samples described in Table 1. [23]. The samples were gently polished by means of fine grain sand paper in order to obtain films of thickness somewhere between 50 to 100 μm. Their IR absorbance spectra were taken from 1030 to 880 cm^{-1} in a Perkin–Elmer dispersive spectrophotometer model 580B, with a resolution of 1.4 cm^{-1} and a relative noise of 0.2. A linear baseline was drawn through 988 and 900 cm^{-1} and subtracted from the spectrum. The resulting spectrum was then decomposed into lorentzian-shaped absorption peaks. Four absorption peaks were found necessary to fit the amorphous sample spectrum between 1030 and 900 cm^{-1}, namely 1011 ± 0.5 cm^{-1}, 967.5 cm^{-1}, 952 ± 0.5 cm^{-1}, and 928 ± 0.5 cm^{-1}. For the semi-crystalline samples, it was necessary to add a fifth peak (previously unmentioned) located at 947 ± 0.5 cm^{-1} in order to correctly fit the recorded spectra which presented a shoulder in the 952 cm^{-1} absorbance peak. The 967.5 cm^{-1} peak location was also found to be sensitive to the degree of crystallinity, since it decreased to 965 cm^{-1} for the highest crystallinities. Apart from the exact peak locations, the fit parameters were the height and half-width of each lorentzian, and a constant shift of the recorded spectra relative to the zero absorbance.

Typical experimental spectra, along with the recalculated spectra, are presented in Fig. 6. One observes clearly on this figure the progressive appearance of the 947 cm^{-1} absorbance peak as crystallinity increases. One can also deduce from this figure that the previously used 965–967.5 cm^{-1} peak is not a crystalline peak, since it is

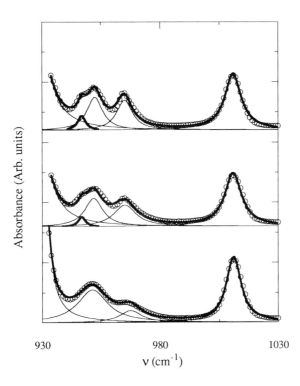

Fig. 3.6 Measured IR absorbance (open symbols), decomposed absorbance peaks and recalculated absorbance (continuous line) for three typical PEEK samples. Lower curve: $\psi = 2.3\%$; mid curve: $\psi = 24.1\%$; upper curve: $\psi = 42.4\%$. The 947 cm^{-1} absorbance peak has been drawn in a thicker line for the sake of clarity.

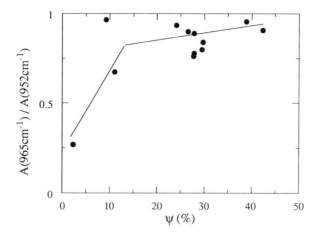

Fig. 3.7 Area ratio of the 965 to the 952 cm^{-1} absorbance bands, as a function of the PEEK weight degree of crystallinity ψ.

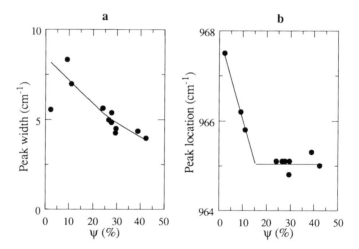

Fig. 3.8 a. Evolution of the half-width of the 965 cm^{-1} peak with the PEEK weight degree of crystallinity ψ. b. Exact wavenumber peak location of the 965 cm^{-1} peak, as a function of the PEEK weight degree of crystallinity ψ.

still present in the amorphous sample. This is further evidenced if one plots the area ratio of this peak to the 952 cm^{-1} reference peak versus ψ_c (Fig. 7). No linear relationship can be observed on this figure. Moreover, the peak half-width (Fig. 8a), and its exact location (Fig. 8b), are strongly dependent on crystallinity.

The previously quoted literature results seem thus not valid when one integrates the absorption peaks after decomposition of the polymer spectrum. However, it is obvious that crystallinity influences in an unnegligible way the intensity of the 965 cm^{-1} peak. One can infer from these observations that this peak must be related to a vibration normal mode associated to a segmental conformation enhanced by the presence of crystallites, i.e. it is a conformational peak.

One could wonder if the 952 cm^{-1} peak is a convenient reference peak. Actually, the same trends can be observed if one makes use of the 1011 cm^{-1} peak as reference peak. We suggested elsewhere [23] that this is due to the fact that both peaks are aromatic C—H vibrations, and may thus equally be chosen as reference since they are proportional to the whole amount of polymer.

Things are getting better if one focusses on the 947 cm^{-1} peak. Firstly, both the peak location and its half-width are independent on crystallinity (Fig. 9). Secondly, the area ratios of this band to the 952 or 1011 cm^{-1} reference bands are linear functions of the degree of crystallinity (Fig. 10), with only a small scattering probably due to the fitting procedure.

Moreover, the linear regression of these ratios versus specific weight converges to an amorphous specific weight comprised between 1.261 and 1.263 g/cm^3, which is in full agreement with the experimental value. These four observations are sufficient to confirm that the 947 cm^{-1} band is a crystalline band (a regularity band or a crystallinity band) which can be used to evaluate crystallinity. Somewhat better results could perhaps be obtained by refining the decomposition procedure.

An examination of the absorption spectra of some highly crystalline PEEK model compounds (i.e. similar to PEEK fragments) and oligomers was undertaken, to check the above assertions. Full description of the synthesis, thermal and structural properties of the PEEK oligomers has been published elsewhere [13, 14]. The spectra were recorded in KBr pellets, so that no attempt was performed to evaluate them in a quantitative way. The frequency positions of the absorption bands of these compounds are presented in Table 2 (between 1020 and 900 cm^{-1}). The 965 cm^{-1} band is observed even for very short PEEK repeating unit fragments; however, the 947 cm^{-1} band can only be detected for the longer oligomers, which have been demonstrated by WAXS to crystallize into a crystallographic lattice similar to the PEEK [14]. This last band is thus actually related to the regular configuration of crystalline PEEK.

However, the 965 cm^{-1} band is associated with a short length segmental conformation, since it is present even for short molecules. As already stated, this conformation is evidently present in the fully amorphous polymer, but its occurrence is increased by the appearance of crystallites, so that one can deduce that this conformation exists abundantly in the PEEK crystalline arrays. This is confirmed by the results obtained on the PEEK oligomers, which are nearly 100% crystalline, and which show clearly this absorption band. The fact that the intensity of this band does not increase linearly with the polymer crystalline content (see Fig. 7), must be due to some

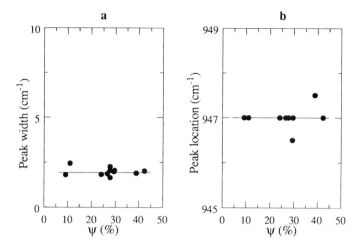

Fig. 3.9 a. Evolution of the half-width of the 947 cm^{-1} peak with the PEEK weight degree of crystallinity ψ. b. Exact wavenumber peak location of the 947 cm^{-1} peak, as a function of the PEEK weight degree of crystallinity ψ. (Same scales as Fig. 8.)

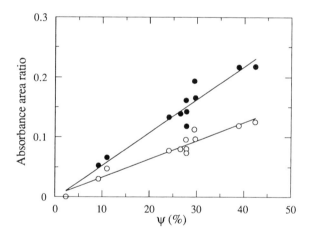

Fig. 3.10 Area ratios of the 947 to the 952 cm^{-1} (●) or 1011 cm^{-1} (○) absorbance bands, as a function of the PEEK weight degree of crystallinity ψ.

Table 3.2 IR Absorption Bands of Some PEEK Model Compounds and Oligomers of High Crystallinity. Symbols between Brackets Represent Qualitative Intensities: s: Strong, m: Medium, w: Weak, vw: Very Weak, sh: Shoulder. (Φ Represents a p-disubstituted Phenyl)

Compound	v_1 (cm^{-1})	v_2 (cm^{-1})	v_3 (cm^{-1})	v_4 (cm^{-1})	v_5 (cm^{-1})	v_6 (cm^{-1})
ΦΟΦΟΦ	1010.5 (m)	976.5(w)	964(w)	954(vw)	–	930(vw)
FΦCOΦF	1011 (w)	–	968(w)	950(w)	–	930(s)
FΦCO(ΦOΦOΦCO)$_1$ΦF	1011 (w)	–	965(w)	952(w)	948(vw)	928(s)
FΦCO(ΦOΦOΦCO)$_2$ΦF	1011 (w)	–	964(w)	951(w)	946(sh)	928(s)

dependence of the probability to find the corresponding conformation in the *amorphous* regions, on the *crystallites* presence and/or nature. It could be for instance due to an increased probability to find this configuration in amorphous zones near the crystallites, i.e. to some degree of coupling between the lamellae surfaces and the neighbouring amorphous phase. With this interpretation in mind, the exact intensity of the 965 cm^{-1} band depends not only on crystallinity but also on other morphological parameters like lamellar thickness, and so on.

In summary, it has been demonstrated that IR spectroscopy is able to evaluate correctly the PEEK degree of crystallinity, provided one performs first a calibration relating crystallinity with the absorbance area ratio of the 947 cm^{-1} band to the 1011 or 952 cm^{-1} reference bands, and provided the spectra are decomposed into lorentzian absorption bands.

5 Limitations of the Two-Phase Crystallinity Concept for PEEK

Each of the crystallinity evaluation techniques described above relies on the validity of a simple two-phase model, in which the properties of the amorphous and crystalline phases are totally decoupled, and which rejects the possibility of existence of an intermediate region joining smoothly the crystalline to the amorphous phase. No doubt this is a simplification of physical reality for every polymer; our present purpose is thus to examine how far from reality this model is in the case of PEEK.

5.1 Influence of Crystallites on Amorphous Regions

As stated above, the existence of so-called rigid amorphous fractions in semi-crystalline PEEK has already been established by *Cheng et al.* [22]. A quantitative evaluation of this fraction is performed by comparing the specific heat jump (Δc_p) measured

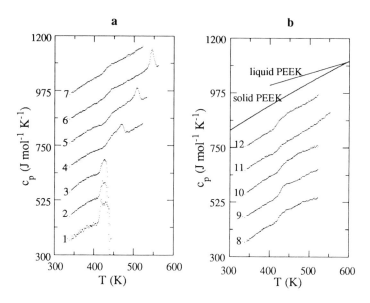

Fig. 3.11 Specific heat of the PEEK samples of this study, versus temperature. Figure a refers to samples isothermally crystallized from the glassy amorphous state, and figure b to samples dynamically crystallized from the molten state. The samples are numbered with reference to Table 1. For clarity, each curve is displaced of 100 J/mol/K upwards from the preceding one. The computed specific heat of solid PEEK, and its liquid c_p, are also shown in fig. b.

at the glass transition (T_g) with the specific heat jump of the pure amorphous sample ($\Delta c_{p,a}$). For an ideal two-phase polymer, the rigid fraction

$$f_r = 1 - \Delta c_p / \Delta c_{p,a} \tag{3}$$

should be equal to the degree of crystallinity of the polymer ψ_c. Some polymers do not obey this law, and show rigid fractions greater than their crystalline fraction [24]. The amorphous rigid fraction, $f_{r,am}$, is the difference between the rigid fraction and the crystallinity by weight, i.e.:

$$f_{r,am} = f_r - \psi_c. \tag{4}$$

The determination of the $f_{r,am}$ of the samples described above has been performed by measuring their specific heat between 50 and 250 or 300°C (*experimental conditions : Perkin-Elmer DSC2, ~15 mg sample, scan rate : 10°C/min, nitrogen gas flow, temperature calibration : Tin and Indium; the energy ordinate was calibrated using the melting enthalpy of Indium; the heat flow of the PEEK samples was compared to the heat flows of the empty pans and of a sapphire reference to obtain specific heat*). The specific heat of the samples is displayed *versus* temperature in Fig. 11. Small hysteresis effects (both negative and positive) are observed for some samples.

The specific heat jump at T_g was calculated as follows. The sample specific heat below T_g was extrapolated upwards through the transition by fitting the low temperature portion of the curve with the calculated solid state specific heat[1], multiplied by an adjustable constant. This scaling constant was comprised between 0.983 and 1.017, and was necessary to take into account small experimental errors (sample weight,...) and the small dependence of the PEEK solid state heat capacity on the rigid fraction, changing from sample to sample [22]. The specific heat above T_g was extrapolated downwards through the transition by fitting it with a linear combination of the computed solid c_p, and of the liquid c_p given by *Cheng et al.* [28]. Typical extrapolations are shown in Fig. 12.

For some samples, the beginning of the reorganization process was far enough from T_g to unambiguously extrapolate the specific heat downwards (samples 1, 6, 7). The end of the glass transition of samples 2 and 3 was found already perturbed by the onset of recrystallization. For the other samples, it was found that the specific heat progressively deviates upwards from the expected (quasi-linear) curve somewhere above T_g (samples 4, 5, 10, 11, 12). This has to be related to melting and/or progressive decrease of the so-called rigid amorphous fraction. In such case, the extrapolated specific heat was drawn through the computed inflexion point (after curve smoothing), as it is supposed to best mark the separation between the end of the glass transition region and the beginning of the reorganization processes. The progressive deviations observed confirm that reorganization is the rule for PEEK during DSC heating scans.

The glass transition was defined as the temperature at which the measured c_p reaches a value corresponding to half the specific heat jump between the 'baselines' at the same temperature (Δc_p). The rigid fraction was obtained by dividing Δc_p with the value $\Delta c_{p,a}$ for the pure amorphous sample, computed at the measured T_g by subtracting the c_p of the liquid polymer and of the solid polymer at this T_g. Results are given in Table 3. No values of $f_{r,am}$ were computed for samples 8 and 9, because their glass transitions look perturbed, probably due to their very low crystallization temperature. Because of uncertainty in 'baselines' choice, together with potential error in the computation of ψ_c and of $\Delta c_{p,a}$ (which is the result of the subtraction of two great numbers), a precision of only $\sim 20\%$ (relative) is expected on $f_{r,am}$; one should thus not be overconfident concerning the absolute figures given in Table 3.

The measured $f_{r,am}$ varies between 0 and 40%. This is of the order of the values found for other polymers (PET: up to 45% [37]; poly(phenyleneoxide): up to 38% [38]). However, this is much higher than the previously published values for PEEK [22], which were ranging from 0 to 14%. To explain this difference, one must compare our crystallization conditions to the conditions used by *Cheng et al.* For PEEK isothermally crystallized from the glass, these authors heated their samples in the DSC furnace during 2 hours, before cooling them at -0.31 K/min to below T_g. The samples of this study were crystallized during 30 min. under pressure (ca. 10 MPa), before being quenched in water. Obviously, our conditions give less chance to the polymer segments to relax, and this is probably the cause of the higher rigid fraction

---

[1] It was found to be represented to better than 0.2%, between T = 300 and 650K, by using:

$$c_{p,sol} (J/mol/K) = 105.8 + 0.853\ T - 3710000\ T^{-2}$$

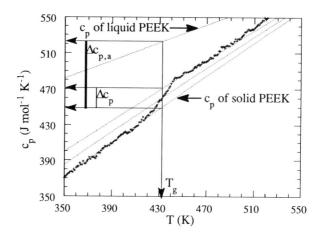

Fig. 3.12 Illustration of the procedure followed to determine T_g and Δc_p at T_g. The sample c_p is extrapolated from above and from below the glass transition using the relations for the solid and liquid c_p of PEEK, as explained in the text. T_g is the temperature where c_p corresponds to half the specific heat jump between the baselines (sample 6 of Table 1).

measured. This explanation is supported by the fact that our sample slowly cooled from the melt has indeed no rigid amorphous fraction.

The quenched sample has a non-zero $f_{r,am}$. This results from overestimation of the $f_{r,am}$ due to hysteresis effects in the glass transition region. The $f_{r,am}$ of samples isothermally crystallized at 180 or 220°C is lower than the $f_{r,am}$ of the samples isothermally crystallized at higher temperatures. This is not considered to be relevant; actually, the low temperature melting endotherm of these samples is very near the glass transition, so that it interferes with it, causing an increase in the measured Δc_p, thus an apparent decrease of the $f_{r,am}$. This explains why the measured $f_{r,am}$ increases as this melting endotherm reaches higher temperatures.

A possible physical meaning for the $f_{r,am}$ is the following. It has been stressed by various authors that a semi-crystalline polymer is comprised of three major regions: the ordered crystalline phase, the disordered, liquid-like, interlamellar phase, and a third interfacial region allowing a progressive match of the configurational differences between the two previous zones [39, 40]. This interface consists in folds (not necessarily sharp nor adjacently reentring), and initial segments of tie molecules bridging the amorphous phase. Obviously, the number of conformations available at a given temperature to the chain segments in this interfacial phase must be quite low, because these segments are tied at least at one side on the crystal surface, and because the steric hindrances between chains are higher than in the less dense liquid phase. Thus the configurational entropy of this zone is much lower than that of the pure liquid phase, at the same temperature. One could talk of an 'immobilization' or 'rigidification' of the amorphous segments in this zone. From this rigidification, two effects result:

The specific heat of the interface zone is much nearer to the solid state c_p than to the liquid state c_p.

- The glass transition temperature of this zone will be much higher than the T_g of the pure liquid phase [41]. In the light of the Gibbs–Di Marzio thermodynamic theory of the glass transition [42], this results simply from the lower configurational entropy of the interfacial region. In simple words, the molecular segments of the interfacial region will request more energy to pass from a conformation to another because many conformations are energetically disfavoured. The onset of these motions in phase space, which characterizes the glass transition, will then occur only at a higher temperature.

The $f_{r,am}$ measured by DSC can thus be partly connected to the interfacial region. It is strongly dependent on the crystallization conditions, since it is associated with the lamellar interface. If enough time is given to the polymer to perfection its interfacial region, i.e. reducing the constraints in it, one expects a decrease of the $f_{r,am}$. This is conform to the above comparison of our measured $f_{r,am}$ with results previously published. Upon heating the polymer above the T_g of the interlamellar regions, modification in the interfacial regions can occur, changing the $f_{r,am}$. Actually, one observes a progressive 'unfreezing' of the $f_{r,am}$ for some polymers [43], in which case one can think of an extremely enlarged glass transition, or no softening at all before the melting temperature for other polymers [44]. For PEEK, we observed an important reorganization somewhere above the T_g of the interlamellar regions; this corresponds also certainly to $f_{r,am}$ changes.

The interfacial thickness for polyethylene was estimated to be 1–3 nm [45]. The $f_{r,am}$ was suggested elsewhere to correspond to nanometer scale micromorphological effects [38]. From the expected lamellar thickness for our isothermally crystallized

Table 3.3 Glass Transition, Specific Heat Jump at T_g, and Rigid Fractions of the Samples of This Study. Full Description of the Samples is Given in Table 1. (d) is Indicated after T_c for Samples Dynamically Crystallized. Italic Characters Have Been Used to Indicate Samples for Which the Upper Baseline Choice Was not Unambiguous

Sample	$T_c(°C)$	$T_g(°C)$	$\Delta c_p(J\ mol^{-1}\ K^{-1})$	$\psi_c(\%)$	$f_r(\%)$	$f_{r,am}(\%)$
1	quenched	141.3	63.3	2.3	17	15
2	150	142.8	45.5	9.2	40	*31*
3	150	144.9	38.3	11.1	49	*37*
4	180	166.5	30.9	24.1	52	*28*
5	220	166.9	24.4	27.7	62	*35*
6	260	158.6	20.8	27.8	70	42
7	300	154.3	21.0	29.7	70	40
8	120 (d)	150.8	29.7	26.6	–	–
9	150 (d)	153.8	35.8	27.8	–	–
10	180 (d)	156.4	28.5	29.5	59	*29*
11	300 (d)	151.3	18.1	31.3	75	*43*
12	slow cool.	149.2	43.9	42.4	39	−3

samples[1], a value between 1 and 2.5 nm could apparently be predicted for the interfacial thickness of these samples, if the $f_{r,am}$ is equated with it. These are not unreasonably high values, if one recalls that the distance between two phenyls is of the order of 0.5 nm in PEEK.

We turn now to the interlamellar regions, i.e. the regions which are responsible of the specific heat increase at T_g. Although these regions have a liquid-like character, they are not fully identical to the pure amorphous phase. This is best seen by noting that the T_g of these regions is higher than that of the fully amorphous polymer (Table 3). The difference amounts up to 25°C. The Gibbs–Di Marzio glass transition theory tells us that this corresponds to a reduction of the configurational entropy S_c of the interlamellar amorphous zones in the semi-crystalline polymer. Such an S_c reduction has been precisely postulated by Robellin-Souffaché and Rault on the basis of the dependence of the long period of polymers crystallized from the amorphous state on the radius (r) of the polymer coils in the liquid state [46]. They suggested that the polymer S_c begins to decrease between two growing lamellae when their mutual distance is of the order of r. This entropy decrease was related to the squeezing of the amorphous tie molecules during the crystalline growth, and to the increase of the entanglement density in the interlamellar regions, due to the existence of tie molecules. From considerations on the free energy balance between the amorphous regions and the growing crystalline lamellae, one can expect that the S_c reduction in the amorphous phase is larger for polymers crystallized at lower temperatures. Thus, the glass transition of the semi-crystalline polymer should be higher when isothermally crystallized at lower temperature. This is exactly the trend observed in Table 3, for samples 4, 5, 6, 7 (other samples are not fully-crystallized, or dynamically crystallized from the molten state).

The interlamellar regions are thus different from the unconfined amorphous polymer, with respect to S_c. This is furthermore expected from theoretical studies on the free energy of polymers confined between two surfaces on which they are attached [47]. If the increase of S_c with temperature is smaller for the confined amorphous regions than for the pure amorphous liquid, the c_p of the amorphous interlamellar regions will be smaller than the c_p of the liquid polymer. The heat capacity jump at T_g will be also smaller. A fraction of the $f_{r,am}$ will then be due to this effect. This means that the $f_{r,am}$ quantifies simultaneously the amount of rigidified matter (the interfacial region), and the entropic modification of the interlamellar regions. The distinction between interfacial and interlamellar regions on the basis of $f_{r,am}$ measurements is thus somewhat artificial.

From the above considerations, the PEEK amorphous regions are seen to be rigidified by the crystallites presence while a configurational entropy decrease is detected in liquid-like interlamellar regions through the increase of the glass transition temperature of these regions. Properties different from the unconfined polymer

---

[1] From the experimental independence of lamellar thickness, l_c, on molecular weight [46], and using data reported in the litterature [3,4,5], an approximate relationship between l_c and crystallization temperature T_c (°C) can be obtained between 180 and 310°C:

$$l_c \text{ (nm)} = -105.8 + 1.739\, T_c - 0.008067\, T_c^2 + 1.302 \cdot 10^{-5}\, T_c^3 \quad (\pm 10\%)$$

glass must exist in the amorphous regions. An illustration of this point was given above, when it was suggested that the conformation giving rise to the 965 cm^{-1} IR absorption band would be favored in amorphous zones near the crystallites[1]. The simple two-phase model finds here its limit; it is but a rough picture of the semi-crystalline PEEK.

5.2 Influence of the Amorphous Regions on the Crystalline Phase

Having determined that the crystalline phase exerts a deep influence on the amorphous regions, it sounds logical to think that the crystalline phase must be somehow influenced in turn by the amorphous regions. Such a coupling could explain the variation of ρ_c with crystallization temperature. We discussed in a previous section of this chapter the two present interpretations of this phenomenon, and we mentioned experimental indications showing that the second interpretation is not able alone to explain the observed density variations. The variation of ρ_c with crystallization temperature would then be due in part to some influence of the lamellar surface on the chain packing into the crystalline lamellae. With increasing crystallization temperature, one should get thicker lamellae, and thus the crystalline packing would be less affected by the coupling of the crystallites with the amorphous zones; the overall result would consist in an increase of the crystalline density. Although this is but an hypothesis at the while, it is conforted by the existence of the reciprocal coupling (crystalline→amorphous). Complementary experimental results in this particular field would be clearly of great scientific interest.

5.3 Practical Conclusions

From the preceding discussions, the simple two-phase model appears to be a very rough approximation of the true physical state of semi-crystalline PEEK. It is thus worth to question the pertinence of the degree of crystallinity one extracts from this model, i.e. its adequacy to describe the evolution of the physical properties of variously crystallized PEEK samples.

For properties like the (short times) elasticity modulus below the glass transition, or probably thermal conductivity (between ~100K and the glass transition), the degree of crystallinity will be a parameter sufficient to predict the value of the property. This can be observed in Fig. 13, where the real part of the traction modulus measured at 60°C is plotted versus crystallinity for some of the film samples of this study (*experimental conditions: Rheometrics Solid Analyzer in the tension mode, static stress around 1.5 MPa, deformation frequency: 1 Hz, oscillating strain around 0.02%, sample thickness of the order of 200 μm, sample width of ca. 3 mm, sample length: 22 mm*). A satisfactory correlation is obtained, the scatter resulting mainly from uncertainties in the sample geometry.

---

[1] However, no correlation was found between this absorption band and the total rigid fraction, or the so-called rigid amorphous fraction.

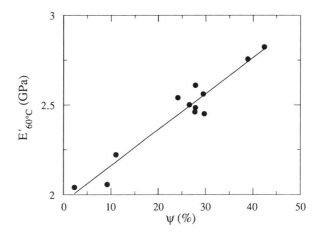

Fig. 3.13 Elasticity modulus, E', of the PEEK samples versus their weight degree of crystallinity (ψ). Measurement temperature: 60°C.

However, for most properties which depends more or less on motions of the chains in the amorphous regions, like for instance the elasticity modulus above T_g or its value below T_g for long times, the impact resistance, the elongation at break, or the specific heat above T_g, one has to expect a marked influence of the so-called rigid amorphous fraction on them, so that a proper prediction of such properties must rely on a more detailed morphological characterization (lamellar thicknesses distribution, width of the transition layer between crystalline lamellae and amorphous regions,...). An illustration of this point can be found in the work of *Cebe et al.* [48], who observed that the degree of crystallinity is not as important as processing history in determining the room temperature mechanical properties of PEEK. Nevertheless, crystallinity allows to have a gross appraisal of the general physical state of the material.

Conclusion

With the notable exception of WAXS, several common techniques to determine the PEEK degree of crystallinity were discussed. From these, the DSC was discarded for routine determinations, since recrystallization of PEEK during heating scans prevents to use simple peak integration techniques, while more sophisticated procedures were found not applicable for theoretical reasons, or very demanding with regard to experimental conditions (perfect baseline flatness). Infrared spectroscopy was shown however to be a valuable tool in this respect, but it needs first a calibration to be performed. It was shown that the 947 cm^{-1} absorbance peak is a regularity or a crystallinity band, sensitive to crystalline fractions only, while the previously used 965 cm^{-1} band is but a conformational band, sensitive to a conformation present not only in crystalline zones, but also in amorphous ones. Gravitometry was also discussed. Its major drawback is the variation of the 100% crystalline PEEK density with crystallization temperature. This problem can be partly circumvented by the choice of $\rho_c = 1.4$ g/cm^3, although some caution has to be taken for samples crystallized at low (<200°C) or high temperatures (>300°C).

The validity of the two-phase simple model, on which all crystallinity determinations are resting, was investigated. Noticeable amounts of a so-called rigid amorphous fraction were shown to sometimes exist in PEEK. Moreover, the thermodynamic properties of the liquid-like interlamellar region are different from the pure amorphous polymer, as examplified by its higher glass transition. Also the variation of ρ_c with crystallization temperature was suggested to be in part the result of a strong coupling between the crystalline and the amorphous regions, whose properties are mutually affected by the presence of each other. As a consequence, the degree of crystallinity can be used only as a rough indicator of the polymer structure, and more sophisticated morphological investigations are required if one tries to understand correctly the relationship between the PEEK structure and its properties.

Acknowledgments

We are indebted to the Belgian National Science Foundation for its financial support of this work. The authors wish also to express their gratitude to M. J.-J. Biebuyck, for its contribution in the critical evaluation of crystallinity by DSC.

References

1. Balta-Calleja, F.J., Vonk, C.G., "X-Ray scattering of synthetic polymers", Elsevier, Amsterdam, 1989.
2. Kumar, S., Anderson, D.P., Wade Adams, W., "Crystallization and morphology of poly(aryl-ether-ether-ketone)", Polymer **27**, 1986, pp 329–336.
3. Blundell, D.J., Osborn, B.N., "The morphology of poly(aryl-ether-ether-ketone)", Polymer **24**, 1983, pp 953–958.
4. Cebe, P., Chung, S.Y., Hong, S.-D., "Effect of thermal history on mechanical properties of polyetheretherketone below the glass transition temperature", J. Appl. Polym. Sci. **33**, 1987, pp 487–503.
5. Blundell, D.J., "On the interpretation of multiple melting peaks in poly(ether ether ketone)", Polymer **28**, 1987, pp 2248–2251.
6. Cebe, P., "Annealing study of poly(etheretherketone)", J. Mater. Sci. **23**, 1988, pp 3721–3731.
7. Nguyen, H.X., Ishida, H.,"Poly(aryl-ether-ether-ketone) and its advanced composites: a review", Polym. Comp. **8**, 1987, pp 57–73.
8. Hay, J.N., Langford, J.I., Lloyd, J.R., "Variation in unit cell parameters of aromatic polymers with crystallization temperature", Polymer **30**, 1989, pp 489–493.
9. Wakelyn, N.T., "Variation of the unit cell parameters of PEEK film with annealing temperature", J. Polym. Sci. C: Polym. Lett. **25**, 1987, pp 25–28.

10. Deslandes, Y., Alva Rosa, E., "Characterization of PEEK films by microindentation", Polym. Comm. **31**, 1990, pp 269–272.
11. Davis, G.T., Eby, R.K., Martin, G.M., "Variations of the unit-cell dimensions of polyethylene: effect of crystallization conditions, annealing, deformation", J. Appl. Phys. **39**, 1968, pp 4973–4981.
12. Abraham, R.J., Haworth, I.S., "Molecular modelling of poly(aryl ether ether ketone): 2. Chain packing in crystalline PEK and PEEK", Polymer **32**, 1991, pp 121–126.
13. Jonas, A., Legras, R., Devaux, J., "PEEK oligomers: a model for the polymer physical behaviour. 1. Synthesis and characterization of linear PEEK monodisperse oligomers", submitted.
   An account of this work has been given at the IUPAC International Symposium on New Polymers, Kyoto, Japan, Nov. 30–Dec. 1, 1991.
14. Jonas, A., Legras, R., Scherrenberg, R., Reynaers, H., "PEEK oligomers: a model for the polymer physical behaviour. 2. Structure and thermal behaviour of linear monodisperse PEEK oligomers", submitted.
   An account of this work has been given at the 2nd Pacific Polymer Conference, Otsu, Japan, Nov.26-29, 1991.
15. Small, R.W.H., "A method for direct assessment of the crystalline/amorphous ratio of PEEK in carbon fibre composites by wide angle X-Ray diffraction", Eur. Polym. J. **22**, 1986, pp 699–701.
16. Wakelyn, N.T., "Resolution of wide-angle X-Ray scattering from a thermoplastic composite", J. Polym. Sci. A : Polym. Chem. **24**, 1986, pp 2101–2105.
17. Cebe, P., Lowry, L., Chung, S.Y., Yavrouian, A., Gupta, A., "WAXS study of heat-treated PEEK and PEEK composite", J. Appl. Polym. Sci. **34**, 1987, pp 2273–2283.
18. Zoller, P., Kehl, T.A., Starkweather, H.W.Jr., Jones, G.A., "The equation of state and heat of fusion of poly(ether ether ketone)", J. Polym. Sci. B: Polym. Phys. **27**, 1989, pp 993–1007.
19. Wunderlich, B., Cheng, S.Z.D., Loufakis, K., in "Encyclopedia of polymer science and engineering", Vol. **16**, J. Wiley & Sons, New York, 1989, pp 767–807.
20. Holdsworth, P.J., Turner-Jones, A., "The melting behaviour of heat crystallized poly(ethylene terephtalate)", Polymer **12**, 1971, pp 195–208.
21. Lee, Y., Porter, R.S., "Double-melting behaviour of poly(ether ether ketone)", Macromolecules **20**, 1987, pp 1336–1341.
22. Cheng, S.Z.D., Cao, M.-Y., Wunderlich, B., "Glass transition and melting behavior of poly(oxy-1,4-phenyleneoxy-1,4-phenylenecarbonyl-1,4-phenylene)", Macromolecules **19**, 1986, pp 1868–1876.
23. Jonas, A., Legras, R., Issi, J.-P., "DSC and infrared crystallinity determinations of poly(aryl-ether-ether-ketone)", Polymer, Polymer **32**, 1991, pp 3365–3370.
24. See for instance Cheng, S.Z.D., "Thermal characterization of macromolecules", J. Appl. Pol. Sci.: Appl. Polym. Symp. **43**, 1989, pp 315–371, *and the references therein*.
25. For a list of baseline choices, see the review of Runt, J.P., in "Encyclopedia of polymer science and engineering", Vol. **4**, J. Wiley & Sons, New York, 1986, pp 482–519.
26. Gray, A.P., "Polymer crystallinity determinations by DSC", Thermochimica Acta 1, 1970, pp 563–579.
27. Blundell, D.J., Beckett, D.R., Willcocks, P.H., "Routine crystallinity measurements of polymers by d.s.c.", Polymer **22**, 1981, pp 704–707.
28. Cheng, S.Z.D., Wunderlich, B., "Heat capacities and entropies of liquid, high-melting-points polymers containing phenylene groups (PEEK, PC, and PET)", J. Polym. Sci.: Part B: Pol. Phys. **24**, 1986, pp 1755–1765.
29. Cheng, S.Z.D., Lim, S., Judovits, L.H., Wunderlich, B., "Heat capacities of high melting polymers containing phenylene groups", Polymer **28**, 1987, pp 10–22.
30. Beattie, J.A., "Six place tables of the Debye energy and specific heat functions", J. Math. Phys. (M.I.T.) **6**, 1926, pp 1–32.
31. Wunderlich, B., "Motions in polyethylene. II. Vibrations in crystalline polyethylene", J. Chem. Phys. **37**, 1962, pp 1207–1216.

32. Grebowicz, J., Wunderlich, B., "On the c_r to c_p conversion for solid linear macromolecules", J. Therm. Anal. **30**, 1985, pp 229–236.
33. Manson, J.-A.E., Seferis, J.C., "Void characterization technique for advanced semicrystalline thermoplastic composites", Sci. & Eng. Comp. Mat. **1**, 1989, pp 75–84.
34. Nguyen, H.X., Ishida, H., "Molecular analysis of the melting behaviour of poly(aryl-ether-ether-ketone)", Polymer **27**, 1986, pp 1400–1405.
35. Zerbi, G., Ciampelli, F., Zamboni, V., "Classification of crystallinity bands in the infrared spectra of polymers", J. Pol. Sci.: Part C: Polym. Symp. **7**, 1964, pp 141–151.
36. Chalmers, J.M., Gaskin, W.F., Mackenzie, M.W., "Crystallinity in poly(aryl-ether-ketone) plaques studied by multiple internal reflection spectroscopy", Polymer Bull. **11**, 1984, pp 433–435.
37. Schick, C., Donth, E., "Characteristic length of glass transition : experimental evidence", Phys. Scr. **43**, 1991, pp 423–429.
38. Cheng, S.Z.D., Wunderlich, B., "Glass transition and melting behavior of poly(oxy-2,6-dimethyl-1,4-phenylene)", Macromolecules. **20**, 1987, pp 1630–1637.
39. Mandelkern, L., "Relation between properties and molecular morphology of semi-crystalline polymers", Farad. Disc. Chem. Soc. **68**, 1979, pp 310–319.
40. Mandelkern L., Alamo, R.G., Kennedy, M.A., "Interphase thickness of linear polyethylene", Macromolecules. **23**, 1990, pp 4721–4723, *and the references therein.*
41. Struik, L.C.E., "Physical aging in amorphous polymers and other materials", Elsevier, Amsterdam, 1978; see also Struik, L.C.E., "The mechanical and physical ageing of semicrystalline polymer", Polymer **28**, 1987, pp 1521–1533; *ibid.* **28**, 1987, pp 1534–1542; *ibid.* **30**, 1989, pp 799–814; *ibid.* **30**, 1989, pp 815–830.
42. Gibbs, J.H., Di Marzio, E.A., "Nature of the glass transition and the glassy state", J. Chem. Phys. **28**, 1958, pp 373–383.
    For a brief review, see Di Marzio, E.A., "Equilibrium theory of glasses", in "Structure and mobility in molecular and atomic glasses", ed.: O'Reilly, J.M., Goldstein, M., Annals of the New York Academy of Sciences 371, New York, 1981.
43. Grebowicz, J., Lau, S.-F., Wunderlich, B., "The thermal properties of polypropylene", J. Polym. Sci. : Polym. Symp. **71**, 1984, pp 19–37.
44. Suzuki, H., Grebowicz, J., Wunderlich, B., "Heat capacity of semi-crystalline, linear poly(oxymethylene) and poly(oxyethylene)", Makromol. Chem. **186**, 1985, pp 1109–1119.
45. Kumar, S.K., Yoon, D.Y., "Lattice model for crystal-amorphous interphases in lamellar semicrystalline polymers: effects of tight-fold energy and chain incidence density", Macromolecules. **22**, 1989, pp 3458–3465.
46. Robelin-Souffaché, E., Rault, J., "Origin of the long-period and crystallinity in quenched semi-crystalline polymers. 1.", Macromolecules. **22**, 1989, pp 3581–3594.
47. Muthukumar, M., Jyh-Shyong Ho, "Self-consistent field theory of surfaces with terminally attached chains", Macromolecules. **22**, 1989, pp 965–973.
48. Cebe, P., Hong, S.-D., Chung, S., Gupta, A., "Mechanical properties and morphology of poly(etheretherketone)", in "Toughened composites", ASTM STP 937, Johnston, N.J. editor, ASTM, Philadelphia, 1987, pp 342–357.

Chapter 4

The Physical Structure and Mechanical Properties of Poly(Ether Ether Ketone)

P.-Y. Jar and Ch.J.G. Plummer

Abstract

Failure in PEEK below T_g appears to arise from intra-spherulitic cracking, which propagates through the spherulitic structure without deviating around regions of locally high crystallinity. This is consistent with thin film observations of crazing in semi-crystalline samples, where scission crazing, argued to be locally favoured over simple shear deformation by the altered constraints in the presence of crystallites, is believed to coexist with shear deformation below T_g.

1 Introduction

The semi-crystalline high temperature-resistant thermoplastic, poly(ether ether ketone) (PEEK), has attracted great interest in both industrial and academic circles since its inception by ICI in the late '70s. There exists, in consequence, a considerable body of published work on PEEK, which places particular emphasis on its mechanical properties and morphology. Since changes in the mechanical properties are generally brought about by varying the thermal history of the samples, which may in turn have a profound effect on their morphology, it is the link between mechanical properties and morphology which has been the focus of much discussion in the last few years. One of the major applications of PEEK is as a matrix for carbon fibre composites, a commercial prepreg of carbon fibre/PEEK being currently available from ICI under the trade name of APC-2 (Aromatic Polymer Composite). This raises the further question of the interaction between the fibres, the matrix morphology and the mechanical properties in PEEK composites.

Here, a brief overview of crystalline structure in PEEK is given, followed by a discussion of the dependence of its mechanical behaviour on such factors as the

---

Pean-Yue Jar, Australian National University, GPO Box 4 ACT2601, Canberra, Australia; Christopher J.G. Plummer, Laboratoire de Polymères, Ecole Polytechnique Fédérale de Lausanne, DMX-D, 1015 Lausanne, Switzerland.

degree of crystallinity and sample morphology, with reference to the role of carbon fibres. Special attention will be given to micro-deformation mechanisms in PEEK and their possible relationship with the behaviour of amorphous glassy thermoplastics.

2 The Microstructure of PEEK

2.1 The Crystalline Structure

2.1.1 The Unit Cell

The crystalline structure of PEEK has been studied using wide-angle X-ray scattering (WAXS) and transmission electron microscopy (TEM) of semi-crystalline samples, obtained by a variety of preparation methods. These include crystallization from the melt [1], crystallization from dilute solution [2, 3] and crystallization by the annealing of amorphous samples above the glass transition temperature T_g at 145°C [4, 5]. It is generally agreed that the PEEK has an orthorhombic unit cell, as shown in Fig. 1. The unit cell parameters are $a = 7.75$–7.83 Å, $b = 5.86$–5.94 Å, and $c = 9.88$–10.06 Å, with the main chain direction along c and the ketone units aligned with b [1–11]. Die-drawn PEEK with a draw ratio of up to 4 has also been investigated [1], the results indicating unit cell parameters consistent with those given above.

2.1.2 Crystal Growth

Crystal growth in PEEK occurs preferentially along the normal to the (110) planes, as indicated by TEM studies of microtomed thin films [7] and solution-cast thin

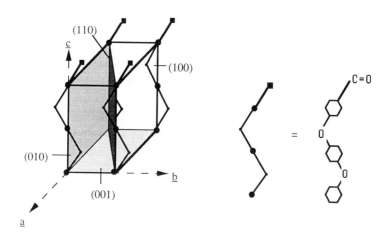

Fig. 4.1 The unit cell of crystalline PEEK: three possible relative positions for the ether and ketone units have been suggested [11]; those given here are taken from reference [10].

films [3]. In the presence of fibres (carbon fibre/PEEK, APC-2), the dominant growth direction is along the fibre surfaces, particularly in slow-cooled specimens, in which *a* tends to lie perpendicular to the fibre direction.

Crystallization rates vary with crystallization temperature. *Blundell* and *Osborn* [3] have reported that the maximum crystallization rate is reached at 230°C, further increases in temperature resulting in a decrease in crystallization rate. Indeed, by 315°C, the crystallization rate has diminished to such an extent that observation of a crystallization peak by DSC requires that the sample be held at this temperature for over 10 minutes.

2.1.3 Determination of Degree of Crystallinity

The degree of crystallinity in PEEK is generally low. Values of between 30 and 35 volume% are commonly reported, although these may be modified by varying the cooling rate or by suitable post-annealing treatments. The various techniques which have been used to measure the degree of cristallinity, including infrared spectrometry (IR), DSC, WAXS and densitometry, have been considered in some detail in chapter 3. The considerable ambiguities associated with many obtained data are basically related to the high chain stiffness which is responsible for the fact that equilibrium conditions are attained relativly slowly. In other words metastable states play an important role; depending on its thermal treatment one and the same material can have a "DSC-degree" of crystallinity of between 2 and 40% (by weight). In chapter 3 it was argued that routine DSC measurements lead to a systematic overestimate of the degree of crystallinity in PEEK by about 10%. DSC nevertheless remains the most popular technique for determining the degree of crystallinity in PEEK, especially for samples containing fibres, although it cannot be used to measure degrees of crystallinity below 22% [12]. Also, with some thermal treatments (for example, the annealing of a semicrystalline sample or the heat treatment of an amorphous sample below 340°C), the interpretation in terms of molecular structure of the low temperature endothermic peak which appears in subsequent DSC scans is somewhat open to question, as will be considered further in 2.3.

Investigations based on these different methods generally indicate a degree of crystallinity in PEEK in the range 0 to 50%. Studies using IR also suggest that the degree of crystallinity in the spherulites may exceed 60% locally, even for samples where the global degree of crystallinity is only 23% [13].

2.2 Spherulitic Structure

2.2.1 Observation of Spherulitic Structure

Polarized-light optical microscopy is one of the commonest techniques used to observe spherulites in semi-crystalline polymers, such as PEEK, for which typical micrographs are given in Fig. 2. Fig. 2(a) is from an injection molded PEEK tensile test bar (Victrex^{TM} 450G from ICI) and Fig. 2(b) is from a melt crystallized PEEK film (Stabar^{TM} K200, also from ICI).

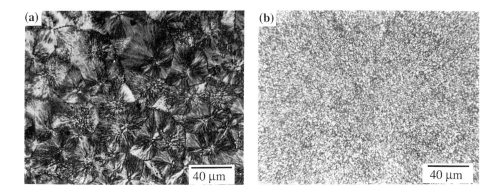

Fig. 4.2 (a) Polarized-light optical micrograph of injection-moulded PEEK (450G); (b) polarized-light optical micrograph of PEEK film (Stabar K200) heated to 400°C.

The effective diameter of spherulites in such samples may vary from less than 1 μm [14] to over 100 μm [7], depending on the preparation conditions, although frequently the fine structures are better characterized in terms of sheaf-like folded-chain lamellar bundles, rather than of well formed spherulites. In either case, b is generally radial, a, tangential and c, perpendicular to the fold plane, with relatively thick amorphous layers separating the lamellae.

There are two disadvantages in using optical microscopy. First, since spherulites in PEEK are generally small, a combination of good resolution with good contrast is often difficult to obtain, particularly where the spherulite diameter is less than 5 μm. Second, sample preparation from bulk specimens for optical microscopy is difficult and time-consuming, not only because the sample has to be sufficiently thin to avoid overlapping of spherulites, but also because most available fast glues do not bond PEEK well to microscope slides, which may engender problems where use is made of petrographic techniques to produce thin sections by polishing.

Because of these disadvantages, certain other methods for observing the spherulitic structure have also been tried, namely chemical etching [14], plasma etching [15] and 'free-surface crystallization' [6, 16, 17]. All these techniques involve examination of the surface morphology of bulk spherulitic samples, usually with SEM, although TEM examination may also be used where details of the sub-spherulitic structure are required [6].

The chemical etch is normally a 2 wt% solution of potassium permanganate in a mixture of 4 parts by volume of orthophosphoric acid to 1 of distilled water. In carbon fibre reinforced PEEK however, the preferred etch is 1 wt% potassium permanganate in a 5:2:2 mixture of sulphuric acid, orthophosphoric acid and distilled water. (Further details of these techniques are given in [14] and [18] respectively.) Fig. 3 shows examples of the surface morphology of a chemically etched carbon fibre/PEEK composite. Fig. 3(a) is from a specimen formed at 400°C and Fig. 3(b) from a specimen formed at 380°C. Plasma etching gives similar surface morphologies to those obtained by chemical etching, but may result in damage to the carbon fibres.

4. Poly(Ether Ether Ketone): Structure and Properties   115

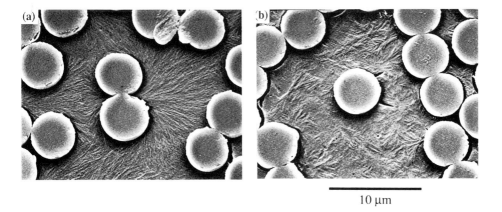

Fig. 4.3 SEM micrographs of chemically etched carbon fibre/PEEK: (a) 400°C; (b) 380°C (the cooling rate was 10 Kmin^{-1} in each case).

'Free-surface crystallization' refers to crystallization in the absence of applied pressure. Because the density of the crystalline phase is higher than that of the amorphous phase, this will result in an apparent volume decrease of the amorphous material remaining after crystallization, which in turn results in a 'mountain-and-valley' morphology at the surface. The 'mountains' are the crystal-rich regions and the 'valleys', the amorphous regions, as shown schematically in Fig. 4. The surface morphology obtained in this way for carbon fibre/PEEK (APC-2) (fibres parallel to the surface) is shown in Fig. 5. Also given for comparison are polarized-light optical micrographs of a specimen with the same thermal history viewed perpendicular to the fibre direction.

2.2.2 Parameters Affecting the Spherulitic Morphology

The Grade of PEEK   Several different grades of PEEK are commercially available from ICI, namely Victrex 450G and P150, and Stabar K200 and XK films. *Chu* and *Schultz* [19] compared the spherulitic morphology of 450G and P150 for a wide range of conditions and found that those for which P150 showed well developed spherulites, resulted in the fine sheaf-like morphology in samples of 450G. This was accounted for in terms of a difference in nucleation density, arising in turn from the differing molecular weights of the two grades (12,000 and 50,000 for P150 and 450G respectively. However, subsequent studies of non-commercial PEEK fractions, with a relatively low nucleation density, have not indicated systematic changes in the nucleation density, although spherulite growth rates did decrease with increasing molecular weight [20].

Thermal History   The spherulitic morphology in PEEK is also influenced by thermal history. The development of spherulitic structure in amorphous PEEK during annealing for example, depends strongly on the annealing temperature. Immediately above the onset of crystallization at 170°C, the fine sheaf-like morphology predominates.

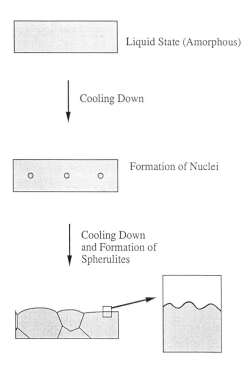

Fig. 4.4 Schematic of formation of surface morphology during free crystallization.

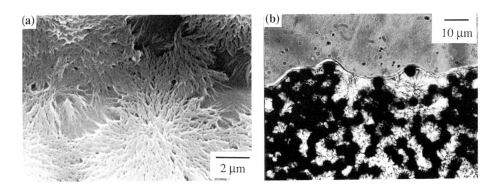

Fig. 4.5 (a) SEM micrograph of the surface of 400°C-treated prepreg prepared by free crystallization (fibre direction horizontal); (b) polarized-light optical micrograph of prepreg using the same thermal treatment (fibre direction out of the surface).

At progressively higher temperatures, the lamellae increase in thickness, but the morphology remains essentially sheaf-like up to 340°C. At this latter temperature, which corresponds to the major DSC endotherm, the bulk of the crystalline material present melts, leading to a loss in birefringence. However, residual crystallinity of relatively high perfection persists to higher temperatures, and acts as nuclei for crystallization during subsequent cooling (self-seeding). Complete melting does not occur until temperatures above the thermal equilibrium melting temperature ($T_{m,o}$), which for PEEK is approximately 395°C [3]. *Blundell et al.* have suggested that self-seeding in PEEK becomes negligible above 370°C [18], in which case well formed spherulites should result. However, our own investigations of the spherulitic structure of carbon fibre/PEEK (APC-2) formed at 380°C with a slow subsequent cooling rate (0.3 Kmin^{-1}), showed the sheaf-like structure to persist under these conditions, as shown in Fig. 6.

The Presence of Fibres Three different morphologies have been reported in the presence of fibres:

(i) single spherulites nucleating from the fibre surface;
(ii) spherulites nucleating between two fibres, where the fibre spacing is low;
(iii) transcrystalline growth.

(i) and (ii) are commonly observed in APC-2. Examples are shown in Fig. 3(b) and Fig. 3(a) respectively.

Spherulite nucleation between fibres may arise from local shearing at the points of contact, which increases the probability of nucleation by orientation of the polymer chains, or else by exposure to specific nucleation sites at the surface of the fibres [18]. This latter effect may also account for the nucleation of single spherulites at the fibre surface.

The existence of transcrystallinity in carbon fibre/PEEK composites is not generally agreed upon, although it is potentially of considerable consequence for the mechanical properties. Whilst it may lead to better matrix-fibre bonding, impingement of transcrystalline regions emanating from adjacent fibres may lead to weaknesses within the matrix [21]. Studies of fibres embedded in certain grades of PEEK [10, 22] have shown that transcrystallinity can occur, but studies of the commercial product, APC-2, are less conclusive. *Tung* and *Dynes* [23] reported that use of a slow cooling rate (1.5 Kmin^{-1}), after heating at 400°C for 10–15 minutes, resulted in clear transcrystallinity. Transcrystallinity was not present, however, in our own samples held at 400°C for 5 minutes and then cooled with a cooling rate of 10 Kmin^{-1}. In this latter case, spherulite nucleation between fibres predominated, as shown in Fig. 3(b). *Peacock et al.* [24], who used etching techniques to investigate the spherulitic morphology in APC-2, also reported an absence of transcrystalline growth in samples heated at 380°C for 5 minutes, with cooling rates of 15 and 470 Kmin^{-1}. The reason for the existence of transcrystallinity for the slightly longer hold-time and slower cooling rate used by *Tung* and *Dynes*, but not in the other cases

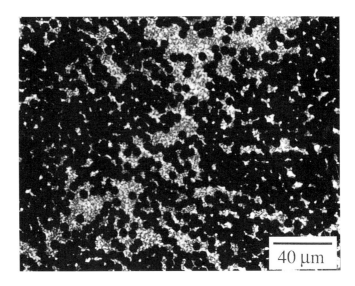

Fig. 4.6 Polarized-light optical micrograph of APC-2 formed at 380°C with a cooling rate of 0.3 Kmin^{-1}.

cited above, is not clear at present. However, since a hold-time of 10–15 minutes at 400°C exceeds that recommended by the supplier, it may well be that transcrystallinity in APC-2 is a consequence of matrix degradation. Further experimental evidence is needed to verify this.

2.3 Secondary Crystallinity

Premature melting of secondary crystals, which have been observed in certain other semi-crystalline polymers [25], may result in the observation of a low-temperature endothermic peak during a DSC scan. However, double melting peaks may also be solely a consequence of melting and recrystallization occurring during the scan itself. In this case, the low temperature endothermic peak would represent the melting of crystals present in the sample prior to scanning and the high temperature peak, the point at which the difference between the endothermic energy of melting and recrystallization passes through a maximum.

For PEEK, a low temperature DSC peak is generally observed in annealed samples whose initial state was either semi-crystalline or amorphous, although it is not generally agreed which of the above mechanisms operates. Here each case will be considered in turn.

2.3.1 Secondary Crystallinity in Annealed Semi-Crystalline Samples

DSC thermograms of semi-crystalline samples annealed at temperatures below 300°C generally show the upper and lower DSC peaks to be well separated. *Jar et al.* [26], in studies of injection moulded PEEK 450G, have shown that whilst the area under the low temperature peak increases with annealing time for an annealing temperature of 250°C, the area under the high temperature peak remains unchanged, as shown in Fig. 7.

A further study of the difference between thermograms obtained prior to and after the annealing treatment was carried out in the following manner. A sample which had been held at 400°C for 2 minutes under nitrogen in the DSC was cooled quickly (320 Kmin^{-1}) to 100°C and re-scanned to 400°C with a scanning rate of 20 Kmin^{-1}. The resulting thermogram is the upper curve in Fig. 8. The same sample was again held at 400°C for 2 minutes and then cooled quickly to 250°C, at which temperature the sample was held for a further 20 minutes, this constituting the annealing treatment. Finally the sample was cooled and re-scanned with a scanning rate of 20 Kmin^{-1} to give the lower curve in Fig. 8.

Comparison of these two curves shows that whilst the annealing treatment at 250°C for 20 minutes results in an increase in the area under the low temperature peak, the remainder of the curve, including the high temperature peak, remains unchanged. Although this observation does not exclude the possibility of melt-recrystallization during the scans, that the low temperature peak appears only in the annealed specimens, and not in the un-annealed specimens, suggests that the microstructure which corresponds to the low temperature endothermic peak, develops during the annealing treatment. It therefore seems reasonable to believe that the low temperature endothermic peak arises from melting of secondary crystals which have grown between the major lamellae during the annealing treatment.

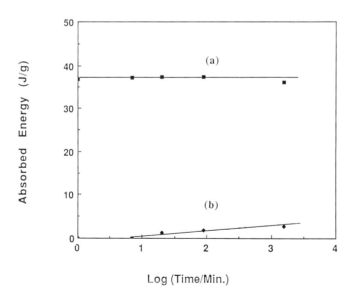

Fig. 4.7 Absorbed energy in DSC thermogram as a function of time (log scale): (a) high temperature endothermic peak; (b) low temperature endothermic peak.

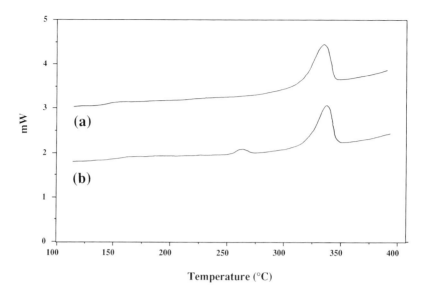

Fig. 4.8 DSC thermnograms of PEEK heated to 400°C: (a) unannealed; (b) annealed at 250°C for 1 hour.

2.3.2 Secondary Crystallinity in Annealed Amorphous Samples

The interpretation of the low temperature endothermic peak in DSC thermograms from annealed amorphous samples remains a contentious issue. Several authors [27–29] argue that the low temperature peak represents the melting of secondary crystallites grown during the annealing treatment, as described above for the case of semi-crystalline PEEK. Others, however, suggest that the low temperature peak represents the melting of crystallites present before the start of the scan, and that the material constituting these then undergoes recrystallization and remelting during the scanning, which gives rise to the upper peak as described at the beginning of this section [3, 31].

Evidence for a single crystal type in annealed PEEK samples and its association with the low temperature melting peak of the DSC thermogram, comes from small angle X-ray (SAXS) analysis, the position of the SAXS peak being found to be independent of the annealing temperature [31]. The evidence for the growth of secondary crystals during annealing comes from DSC analysis of samples subject to different thermal histories [27, 29], TEM [28] and consideration of crystallization kinetics [27].

It is perhaps also significant that, as will be discussed in the next section, annealed PEEK, whether initially semi-crystalline or amorphous, always shows the same behaviour during tensile tests, suggesting similarities in the annealed state in the two cases.

3 The Mechanical Properties of Bulk Samples

In this section it is intended to consider first, in 3.1, some basis aspects of the phenomenology of the mechanical behaviour of PEEK in relationship to its crystalline structure. In 3.2 the interrelationship between deformation and structure at a microscopic level, based on studies of bulk samples and thin films respectively, will be considered.

3.1 Basic Phenomenology

Although some compression tests and shear tests have been carried out on PEEK [32], most of studies of the mechanical properties of PEEK involve either short term tensile tests or compact tension tests. PEEK is generally considered to be highly ductile in tension (draw ratio of about 2), with significant plastic necking persisting up to cross-head speeds in excess of 11 ms^{-1} [33]. Nevertheless, studies of single-edge notched specimens of unreinforced PEEK and short glass fibre reinforced PEEK have indicated the occurrence of a ductile-brittle transition at cross-head speeds between 1 and 1000 mm/min. at temperatures up to 115°C [34].

Typical room temperature values for the tensile strength, modulus and Izod impact toughness of the unreinforced resin are 92 MPa, 3.6 GPa and 83 Jm^{-1} respectively, whence PEEK may be seen to combine high strength and stiffness with a reasonable degree of toughness, bearing in mind the notch sensitivity. As with many semi-crystalline thermoplastics, it has good creep properties and in particular, and an outstanding, albeit again notch sensitive, performance in fatigue [26, 35–39]. With the addition, for example, of short carbon fibres the tensile strength and modulus of injection mouldings rise to around 200 MPa and 13 GPa.

Strain induced crystallization may also improve the strength and modulus of PEEK films, given an appropriate post-drawing annealing treatment (zone-drawing/annealing) [40]. The dynamic mechanical modulus after such treatments may be up to 4.4 times that of the commercially available films. It is believed that the zone-drawing causes strain induced crystallization [40–42].

More details of these basic properties and those of continuous fibre composites are considered in Part II of this book. Here emphasis will be placed on the effects of the degree of crystallinity, annealing treatments, physical aging and the spherulite size.

3.1.1 The Degree of Crystallinity

In spite of the uncertainty which persists as far as determination of the absolute degree of crystallinity is concerned, it is generally agreed that the tensile strength of PEEK increases with the degree of crystallinity. *Talbott et al.* [32] suggest the tensile strength of PEEK (150P) to be described by the expression

$$S_t = 57.65 + 1.20 \; c \tag{1}$$

where S_t is the tensile strength of PEEK (in MPa), and c is the degree of crystallinity (expressed as a percentage by volume). This expression also fits the earlier results of Seferis [43].

However, *Lee et al.* [44] tested PEEK films and reported tensile strengths some 20% greater than would be consistent with the above expression for a sample with 28% crystallinity. *Cebe et al.* [45] also performed tensile tests on PEEK films and suggested an initial dependence of the mechanical behaviour on the thermal history of the specimens. For a given degree of crystallinity, they reported higher yield stresses for fast cooling in air than for slow cooling. Notwithstanding the possibility of discrepancies arising from the use of different grades of PEEK for such tests, other factors may also affect the tensile strength independently of the degree of crystallinity, such as the nature of annealing and physical aging treatments, and sample degradation. Annealing and aging will be discussed in what follows and degradation is considered in chapter 2 of this book.

3.1.2 Annealing

Whilst there is disagreement over the effect of annealing treatments on the crystalline structure, it is well known that annealing increases the tensile strength of both semi-crystalline and amorphous PEEK [26, 32, 44, 46]. *Talbott et al.* [32], for example, have reported increases of up to 20% on annealing, in spite of very small changes in the degree of crystallinity.

The effect of annealing time and temperature has been studied by *Lee et al.* [44] and *Jar et al.* [26]. Both groups reported improvements in tensile strength. *Lee et al.* also observed that the fracture strain of almost amorphous PEEK (0.2% crystallinity) increases on annealing at temperatures ranging from 210 to 250°C. However, *Jar et al.* reported that the fracture strain decreases with increasing annealing time for semi-crystalline PEEK (450G) at 250°C. Significantly, the annealed specimens in [44] fractured during work hardening after mechanical drawing, but the annealed specimens in [26] fractured during mechanical drawing. Since in the later case, the specimens were injection-moulded, the release of internal stresses in the sample surface during annealing may have resulted in the appearance of surface flaws, which would induce premature fracture of the specimens during tensile tests [26]. However, bearing in mind that these samples had a cross section of 3 mm × 6 mm, whereas those used in [44] where film samples with a thickness of 100 μm, the discrepancy between the two sets of results may have been due to size effects.

Work hardening behaviour immediately after yielding of annealed specimens was observed by *Jar et al.* [26] in stress-strain curves obtained at and above 80°C, as shown in Figure 9 for specimens tested at 120°C. Similar phenomena have also been reported by *Cottenot et al.* [46]. This post-yield hardening behaviour was correlated with the growth of secondary crystals during the annealing treatment by *Jar et al.* but was considered to be due to a change in deformation mechanism by *Cottenot et al.*. Since this work hardening behaviour does not occur in un-annealed specimens, as also shown in Fig. 9, the possibility of a change in deformation mechanism seems unlikely. The results of *Cebe et al.* [47] also indicated post-yield hardening in slow-cooled specimens tested at 125°C, although this was not discussed.

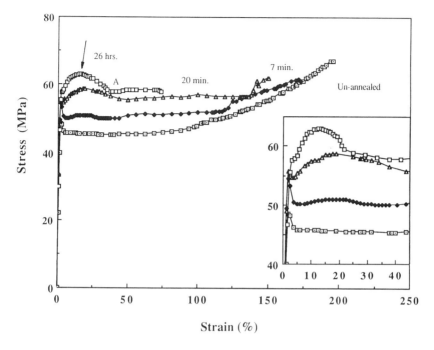

Fig. 4.9 Stress-strain curves at 120°C of injection-moulded PEEK with different annealing times; annealing temperature was 250°C.

The onset of post-yield work hardening on annealing may occur in both amorphous and semi-crystalline samples (these latter obtained by heating amorphous PEEK films (Stabar K200) at 400°C for 5 minutes and then cooling in air), as shown in Figure 10, along with results for an un-annealed sample for comparison. In this case, the samples were annealed at 250°C for 1 hour.

3.1.3 Physical Aging

The effect of physical aging, that is aging just below T_g, on the tensile behaviour of amorphous PEEK has been discussed by *Kemmish* and *Hay* [48], *Wang* and *Ogale* [49] and by *Ogale* and *McCullough* [30], who considered its effect on creep. The results of Kemmish and Hay indicated similar changes in mechanical and fracture properties on aging of amorphous PEEK to those observed in polycarbonate (PC) and poly ethylene terephthalate (PET), with a ductile-brittle transition being induced by the aging treatments.

3.1.4 Spherulite Size

Spherulite size often plays an important role in the mechanical properties of semi-crystalline polymers such as the polypropylene (PP), where large spherulites may be associated with inter-spherulitic fracture and consequent embrittlement [25].

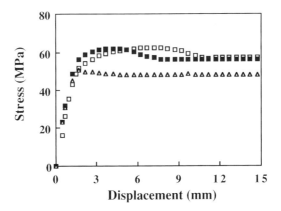

Fig. 4.10 Stress-displacement curves of PEEK films (Stabar K200) tested at 120°C: open squares - heat treated at 250°C for 1 hour; filled squares - heat-treated at 400°C and then annealed at 250°C for 1 hour; open triangles - heat-treated at 400°C without annealing at 250°C (all specimens had a gauge length of 50 mm).

The effect of spherulite size in PEEK has been studied using Stabar K200 films of 100 μm in thickness. These films were crystallized by heating to a temperature between 360°C and 400°C and cooling in air. The specimens heated to 360°C showed a fine sheaf-like structure and a degree of crystallinity of 27 wt% (from DSC) whereas those heated to 400°C showed well developed spherulites (cf 2.2.2) and a degree of crystallinity of 25 wt%. The corresponding surface morphologies are shown in Fig. 11.

The room temperature tensile yield strength was 95 MPa in both cases and the stress-displacement curves were similar in shape. Whilst as discussed in 3.1.1, the lower degree of crystallinity in the 400°C annealed samples taken in isolation should have resulted in a slightly lower yield strength, this was compensated by crosslinking effects. More importantly, these observations suggest that spherulite size does not significantly affect the tensile behaviour, whence it may be concluded that observed variations in the mechanical properties for a given degree of crystallinity result from changes at the molecular level rather than at the spherulitic level.

3.1.5 Long Term Mechanical Behaviour

Karger-Kocsis et al. [35] have studied the effect of annealing on the fatigue crack propagation (FCP) of notched injection moulded specimens, and found that annealing at 150°C for 120 hours had no effect on unreinforced PEEK, but that the FCP resistance decreased on annealing at 325°C for 120 hours. However, *Carfagna et al.* showed that the creep behaviour of amorphous PEEK films could be significantly improved by annealing at 135°C for 170 hours [50]. A similar improvement in creep resistance was also found by Jar *et al.* when semi-crystalline PEEK was annealed at 250°C for up to 26 hours [26].

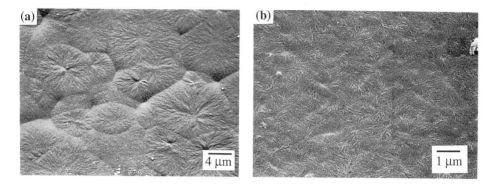

Fig. 4.11 SEM micrographs of surface morphology of PEEK (Stabar K200) prepared by free crystallization: (a) heated to 400°C; (b) heated to 360°C.

Annealing has several effects on PEEK injection mouldings, which may have consequences for long term behaviour: it causes crystallization of the amorphous skin region present in as-molded specimens; it enhances the growth of secondary crystals (see 2.3); it may cause molecular degradation in the skin for long times and high temperatures. The effect of annealing on the mechanical behaviour of PEEK depends on a combination of these factors, weighted according to the annealing conditions. The annealing temperatures used by *Karger-Kocsis et al.* were either relatively low (150°C) or relatively high (325°C), and the annealing times were long (120 hours). This contrasts with the annealing treatments of *Jar et al.* in which the samples were held for a much shorter time (26 hours) at an intermediate temperature (250°C). Thus one might speculate that 150°C is too low a temperature to induce significant change in the mechanical properties, and that at 325°C, molecular degradation cancelled out any possible beneficial effect of annealing, such as the development of secondary crystallinity, argued to result in the increases in the creep resistance in the studies of *Jar et al.*

3.2 Deformation Behaviour in Bulk Samples

3.2.1 Methods of Observation

To observe the deformation behaviour in bulk semi-crystalline PEEK, the same methods were used as described in 2.2.1 for the observation of spherulitic structure, that is: polarized-light optical microscopy and SEM of surface morphology and fracture surfaces.

These have been applied both to unreinforced PEEK and to carbon fibre/PEEK prepreg and bulk composites.

3.2.2 Unreinforced PEEK

Consistent with the discussion in 3.1.4, direct observation appears to confirm the absence of inter-spherulitic weak regions in PEEK. By way of comparison, Fig. 12 shows the inter-spherulitic regions of PP (heated to 210°C for 30 minutes and air-cooled), which are generally associated with crack initiation when the sample is stressed. For PEEK specimens heated to 400°C for 30 minutes and then air-cooled, Fig. 11(a), no inter-spherulitic weak regions were observed, although the spherulites were clearly well developed.

The reason for the absence of inter-spherulitic weak regions in the PEEK is not clear at present. However, early work on carbon fibre/PEEK [51] suggested that slow-cooled specimens did contain such regions, this resulting in turn in a decrease in the delamination toughness as determined from crack propagation measurements. Such a decrease was not observed in later generation carbon fibre/PEEK composites.

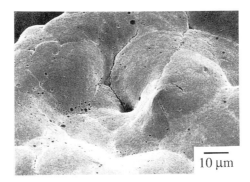

Fig. 4.12 SEM micrograph of PP prepared by free crystallization at 210°C and cooled in air.

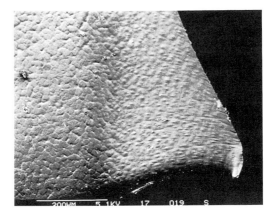

Fig. 4.13 SEM micrograph of the tensile fracture surface of PEEK (Stabar K200) prepared by free crystallization.

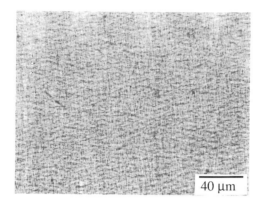

Fig. 4.14 Polarized-light optical micrograph of necked region of injection-moulded PEEK (450G) (draw direction vertical).

SEM observations of the surface of free-crystallized film deformed in tension are shown in Fig. 13, the spherulites being extended in the draw direction (horizontal in the Figure). Final fracture occurred in the neck and the crack path did not appear to deviate around the spherulites, consistent with the absence of inter-spherulitic weak regions.

Deformation behaviour of PEEK with the fine sheaf-like structure was also investigated using polarized-light optical microscopy of thin sections as shown in Fig. 14 (draw direction vertical). The observations suggest stretching of the sheaves in the draw direction.

3.2.3 Carbon Fibre/PEEK Prepreg and Bulk Composites

The examination of etched surfaces of bulk composite samples tested in delamination showed inter-fibre cracks beneath the fracture surface (Fig. 15). Cracks are visible which penetrate through the spherulites. This phenomenon has also been reported by *Crick et al.* [52] and provides an indication of the deformation pattern of the spherulites, which in this case would be consistent with that observed for unreinforced PEEK as described in 3.2.2. However, since these cracks do not exist prior to etching, they should only be considered to indicate the distribution of stress concentrations in the spherulites (which are more easily attacked by the etching solution). It is nevertheless reasonable to believe these to represent eventual crack propagation paths during fracture.

Prepregs were also prepared by the 'free crystallization' method and stretched perpendicular to the fibre direction. The surface morphology of the fractured specimens is shown in Fig. 16, providing further evidence in support of the above interpretation.

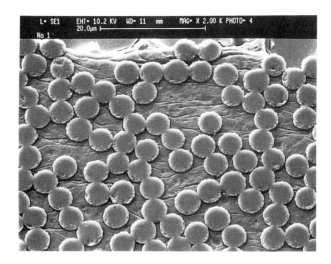

Fig. 4.15 SEM micrograph of chemically etched surface of APC-2 after mode I delamination test; the fracture surface generated by the delamination test is uppermost in the micrograph.

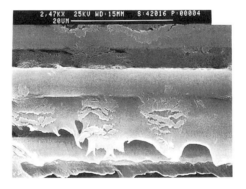

Fig. 4.16 SEM micrograph of the surface of prepreg formed at 400°C by free crystallization; the specimen was tested in tension at 90° to the fibre direction.

4 Microdeformation Behaviour in Thin Films

The additional levels of microstructural complexity present in semicrystalline materials have to date made them less attractive as subjects for fundamental research into deformation mechanisms than intrinsically amorphous polymers. Thus the 'copper grid' technique for TEM observation of tensile deformation in thin films, to be described in what follows, whilst having proven a very powerful means of investigation, has to date, been almost exclusively applied to amorphous polymers and polymer blends [53–59].

The initial aim for this work was to establish whether the behaviour of PEEK in the amorphous glassy state (as induced by rapid quenching) fits the general pattern found for non-crystallizable amorphous polymers, to be described in 4.1, in which entanglement appears to play a crucial role. Such studies of deformation in the amorphous phase of semi-crystalline polymers may well be a useful way of bridging the gap between developments in the understanding of intrinsically amorphous and of semi-crystalline systems.

4.1 Microdeformation in Amorphous Polymers

Most of the basic concepts pertaining to the following discussion are introduced in chapter 5, so that it is only necessary to recall some of the essential points here.

First, amorphous polymers below T_g may be modelled in terms of a network of entanglement points, of entanglement density v_e, a spatial separation d_e and a corresponding polymer strand contour length l_e. Thus in tension, deformation is characterized by a fixed maximum extension ratio λ_{max} of the order

$$\lambda_{max} = l_e/d_e \tag{2}$$

Second, it is found that crazing is favoured over simple shear deformation in amorphous polymers by low v_e [57, 58]. The key to the understanding of this, is the presence of voids in the craze body. For a void to grow, and hence for a craze to propagate, new surfaces must be created within the material. Hence either a certain number of entangled strands must break, or entanglement must be lost in some other way.

Where entanglement loss is mediated by scission, *Kramer* has argued that the crazing stress will increase as $v_e^{1/2}$ since at higher v_e, more entangled strands must break to create unit area of void surface [57] (see also chapter 5 of this book, equations 3 and 4). Thus for large v_e the crazing stress becomes much larger than the shear stress and simple shear deformation dominates. Kramer's model suggests a weak temperature and strain rate dependence of the crazing stress, consistent with the observation of a transition from crazing to shear as the temperature is increased or the strain rate decreased in low ne polymers such as polystyrene (PS, entanglement density $\sim 4 \times 10^{25}$ m^{-3}) which craze at room temperature [53, 54].

What this approach fails to account for however, is the transition from shear to crazing with increasing temperature observed in thin films of high ne polymers such as polyethersulphone (PES) and PC (v_e in PES is of the order of $30 \times 10^{25}\,\mathrm{m}^{-3}$) [55]. Explanation of this requires an alternative view of entanglement to the network model described above, that is, the *tube model* [60, 61], in which entanglement is viewed as a non-local tube-like constraint on a given chain, representing its interactions with the surrounding chains.

During high temperature disentanglement crazing in PES and PC, chains are 'pulled' out of their tubes as the voids advance, which has led to the adoption of the term *forced reptation* to describe this mechanism for disentanglement [62]. The frictional forces opposing forced reptation will increase strongly with the molecular weight M of the chains in question, consistent with the observed M dependence of high temperature crazing in PES and PC [55].

4.2 Implications for PEEK and Other Semi-Crystalline Thermoplastics

In the case of PEEK, the extension ratios of approximately 2 observed during room temperature shear deformation in amorphous bulk samples, suggest a similar entanglement density to PES and PC. This is reasonable in view of the structural similarities of all three materials. The grade of PEEK used here to fabricate thin films (supplied by ICI) has a molecular weight average M_w of the order of 50,000. For PES, a clear transition from shear to crazing has been observed in thin films for a range of M_w, occurring at approximately 100 K below T_g (218°C) for $M_w = 47,000$ and a strain rate of $10^{-2}\,\mathrm{s}^{-1}$. Given the increased tendency for disentanglement crazing with reduced M, one might therefore reasonably expect to observe a transition from shear to crazing in amorphous PEEK as the temperature is raised towards its T_g.

The phenomenology of both crazing and shear is widely documented [63, 64], and several qualitative explanations have been advanced both for crazing below T_g [63], and for coarse fibrillation above T_g [65–69] (also referred to as crazing in the literature). It is certainly reasonable that coarse fibrillation should occur during drawing above T_g of semi-crystalline polymers following break-up of the crystalline lamellae, and one might expect similar effects during the latter stages of plastic deformation below T_g [65]. However, below T_g, the size and spacing of fibrils in crazes in semi-crystalline polymers may be similar to those observed in those observed in amorphous polymers [64, 71, 72]. This suggests that the morphology of the former is not strongly influenced by the presence of the lamellae. Moreover, since the magnitude of the observed fibril spacings in amorphous polymers is accounted for by the surface drawing model on which Kramer's model is based [56], one may argue that the basic mechanism for crazing is the same in both cases. There is certainly strong evidence from previous use of the copper grid technique to investigate deformation in amorphous and semi-crystalline isotactic polystyrene (iPS), that entanglement concepts remain highly relevant to crazing in this latter [72]. Interpretation of thin film deformation of PEEK in terms of such ideas may thus provide a means of rationalizing the deformation behaviour of bulk samples described in the previous section.

4.3 Preparation of PEEK Films

The basic experimental method involves the solution casting of a uniform thin film of about 0.5 µm in thickness by drawing a glass slide from a suitable solution of the polymer in question at a constant rate. The film left behind on the glass slide after evaporation of the excess solvent is then floated off on a water bath and picked up on an annealed copper grid which has been previously coated with the same polymer. A short exposure to the solvent vapour then serves to bond the film to the copper grid, and to remove any wrinkles. The copper grid acts as a support to the film during deformation, and prevents craze closure during subsequent mounting of individual grid squares for TEM observation [53–59].

Here, PEEK films were cast from boiling 1-chloronaphthalene (boiling point 230°C) immediately after dissolution. Since exposure to the solvent vapour was impractical, it was necessary to melt the films in order to bond them to the copper grids (at just above 350°C, that is, the temperature of the main DSC endotherm). Amorphous samples were prepared by quenching from 400°C, and crystallinity could then be induced in a relatively controlled fashion by suitable heat treatment above T_g. Fig. 17 shows a micrograph of a film which had been crystallized from the amorphous state by holding it at 300°C for five minutes, showing the fine sheaf-like structure described in 2.2.2.

To obtain coarse spherulitic structures, longer heat treatments were required. To avoid damaging the films, the samples were placed face-down on single salt crystals (NaCl), that is with the film in full contact with the surface of the crystal. The salt ensured sufficient mechanical support to the molten film when the assembly was heated to prevent damage to the film both during heating to 400°C and during subsequent isothermal crystallization, which was monitored using a hot stage

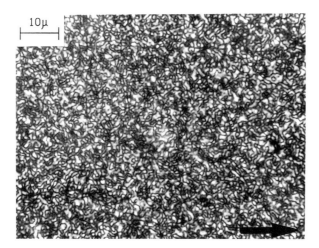

Fig. 4.17 Optical micrograph under crossed polarizers of a PEEK film crystallized from the amorphous state at 300°C (analyzer direction indicated by arrow).

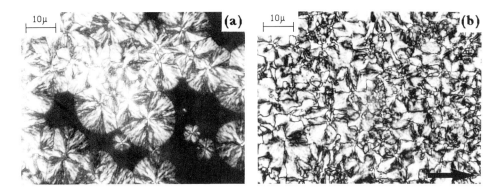

Fig. 4.18 Optical micrograph under crossed polarizers of a spherulitic PEEK film: (a) during crystallization; (b) after completion of crystallization.

mounted on the optical microscope. After subsequent cooling to room temperature, the salt, at this stage bonded to the film, was dissolved in water. Fig. 18 shows a film which had been prepared in this way, as well as the microstructure at an intermediate stage of crystallization, during which the spherulites were most clearly visible.

The films, mounted on the copper grids, were finally heat treated overnight in vacuo at 110°C, that is, below T_g. Given the comments in 3.1.3, such a physical aging treatment might be expected to encourage crazing, as observed in PES and PC [55]. Deformation was carried out using tensile test machine equipped with an environmental chamber. Since direct observation was not possible during deformation, the copper grids were all strained to approximately 5%, which was generally sufficient to ensure the onset of irreversible deformation mechanisms such as crazing. Subsequent to deformation, individual grid squares were cut from the grids for TEM observation. The constraints imposed by the sample preparation technique described above meant that the resulting samples were somewhat thicker than is ideal for TEM, and problems with sample transparency were compounded in the presence of crystallites. Although deformation could be imaged, the samples were not suitable for electron diffraction.

4.4 Deformation of PEEK Films

4.4.1 Deformation of Amorphous Films

Deformation of the amorphous films was carried out in the range of temperatures up to T_g, and in the first series of tests, at a strain rate of approximately $2 \times 10^{-2} \, \text{s}^{-1}$. Viewed under cross-polars, crack-like regions of localized shear deformation, resulting in local necking down of the film thickness are visible. Such features are known as *deformation zones* (DZs) and are characteristic of shear deformation in thin films, being the analogue of macroscopic ductile necking in the bulk samples [55, 72]. TEM shows these regions to contain no fibrillation (Fig. 19). DZs were observed at all

4. Poly(Ether Ether Ketone): Structure and Properties 133

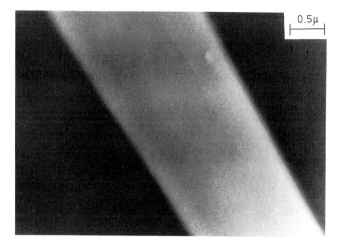

Fig. 4.19 TEM of a deformation zone (DZ) in an amorphous film deformed at room temperature.

temperatures below T_g at the usual strain rate of $2 \times 10^{-2} s^{-1}$ and indeed reducing the strain rate by three decades did not induce any apparent change in deformation mechanism, even at temperatures close to T_g. Although the DZs remain clearly visible under such conditions, they are somewhat more diffuse than for low temperatures/ high strain rates, as expected given the general phenomenology of DZs in other materials and also the general tendency for shear necks in bulk samples to become more diffuse as the temperature is increased [55, 72].

4.4.2 Deformation of Semi-Crystalline Films

Deformation of the semi-crystalline films with sheaf-like microstructures was again carried out at $2 \times 10^{-2} s^{-1}$, and in this case, the temperature range was increased to 200°C, that is, above T_g. Fig. 20 shows optical micrographs of films deformed at room temperature. Most striking is the continued presence of DZs, but as shown in Fig. 20a, in contrast with the amorphous case, these DZs contain a faint texture when viewed under crossed polarizers, corresponding to the stretching of the sheaves in the draw direction as described in 3.2.2 (cf. Fig. 14). Fig. 20b shows a micrograph of a different region of the same film, without crossed polarizers. Here a second type of deformation is visible (no longer being masked by the birefringence contrast), consisting of many fine crack-like defects running perpendicular to the tensile axis, similar in optical appearance to crazes in thin films of amorphous polymers. These features do not appear to deviate around the sheaves, consistent with the observations of crack paths in 3.2.2.

Figs. 21a and 21b show TEM micrographs of the features in Fig. 20a and 20b respectively. The structure within the DZs (Fig. 21a) is highly inhomogeneous, with regions of high crystallinity interspersed with more highly drawn amorphous regions.

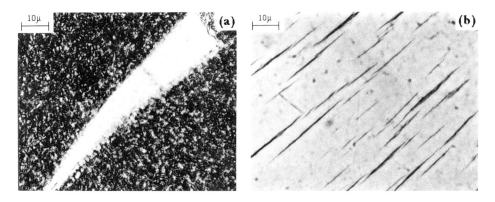

Fig. 4.20 Optical micrographs of deformation in a crystalline film at room temperature: (a) under crossed polarizers; (b) a different region without crossed polarizers.

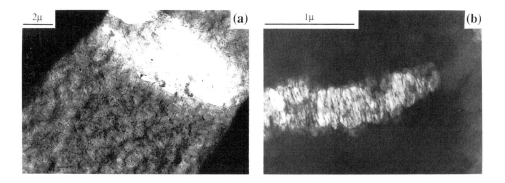

Fig. 4.21 TEM micrographs of deformation in a crystalline film at room temperature: (a) deformation zone, (b) crazing.

Also visible is a faint banded texture perpendicular to the tensile axis, which resembles what is referred to in the literature as a "string of pearl" structure, that is, the break-up of the crystalline lamellae into blocks, separated by more highly drawn material [63-69]. Although the texture of these DZs has a generally fibrillar appearance it is difficult to distinguish true voids.

The deformation shown in Fig. 20b and 21b, and at higher magnification in Fig. 21c, on the other hand, shows a far clearer void/fibril structure, with an apparent fibril spacing of the order of 40 nm. This is typical of crazes in amorphous polymers (although with somewhat irregular boundaries and less coherent fibrillar structures, corresponding to the inhomogeneity of the surrounding material).

There appears little change in the optical appearance of the DZs in the semi-crystalline samples with increasing temperature, even beyond T_g, although the DZs become gradually more diffuse above 160°C. Changes are apparent in the crazing behaviour however. The density of crazing decreases with temperature, and the individual crazes appear longer than those observed at room temperature. Eventually, within approximately 40 K of T_g crazing is replaced entirely by shear deformation in the majority of the grid squares (some residual crazing persists in isolated grid squares to within 20 K of T_g but there is little doubt as to the overall tendency). No crazing is seen above T_g.

4.4.3 Spherulitic Films

DZs in these coarse spherulitic films (Fig. 22a) show much less well defined boundaries than for either the amorphous or the non-spherulitic semi-crystalline films, and a marked tendency towards branching. From the optical micrographs, initial deformation appears to take place predominantly along the spherulite diameters perpendicular to the tensile axis, developing into the highly birefringent 'butterfly-shaped' regions seen to the left of the main DZ in Fig. 22a. As the TEM micrograph in

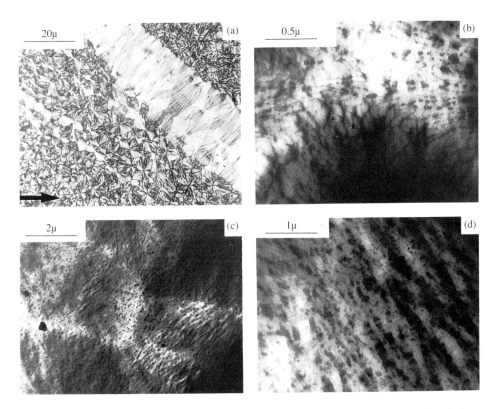

Fig. 4.22 Deformation in a spherulitic film: (a) optical micrograph under crossed polarizers of a deformation zone (DZ) (b) TEM of localized shear deformation between three spherulites; (c) and (d) TEM of part of the DZ in (a).

Fig. 22b shows, localized shear can also occur in inter-spherulitic regions (in this case, at the point of intersection of three adjacent spherulites), but no cracking or crazing appears to be associated with such regions.

Within the DZ, it appears that the spherulite diameters perpendicular to the tensile axis remain relatively intact, with large deformation being localized to regions of the spherulites where the spherulite arms lie along, or at low angles to the tensile axis. Further, well defined angled boundaries appear to separate regions of low and high deformation in adjacent spherulites, which are most clear under TEM (Fig. 22c).

Although yielding of the amorphous material between the spherulite arms is easiest where it is in series with the crystalline material, that is, where the spherulite arms are perpendicular to the tensile axis, the crystallites themselves are presumably unfavourably oriented for further deformation in such regions. Thus, on initial work hardening, deformation will transfer to regions where the spherulite arms lie along the tensile axis, and hence the orientation in the crystalline material is perpendicular to the tensile direction. In such regions, as with the DZs in the non-spherulitic semi-crystalline films, there is a banded structure perpendicular to the tensile axis corresponding to break-up of the crystalline lamellae (Fig. 22d).

It proved difficult to image crazes and the structure of the surrounding undeformed spherulites simultaneously, the former tending to break down at beam intensities sufficient to show the latter in films of this thickness. Small crazes were however found below T_g, nucleating predominantly in intra-spherulitic regions, and growing along the spherulite diameters perpendicular to the tensile axis (which present a continuous path of amorphous inter-lamellar material to a craze propagating in this direction). However, these crazes were restricted to individual spherulites, presumably because in these samples the discontinuous change in crystallite orientation between spherulites means that paths favourable to craze propagation are generally discontinuous at the spherulite boundaries.

4.5 Comparison with Crazing in Amorphous Polymers

The observation that thin films of amorphous PEEK show no evidence of disentanglement crazing for conditions under which it is expected for similar, but non-crystallizable polymers, may be linked to strain induced crystallization. During craze propagation, a region of highly strain softened and deformed polymer exists ahead of the advancing void tips, in which, in the case of disentanglement crazing, chains are gradually being pulled out of their tubes. Indeed it is likely that as the void tips advance, there will be a build-up of highly stretched mobile chains in this region, and thus locally a strong tendency for crystallization to occur. Such crystallization may then act to 'pin' the chains in their tubes, and hence render continued disentanglement progressively more difficult. (The analogous effect of chemical cross-linking has been observed in PES, where it has been found to suppress disentanglement crazing in the high T regime [73].)

The absence of scission crazing in the amorphous films is likely to be a consequence of the high entanglement density in PEEK, as discussed in the introduction to this section. Crazes have been seen, however, in the amorphous bulk samples of PEEK subject to room temperature tensile deformation. The reason appears to be that in bulk samples there is a high degree of constraint on the lateral contractions resulting from homogeneous drawing, resulting in an increase in the hydrostatic tension at the expense of the shear stress driving the deformation. On the other hand, the presence of voids in the craze body means that crazing can result in local axial deformation without the need for lateral contraction. Thus a transition from shear to crazing can result from increased sample thickness. Earlier investigations of amorphous PEEK suggested that this occurs at film thicknesses of the order of 10 µm [74].

Qualitative models for crazing in semi-crystalline thermoplastics in the literature have been based on the idea of coalescence of micro-voiding, arising from stress concentrations subsequent to break-up of the crystalline lamellae [63]. This mechanism would seem most appropriate to what is referred to here as a DZ (although from the present TEM observations it is difficult to confirm whether true voiding is in fact present in such structures). We believe instead that the appearance of crazing in the semi-crystalline films is linked to changes in the degree of constraint on shear deformation in the presence of crystallites, and that the crazing mechanism is the same as that for scission crazing in amorphous polymers.

Strain induced crystallization is unlikely to be an important factor in scission crazing. Indeed there is some evidence that strain induce crystallization accompanies low strain rate scission crazing in amorphous iPS [75]. Certainly the phenomenology of the crazing seen in the crystalline films of PEEK is consistent with the scission crazing mechanism, that is, the competition between crazing and shear is strongest at room temperature, with a tendency for shear to dominate closer to T_g. A gradual transition to shear with increasing temperature is also suggested by Narisawa and Ishikawa in their fracture tests on semi-crystalline PEEK below T_g [64].

Conclusions

One of the key advantages of PEEK is its ductility, whence one may infer that in simple tension, shear deformation dominates other possible modes of failure such as inter-spherulitic failure or crazing/craze breakdown. This contrasts with the behaviour of polymers such as PP and PS, and PES at high temperature.

There appears little evidence either from direct observation, or from attempts to correlate tensile strength with spherulite size, for inter-spherulitic failure in PEEK, at least below T_g. Ultimate failure appears to arise from intra-spherulitic cracking, which propagates through the spherulitic structure without deviating around regions of locally high crystallinity. This is consistent with thin film observations of crazing in semi-crystalline samples, where scission crazing, argued to be locally favoured over simple shear deformation by the altered constraints in the presence of crystallites, is observed to coexist with shear deformation below T_g. Further, treatments such as physical aging which raise the yield stress, tend to result in embrittlement, and similar effects seen in amorphous thermoplastics such as PC, have been accounted for in terms of the increased tendency to craze of aged samples.

However, the transition from shear or scission crazing to disentanglement crazing as the temperature approaches T_g shown by amorphous polymers of similar entanglement density is absent from amorphous quenched films of PEEK, shear being observed at all temperatures. It is suggested that this might be due to strain induced crystallization disfavouring craze propagation owing to its effective pinning of the chains in their tubes. Whatever its origins, such an absence of disentanglement crazing is likely to be reflected by improved high temperature and long term properties.

It is as well to emphasize the clear distinction made here between the two types of deformation in the semi-crystalline films of PEEK, since it has not always been clearly made in the literature, owing to the general fibrillar appearance of highly drawn structures in semi-crystalline polymers. Nevertheless, it is important to recognize that the crazing mechanisms described in section 3 are unlikely to be directly relevant to fibrillar or quasi-fibrillar structures which result from the break-up of crystalline lamellae during shear, for which the term 'fibrillar shear' has recently been suggested [76].

The thin film technique offers promise for future work, regarding the elucidation of the mechanisms of deformation in PEEK, and in particular the role of secondary crystallinity, and the relationship between crazing and craze stability and macroscopic ductile-brittle behaviour (cf. chapter 5). It should nevertheless be kept in mind that: (i) it is not suited for quantitative investigation of mechanical properties; (ii) where detailed correlation with bulk tests is to be made, careful control of the sample preparation is necessary to assure an identical physical state with the bulk samples (factors to be considered include not only the degree and morphology of the crystallinity, but also degradation, the loss of additives during dissolution and so forth).

Acknowledgments

We wish to acknowledge the valuable help and advice of B. Senior (SEM operation), J.W. Smith and A.M. Donald during the course of this work, some of which was carried out as part of the Materials Programme NFP19 of the Swiss National Research Foundation. CJGP is supported by the Swiss Committee for the Encouragement of Scientific Research (CERS).

References

1. D.R. Rueda, F. Ania, A. Richardson and I.M. Ward, Polym. Comm. **24**, 258 (1983).
2. A.J. Lovinger and D.D. Davis, Bull. Am. Phys. Soc. **30**, 249 (1985).
3. D.J. Blundell and B.N. Osborn, Polymer **24**, 953 (1983).
4. J.N. Hay, D.J. Kemmish, J.I. Langford and I.M. Rae, Polym. Comm. **25**, 175 (1983).
5. N.T. Wakelyn, Polym. Comm. **25**, 306 (1984).
6. A.J. Lovinger and D.D. Davis, J. Appl. Phys. **58**, 2843 (1985).
7. S. Kumar, D.P. Anderson and W.W. Adams, Polymer **27**, 329 (1985).

8. P.C. Dawson and D.J. Blundell, Polymer **21**, 577 (1980).
9. J.M. Chalmers, W.F. Gaskin and M.W. Mackenzie, Polym. Bull. **11**, 443 (1984).
10. A.J. Waddon, M.J. Hill, A. Keller and D.J. Blundell, J. Mat. Sci. **22**, 1773 (1987).
11. R.J. Abraham and I.S. Howarth, Polymer **32**, 121 (1991).
12. J.-A. E. Månson and J.C. Seferis, Sci. Eng. Comp. Mater. **1**, 75 (1989).
13. H.X. Nguyen and H. Ishida, Polymer **27**, 1400 (1986).
14. R.H. Olley, D.C. Bassett and D.J. Blundell, Polymer **27**, 344 (1986).
15. A. Lustiger, F.S. Uralil and G.M. Newaz, Polym. Comp. **11**, 65 (1990).
16. A Lustiger, ANTEC '90, 1271 (1990).
17. P.-Y. Jar, W.J. Cantwell and H.-H. Kausch, Comp. Sci. & Eng., to be published.
18. D.J. Blundell, R.A. Crick, B. Fife, J. Peacock, A. Keller and A. Waddon, J. Mat. Sci. **24**, 2057 (1989).
19. J.-N. Chu and J.M. Schultz, J. Mat. Sci. **24**, 4538 (1989).
20. Y. Deslandes, F.-N. Sabir, J. Roovers, Polymer **32**, 1267 (1991).
21. J.T. Hartness, SAMPE J. **20**, 26 (1984).
22. Y. Lee and R.S. Porter, Polym. Eng. & Sci. **26**, 633 (1986).
23. C.M. Tung and P.J. Dynes, J. Appl. Polym. Sci. **33**, 505 (1987).
24. J.A. Peacock, B. Fife, E. Nield and R.A. Crick, "Examination of the morphology of the aromatic polymer composite (APC-2) using an etching technique", presented at the 1st International Conference on Composite Interfaces, Cleveland Ohio, 27-30 May 1986.
25. P. Davies and J.-A. E. Månson, J. Thermoplastic composites, to be published.
26. P.-Y. Jar, H.-H. Kausch, W.J. Cantwell, P. Davies and H. Richard, Polymer Bulletin **24**, 657 (1990).
27. P. Cebe and S.-D. Hong, Polymer **27**, 1183 (1986).
28. D.C. Bassett, R.H. Olley and I.A.M. Al Raheil, Polymer **29**, 1745 (1988).
29. S.Z.D. Cheng, M.-Y. Cao and B. Wunderlich, Macromolecules **19**, 1868 (1986).
30. A.A. Ogale and R.L. McCullough, Comp. Sci. Tech. **30**, 137 (1987).
31. D.J. Blundell, Polymer **28**, 2248 (1987).
32. M.F. Talbott, G.S. Springer and L.A. Berglund, J. Comp. Mater. **21**, 1056 (1987).
33. Ph. Béguelin, "Current activities of the Laboratoire de Polymères in the field of high speed deformation of polymers and composites", 5th Lausanne Polymer Meeting, Lausanne July 2–3, 1990.
34. J. Karger-Kocsis and K. Friedrich, Polymer **27**, 1753 (1986).
35. J. Karger-Kocsis, R. Walter and K. Friedrich, J. Polym. Eng. **8** (1988).
36. K. Friedrich, R. Walter, H. Voss and J. Karger-Kocsis, Composites **17**, 205 (1986).
37. P.S. Pao, J.E. O'Neal and C.J. Wolf, Polym. Mater. Sci. Eng. **53**, 677 (1985).
38. J.F. Mandell, F.J. McGarry, D.D. Huang and C.G. Li, Polym. Compos. **4**, 32 (1983).
39. D.P. Jones, D.C. Leach, D.R. Moore, Polymer **26**, 1385 (1985).
40. T. Kunugi, T. Hayakawa and A. Mizushima, Polymer **32**, 808 (1991).
41. M.C. Chien and R.A. Weiss, Polym. Eng. & Sci. **26**, 6 (1988).
42. Y. Lee and R.S. Porter, Macromolecules **24**, 3537 (1991).
43. J.C. Seferis, Polym. Comp. **7**, 158 (1986).
44. L.H. Lee, J. Vanselow and N.S. Schneider, Polym. Eng. & Sci. **28**, 181 (1988).
45. P. Cebe, S.-D. Hong, S. Chung and A. Gupta, ASTM STP **937**, Ed. N.J. Johnston, ASTM, Philadelphia, 342 (1987).
46. P. Cebe, S. Y. Chung and S.-D. Hong, J. Appl. Polym. Sci. **33**, 487 (1987).
47. C. Cottenot, C. G'Sell, J.-M. Hiver and P. Sigety, 7° Journées Nationales sur les Matériaux Composites (JNC 7), Lyon, Nov. (1990).
48. D.J. Kemmish and J.N. Hay, Polymer **26**, 905 (1985).
49. S.F. Wang and A.A. Ogale, Polym. Eng. & Sci. **29**, 1273 (1989).
50. C. Carfagna, E. Amendola, A. D'Amore and L. Nicolais, Polm. Eng. & Sci. **28**, 1203 (1988).
51. P.T. Curtis, P. Davies, I.K. Partridge and J.-P. Sainty, Proc. ICCM6-ECCM2, Elsevier Applied Science, London & New York, **4**, 401 (1987).

52. R.A. Crick, D.C. Leach, P.J. Meakin and D.R. Moore, J. Mater. Sci. **22**, 2094 (1987).
53. E.J. Kramer and L.L. Berger, Ch. 1, Adv. Poly. Sci. **91/92**, Ed. H.-H. Kausch, Springer-Verlag (1990).
54. A.M. Donald, J. Mat. Sci. **20**, 2630 (1985).
55. C.J.G. Plummer and A.M. Donald, J. Poly. Sci. - Poly. Phys. Ed. **27**, 327 (1989).
56. L.L. Berger, and E.J. Kramer, Macromolecules **20**, 1980 (1987).
57. E.J. Kramer, Ch. 1, Adv. Poly. Sci. **52/53**, Ed. H.-H. Kausch, Springer-Verlag (1983).
58. C.J. Henkee and E.J. Kramer, J. Poly. Sci. - Poly. Phys. Ed. **22**, 727 (1985).
59. B.D. Lauterwasser and E.J. Kramer, Phil. Mag. **A39**, 469 (1979).
60. P.G. DeGennes, J. Chem. Phys. **55**, 572 (1971).
61. M. Doi and S.F. Edwards, J. Chem. Soc. Far. Trans. **74**, 1789, 1802 (1978).
62. T.C.B. McLeish, C.J.G. Plummer and A.M. Donald, Polymer **30**, 1651 (1989).
63. K. Friedrich, Ch. 5, Adv. Poly. Sci. **52/53**, Ed. H.-H. Kausch, Springer Verlag (1983).
64. I. Narisawa and M. Ishakawa, Ch. 8, Adv. Poly. Sci. **91/92**, Ed. H.-H. Kausch, Springer-Verlag (1990).
65. A. Peterlin, J. Mat. Sci. **6**, 490 (1971).
66. J. Petermann and H. Gleiter, J. Poly. Sci. - Poly. Phys. Ed. **10**, 2333 (1972).
67. W.W. Adams, D. Yang, and E.L. Thomas, J. Mat. Sci. **21**, 2239 (1986).
68. H.G. Olf and A.J. Peterlin, J. Poly. Sci. - Poly. Phys. Ed. **12**, 2209 (1974).
69. J.H. Harris and I.M. Ward, J. Mat. Sci. **5**, 573 (1970).
70. T.E. Brady and G.S.Y. Yeh, J. Mat Sci. **8**, 1083 (1973).
71. D.E. Morel and D.T. Grubb, Polymer **25**, 417 (1984).
72. A.M. Donald and E.J. Kramer, Polymer **23**, 461 (1982).
73. C.J.G. Plummer and A.M. Donald, J. Mat. Sci. **26**, 1165 (1991).
74. C.J.G. Plummer, Unpublished results, Cambridge (1990).
75. D.E. Morel and D.T. Grubb, J. Mat. Sci. Lett. **3**, 5 (1984).
76. A.P. More and A.M. Donald, Paper 11., Proc. 8th International Conference on Deformation, Yield and Fracture of Polymers, Cambridge, April (1991).

Chapter 5

Structure and Mechanical Properties of Other Advanced Thermoplastic Matrices and Their Composites

P. Davies and Ch.J.G. Plummer

Abstract

The basic properties of potential matrices for polymeric composites other than poly (ether ether ketone) (PEEK) have been reviewed. More detailed discussion is given of two polymers, polyphenylene sulphide (PPS) and polyethersulphone (PES), which are representative of high performance semi-crystalline and amorphous polymers respectively. Finally some consideration is given of the properties and the suitability for composite applications of thermoplastic (thermotropic) liquid crystalline polymers (LCPs).

1 Introduction

High performance polymers at present account for around 1% of the world market for engineering polymers (currently dominated by such as polyamide (PA), polycarbonate (PC) etc.). Nevertheless, this balance is likely to shift markedly in their favour as they become increasingly competitive with traditional materials such as light metals, glasses and ceramics [1].

As far as thermoplastic polymers for composite matrices are concerned, polyetheretherketone (PEEK) has dominated the scientific literature in the last five years, but a large and increasing number of other polymers are being considered as matrices for high performance applications. Indeed, as will be discussed below, much initial development work on thermoplastic composites was carried out on carbon fibre reinforced polysulphone (PSU). It is the aim of this chapter to give a brief overview of these other options available. This will then be followed by a more detailed discussion, in which three examples will be treated:

Peter Davies, Laboratoire de Matériaux, Insitut Français de Recherche pour l'Exploitation de la Mer, (IFREMER), Centre de Brest, BP70, 29280 Plouzané, France; Christopher J.G. Plummer, Laboratoire de Polymères, Ecole Polytechnique Fédérale de Lausanne, DMX-D, 1015 Lausanne, Switzerland

- Polyethersulphone (PES)
- Polyphenylene sulfide (PPS)
- Liquid crystalline polymers (LCPs).

These are intended to represent the three classes of polymers currently under active consideration. The first two are amorphous and semi-crystalline respectively and have been widely studied. Discussion of the third, the liquid crystalline polymers, will be more speculative as their oriented nature has to date thwarted attempts to produce satisfactory fibre reinforced composites.

The chapter is not intended to be a comprehensive review; the reader is referred to recent authoritative publications by *Leach* [2] and *Carlsson* [3] for more complete summaries of the commercially available materials and their properties. The intention here is rather to provide an introduction to the subject and to indicate areas where problems exist with currently available matrix materials.

1.1 Historical

Although the commercial development of thermoplastic matrix composites dates from the early 1980's, the potential for such materials was recognized much earlier. For example, studies by *Hoggatt* [4] and *Maximovich* [5, 6] demonstrated the attractive properties of laboratory samples of carbon and aramid fibre reinforced PSU and polyarylsulphone (PAS) while *Husman* and *Hartness* described mechanical properties of carbon reinforced polyphenylsulphone (PPSU) [7].

However, a major problem slowed further development of these composites for aerospace applications; the amorphous matrices were susceptible to solvent attack. A NASA programme directed towards improving the solvent resistance of PSU by introducing reactive groups at the ends of molecules met with some success [8], but it was not until the introduction of high performance semi-crystalline thermoplastics, and in particular, the development of fibre impregnated PEEK, that the thermoplastic matrix composites emerged as serious competitors for structural aerospace components. Prepregs of carbon fibres in PEEK and PPS were commercially available from the early 1980's, and now in the early 1990's, they are being considered for primary structures in military aircraft.

1.2 Polymer Matrix Materials and Properties

A selection of polymers susceptible to be used as matrices for advanced composites is presented in Table 1. Three groups may be identified, amorphous, semi-crystalline and "aligned molecular" polymers. Materials are presented with trade names, corresponding composite designations, glass transitions and melting temperatures (where appropriate).

It would be useful to be able to list the properties of these polymers and then use such a list to comment on their potential as composite matrices. Unfortunately this is an extremely hazardous exercise. Property values are quoted by manufacturers,

5. Structure and Properties of Matrices

Table 5.1 High Performance Matrix Materials[1]

Class	Polymer	Trade names, Supplier	T_g °C	T_m °C
Amorphous	PES	*Victrex*, ICI	230	
		Ultrason E, BASF		
	Polyetherimide (PEI)	*Ultem*, GE	215	
	Polyamideimide (PAI)	*Torlon*, Amoco	260	
	Polyarylene sulfide (PAS)	*PAS2*, Phillips Petroleum	215	
	Polyimide	*Avimid K-polymer*, Du Pont	250	
	PSU	*Udel*, Amoco	190	
		Ultrason S, BASF		
Semicrystalline	PEEK and similar	*Victrex PEEK*, ICI,	140	345
		Victrex PEK, ICI	160	365
		Hostatec PEEKK, Hoechst	165	360
		Ultrapek PEKEKK, BASF	175	375
		Declar PEKK, Du Pont		
	PPS	*Ryton, Avtel*, Phillips Petroleum	85	285
		Fortron, Hoechst		
		Primef, Solvay		
		Tedur, Bayer		
		Supec, GE		
	PA	*J-polymer*, Du Pont	160	
Aligned Molecular Polymers	LCP	*Vectra*, Hoechst		285
		Xydar, Amoco		420
		Ultrax, BASF		
		Rhodester, Rhône Poulenc		310
		HX, Du Pont		
		LCC, Eastman Kodak		
		and many others in development		

[1] trade names in italics

but generally for injection moulding grades. For example, the tensile moduli of most of the amorphous and semi-crystalline polymers are in the range 2.5 to 3.5 GPa. However, other properties of interest such as failure strains and toughnesses are very dependent on molecular weight and crystalline structure, two parameters which are in turn determined by the impregnation technique and processing route selected to produce the composites. Such data for unreinforced polymers will therefore be of limited relevance to the final composite behaviour. In order to appreciate the constraints imposed by the need to impregnate fibres it is necessary to consider the techniques available.

1.3 Impregnation

Most of the early work on thermoplastic matrix composites was performed on panels produced by the film stacking method. This is a patented approach allowing the impregnation and forming of parts, provided that films of the matrix polymer are available [9, 10]. It is a discontinuous process but has been employed in the commercial production of large (0.9 m diameter) glass fibre reinforced PES radomes [11].

A second variation of this method uses polymer film and pre-impregnated fibres [9]. For amorphous polymers such as PES, a dichloromethane solvent has been used to impregnate fibres from solution. Park has patented a method to produce a thermoplastic composite by solution coating [12]. The production of prepreg from solution can be applied to most amorphous polymers, although the quality of the prepreg produced after removal of solvent is variable (Fig. 1). However, as the prepreg undergoes a second thermal cycle during moulding, defects such as those in Fig. 1 may not be critical.

While production of prepreg from amorphous polymers is relatively straightforward, the impregnation of fibres with semi-crystalline polymers, which are only soluble in products such as hot, concentrated acids, has led to the development of several novel techniques.

An attractive choice for continuous production of preforms is hot melt impregnation. Unfortunately the high melt viscosity of the technical semicrystalline thermoplastics will lead to poor wetting. However, if the melt viscosity can be reduced then this problem can be avoided and a patent application from ICI proposes a number of ways in which good wetting can be achieved [13]. A melt viscosity of between 1 and 10 Ns/m^2 is required, but results suggest that such viscosities can be achieved by lowering molecular weight, without seriously reducing properties. Pultrusion processing with local heating can therefore be used to produce tapes or sheet material.

A number of other ingenious impregnation methods have also been suggested. For example, an Atochem patent developed by Ganga describes the production of sheaths of polymer containing fibres surrounded by polymer in powder form (Fig. 2) [14]. Such prepregs are very easy to drape and can be filament wound and Ciba Geigy have taken out a licence for this process in the USA. Powder technology routes are also being actively studied by BASF [15]. Other patents involving matrix polymer in powder form are those of *Spie-Batignolles* [16] and *de Charentenay* and *Robin* [17]. The latter used the Joule effect to heat carbon fibres as they were passed through a fluidized bed.

Finally, another approach developed by BASF has been to co-mingle carbon fibres with fibres of matrix material. Such co-mingled yarns can then be used for weaving or filament winding [18].

Impregnation difficulties can thus impose significant restrictions on the range of properties available for composite matrices and the memory of the impregnation route may not be eliminated by subsequent processing. Given that a prepreg has been produced however, a second parameter to be considered is the fibre-matrix interface.

1.4 The Fibre-Matrix Interface

One of the most difficult properties of a composite to quantify is the strength of the fibre-matrix interface. Part of the reason for this is that the interface is not a clearly defined boundary but rather a region which may extend over a significant distance from the fibre, frequently described as an interphase [19]. While such interphases in epoxy composites may include regions where fibre sizing (the fibre coating added to

5. Structure and Properties of Matrices   145

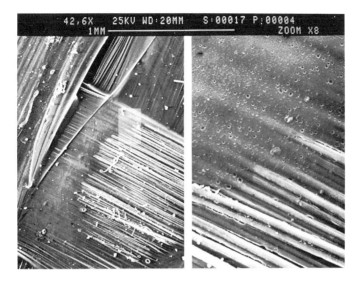

Fig. 5.1 Prepreg of woven T300 carbon fibres in a PEI matrix, showing regions of poor fibre wetting and traces left by bubbles after solvent removal.

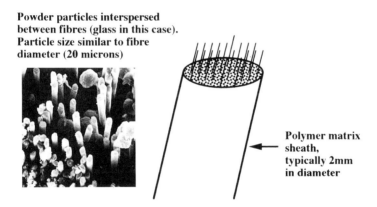

Fig. 5.2 Fibre impregnated thermoplastic (FIT). Carbon and glass fibres can be mixed with powdered matrix in a sheath formed of the matrix material.

improve handling and bonding) and matrix interact they are likely to be limited in size. In some thermoplastic composites with high fibre contents however, the influence of the fibres on crystallite nucleation in the matrix may be such that many properties of the composite are controlled by this interphase as crystals growing from adjacent fibres impinge. The influence of interphase regions on different mechanical properties has been discussed recently [20] but the incorporation of such a region in predictive models is limited by difficulties in measuring interphasial properties.

A range of tests are available to quantify the shear strengths of fibre-matrix interfaces [21] but many of these are based on specimens containing single fibres and are of doubtful application to real composites where both the influence of adjacent fibres on matrix structure and the stress state around fibres may be completely different [22]. Nevertheless, some work on fibres surrounded by thermoplastic matrices has shown that it is possible to obtain good adhesion of thermoplastics to some types of fibre, e.g XAS [23]. Interestingly, in that study AS4 fibres could not be made to show good adhesion with any of a range of thermoplastics (not including PEEK or PPS), and the AS4 fibres are those most frequently used in PEEK and PPS composites.

The interactions between fibres and semicrystalline thermoplastics have been the subject of a very large number of publications. The effect of fibres on PEEK is discussed elsewhere in this book while some comments on PPS are also given below. It is worth noting here, however, that one of the most interesting aspects of semicrystalline thermoplastics is the apparent increase in deflection temperature under load (DTUL) when fibres are present, resulting in useful mechanical properties well above T_g [24]. Finally, the optimization of fibre sizings for thermoplastic composites is not a subject often discussed in the open literature but, as will be seen later for PPS, it can be a major source of concern in the development of a commercial product.

2 Case Studies

Three examples will now be considered, which correspond to the three principal types of thermoplastic matrix, involving amorphous, semi-crystalline and aligned molecular polymers. The first case to be considered is that of the amorphous thermoplastic PES. Here recently established molecular models allow discussion of mechanical properties in microscopic terms, an approach which may provide valuable pointers for its successful application as a composite matrix. In the second case, that of the semicrystalline polymer PPS, the discussion will centre on the influence of crystallinity on mechanical properties, and its role in composite behaviour, areas which have received most attention in the past. This has perhaps been to the detriment of fundamental studies, and indeed no firm basis exists to discuss PPS or indeed any other semicrystalline polymer in terms of the molecular ideas advanced for the amorphous thermoplastics. More discussion of this question in the context of PEEK is given in chapters 3 and 4. Finally some comments will be made on the possibilities of using aligned molecular thermoplastics, that is LCPs, both to reinforce composites and as composite matrices.

2.1 Polyethersulphone (PES)

Of the amorphous thermoplastics, PES appears particularly promising as a matrix for advanced composites in view of its elevated T_g (approximately 230°C). It has an outstanding creep performance with a heat deflection temperature in the region of 203°C, which rises to 216°C on glass fibre addition. Disadvantages include relatively high processing temperatures (340 to 390°C), a tendency for water absorption and lack of resistance to certain organic solvents.

Both the basic resins and short fibre filled grades, supplied by ICI until August 1991 (Victrex PES^{TM}) and BASF (Ultrason E^{TM}) currently find wide application for moulded parts in the electrical and electronics industry, and also increasingly in the automotive and aeronautics industries.

Since its inception as a resin for continuous fibre composites [6], PES has received far less attention than its semi-crystalline counterpart PEEK. High matrix ductility is certainly a key factor in the superior mechanical performance of both PES and PEEK composites. However, the tendency in the literature to refer to PEEK as a 'typical' thermoplastic matrix may be misleading. Indeed the detailed behaviour of the amorphous PES matrix needs to be considered from a distinct viewpoint, not only from that of semi-crystalline thermoplastics, but also from that of what might be termed "conventional" low performance amorphous thermoplastics (polystyrene (PS) for example).

A particularly important aspect of the mechanical behaviour of amorphous polymers is the existence of ductile-brittle transitions. These may depend not only on length scales associated with sample geometry and extrinsic defects, but also on the competition between the two principal micro-deformation mechanisms, shear and crazing. These in turn are linked to molecular parameters such as molecular weight and entanglement density. In what follows some recent ideas concerning the molecular mechanisms of deformation in amorphous polymers will be reviewed with specific reference to PES. Where possible, these will be related to the macroscopic behaviour. Finally a brief assessment of PES as a composite matrix will be given.

2.1.1 Amorphous Thermoplastics; General Considerations

At room temperature amorphous polymers generally take the form of an isotropic glass, such that for sufficiently high molecular weights, relative motion of the constituent chains is held to be "frozen out" below the T_g. Thus, while local rearrangements of the chain conformations are possible above a certain yield point, leading to a locally ductile response to applied stresses, such deformation should remain stable, with work hardening setting in as the chains become highly stretched.

The similarity of the rheological behaviour of high molecular weight amorphous polymers to that of a physically cross-linked rubber has led to the concept of the entanglement network. The amorphous polymer in the glassy state is viewed as a network of polymer strands of mean molecular weight M_e. These are linked at so-called entanglement points, whose density in space, v_e, is given approximately by $6 \times 10^{26} \rho/M_e$, where ρ is the density. An effective M_e may be determined from the short-term shear modulus G' in the temperature regime just above T_g, where it displays a "rubbery plateau". In this regime, conventional rubber elasticity theory gives [25]

$$M_e = 10^3 \frac{\rho RT}{G'}. \qquad (1)$$

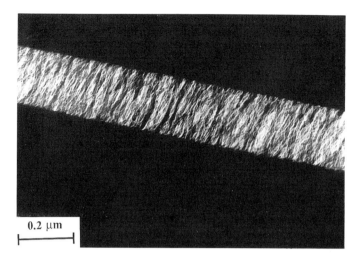

Fig. 5.3 A TEM micrograph of a craze in a thin film of PES deformed in tension at 170°C.

Based on a knowledge of the relevant bond lengths and angles, one can thus calculate the corresponding entanglement strand contour length λ_e. Assuming such strands to adopt a Gaussian conformation in the undeformed state, one may also estimate the spatial separation of topologically linked entanglement points, d_e.

The theoretical maximum extension ratio of the polymer in the glassy state, λ_{max}, corresponding to full extension of the entanglement strands, may then be estimated from

$$\lambda_{max} = \frac{\lambda_e}{d_e}. \qquad (2)$$

For PES, v_e is $50 \times 10^{25}\,m^{-3}$ and λ_{max} is approximately 1.5 [26]. In view of the approximations involved in such calculations, this value of λ_{max} is remarkably close to the extension ratios characteristic of ductile shear necking in PES at room temperature (between 1.4 and 1.7), and indeed it is one of the strengths of the entanglement network model that experimental and theoretical extension ratios are in good agreement for a wide range of amorphous polymers [27].

2.1.2 Crazing

It is also found, again based on observations of a wide range of amorphous polymers with different v_e, and also model systems such as polyphenylene oxide (PPO)-PS blends, in which v_e may be varied by varying the composition [28], that there is a tendency towards brittle failure at room temperature with decreasing v_e. This is characterized by the appearance of widespread crazing under tensile loads [29]. Although crazing is essentially a ductile deformation mechanism, it differs from simple shear necking in that deformation is highly localized, and characterized by a dense network of interconnected voids and highly drawn fibrils as illustrated in Fig. 3. Subsequent breakdown of the craze fibrils leads readily to crack nucleation and consequent embrittlement in bulk samples where crazing dominates.

The presence of voids within the craze body provides an explanation for the link between the onset of crazing and v_e. The energy required to create new regions of void surface in an entangled polymer comprises not only the Van der Waals surface energy, but also a term relating to disruption of the entanglement network. For low v_e polymers such as PS which craze readily at room temperature, *Kramer* [29] assumed that all entangled strands crossing unit area must undergo scission in order for unit area of new void surface to be created. The effective void surface energy is then given by

$$\Gamma = \gamma + \tfrac{1}{4} d_e v_e \quad (3)$$

where γ is the Van der Waals surface energy and U is the energy needed to break a primary chain. For PS the two contributions are approximately equal, but for high entanglement density polymers such as PES, the second term may be increased by up to an order of magnitude.

The crazing stress S_c in Kramer's model scales as

$$S_c \sim \left(\frac{\sigma_f \Gamma}{h}\right)^{1/2} \quad (4)$$

where σ_f is the flow stress for simple shear deformation and h is the width of the strain softened layer at the craze-bulk interface (from which the craze fibrils are formed by a surface drawing mechanism). It is thus easy to see why crazing should be favoured over shear deformation in PS at room temperature, but not in PES and other high v_e amorphous polymers [29, 30].

Such theoretical ideas are principally based on the results of transmission electron microscopy (TEM) studies of the tensile deformation of thin films. For film thicknesses of the order of 1 mm and for strain rates between $10^{-6} s^{-1}$ and $10^{-2} s^{-1}$, PES shows only shear deformation at room temperature, whereas PS shows only crazing, consistent with the above discussion. However, it is found that on heating there is a transition from shear to crazing in physically aged PES at a temperature which decreases with decreasing molecular weight and strain rate [30]. Physical aging refers here to heat treatment just below T_g, which has the effect of raising the yield stress. Since physical aging generally has no effect on the crazing stresses, it is a useful way of promoting crazing in thin films in experimentally accessible regimes, without altering its basic phenomenology. In high molecular weight PS on the other hand, crazing is favoured by high strain rates and low temperatures, as would be expected on the basis of Kramer's scission model for crazing. Moreover, since neither scission nor shear deformation is associated with any molecular weight dependence, it is clear that an alternative explanation must be sought for the crazing behaviour of PES.

Similar anomalies in the high temperature behaviour of low molecular weight PS [31] led to the suggestion that under certain conditions, scission could be replaced by disentanglement as the mechanism for entanglement loss during crazing. In this view, entanglement is no longer a permanent constraint in the glassy state. Rather, for sufficiently high temperatures and/or low strain rates, individual polymer chains may be 'pulled' out of the entanglement network under the influence of local stresses. Indeed the whole concept of the entanglement network is discarded in favour of the 'tunnel' model [32,33]. Here, the individual chains are considered to be trapped within a notional tube, representing the effect of entanglement, and whose length is proportional to the chain molecular weight. Thus when a chain of molecular weight M is pulled out of its confining tube at some velocity v (forced reptation [34]), it experiences a frictional force, f_d, which scales as

$$f_d \sim \zeta_o v(M/M_o) \qquad (5)$$

where ζ_o is the monomeric friction coefficient and M_o is the monomer molecular weight. Where f_d is less than the force required for scission, then clearly disentanglement will be the preferred mechanism.

Hence, since ζ_o is a strongly decreasing function of temperature and strain rate, disentanglement crazing is expected to be seen as the temperature approaches T_g, and to be favoured by low strain rates. Recent work [34–36] has linked S_c explicitly to the frictional force and suggests that thin films of glassy polymers should show the following regimes of behaviour:

(i) at low temperature, scission crazing in low entanglement density polymers (PS) and shear deformation in high entanglement density polymers (PES);

(ii) above some transition temperature a steep drop in the crazing stress as disentanglement progressively replaces scission, whose strong temperature dependence reflects that of ζ_o; for high entanglement density polymers this is seen as a shear to craze transition;

(iii) as f_d continues to decrease, contributions from disentanglement to the void surface energy Γ become negligible and there is no longer a strong temperature dependence in the crazing stress; this is called the 'Van der Waals' regime.

Fig. 4 gives a 'deformation map' [37] of the behaviour expected for PES aged at 200°C for 70 hours, based on thin film data for different strain rates, which reflect well these ideas. Evidence for the low temperature/high strain rate regime of scission crazing has not been obtained directly, although its existence is implied by the model. Similarly, the high temperature/low strain rate regime in which shear deformation replaces Van der Waals crazing has been inferred from experiments on cross-linked PES thin films at relatively high strain rates, in which the crazing stress in (ii) and (iii) was raised [38].

It should however be emphasized that the competition between shear and crazing depends not only on the molecular weight and physical ageing, but also on the local sample stress state, and consequently on the sample thickness. Crazing is promoted over shear in the presence of a large dilatational stress component such as is normally found at notch tips in plane strain samples under tension, but not, for example, by compression. Thus, although not observed in thin films, scission crazing can and does play an important role in the room temperature fracture behaviour of bulk PES

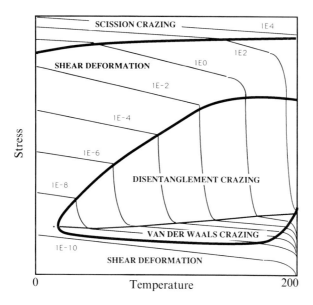

Fig. 5.4 Schematic of the deformation behaviour of physically aged PES films for different strain rates and temperature, based on data for strain rates in the range 10^{-2} to 10^{-6} s^{-1} and a molecular weight average of 47,000. Increasing M will move the boundary between shear deformation and disentanglement crazing to higher temperatures.

specimens. Fig. 4 may nevertheless be considered as a qualitative indication of the general trends implied by the above molecular interpretation.

2.1.3 Crazing and Toughness

Craze break-down is of crucial importance since it potentially provides a direct link between the microscopic behaviour and crack nucleation and the critical strain energy release rate G_c. There has been much recent discussion of this question in the literature for a variety of polymers, although it remains substantially unresolved.

Craze Stability In PS thin films Yang et al. [39] have provided compelling evidence that crack *nucleation* within crazes at low strain rates arises from disentanglement at local weak spots at the craze-bulk interface. Consequently high M increases the resistance to crack nucleation within crazes. However, although correlated with the bulk value for M, at room temperature the effective molecular weight distribution in PS crazes is considerably broadened owing to the predominance of scission during craze formation. In clean samples, the weak spots thus represent the extreme low molecular weight tail of this distribution.[1]

---

[1] In samples containing significant amounts of dust or poorly bonded second phase particles, craze stability may be substantially reduced.

As the temperature is raised, the mean craze molecular weight will gradually increase, and its distribution sharpen as scission is replaced by disentanglement. Thus, although a disentanglement mechanism would imply a strong decrease in stability with increasing temperature for a fixed craze molecular weight distribution, the stability may in fact increase with temperature in the immediate vicinity of the scission-disentanglement transition temperature, as the effective molecular weight increases [40]. Nevertheless, when the temperature exceeds the scission-disentanglement transition temperature there will be a rapid decrease in craze stability.

Craze Breakdown During Crack Propagation   Whilst crack nucleation as described above occurs at the craze-bulk interface, during crack propagation the crack advances through the so-called 'mid-rib' of the craze ahead of the crack tip. This process has recently been considered at a molecular level by *Brown* [41], who has argued the case for a mixed scission/disentanglement mechanism during mode I crack propagation through a craze. That the local stresses at the crack tip can be sufficiently high to cause fibril breakdown by scission at the craze mid-rib is argued to be a consequence of coupling of adjacent fibrils via 'cross-tie fibrils' (whose existence has been confirmed by TEM [42]).

Even at low temperature, G_{Ic} is well known to decrease with decreasing M, and also with increased polydispersity, since the increasing chain end density will have a weakening effect on the fibrils. Eventually however, as the temperature is raised, the fibril strength might reasonably be expected to become dependent on the competition between scission and disentanglement much as does the crazing stress, and in particular, G_{Ic} should decrease strongly with temperature above the transition from scission to disentanglement crazing. In intermediate regimes the situation may be complicated by changes in the effective molecular weight distribution in the crazes and the density of cross-tie fibrils with temperature. G_{Ic} is also approximately inversely dependent on the craze stress itself (whence the toughness of high impact polystyrene (HIPS) where the local stress fields adjacent to rubber particles result in a reduction in the applied stress level required for crazing, compared with that in pure PS). At present, lack of experimental data severely limits the extent to which such effects can be quantified.

These ideas also diverge somewhat from earlier conclusions based on extensive characterization by optical interferometry of bulk crazing in PMMA [43, 44], which stated that craze break-down is due to fibril creep at the craze mid-rib and that this is in turn controlled by the β relaxation. The main objections raised to this have centred around the insensitivity of the craze width at the crack-tip to crack speed [41], and the absence of systematic extension ratio variations within the craze bulk in thin films [35].

Further, whilst it is clear that the β process must control many aspects of the phenomenology of scission crazing in PMMA, for which the kinetics are essentially determined by plastic flow, it cannot, taken in isolation, explain the molecular weight dependence of G_{Ic}. Nevertheless, the creep model does retain considerable support, and it is possible that future interpretations will combine elements of both approaches.

2.1.4 Macroscopic Failure

As previously mentioned, the crucial difference between thin films and bulk specimens is the possibility of a macroscopic plane strain state in the sample interior. In the presence of a notch this may lead to elevated triaxial stresses. Similarly, local shear yielding in the bulk will tend to result in a decrease in the deviatoric component of the stress tensor owing to the constraints imposed by the surrounding undeformed material, and again, to locally elevated triaxial stresses, which will tend to mitigate against propagation of shear deformation.

On the other hand, craze nucleation is widely held to be favoured by hydrostatic tension (which promotes the formation of voids) and to depend on the maximum principal stress rather than the deviatoric stress. Likewise, craze propagation is dependent on the maximum principal stress. Thus plane strain will tend to favour crazing, while plane stress will favour shear.

Simple Tension; The High Temperature Ductile-Brittle Transition Simple tensile tests on aged and unaged injection moulded PES bars of various molecular weights [45] have shown similar behaviour to that of thin films, in spite of the thickness differences. Fig. 5 shows results for aged samples tested at 1mm/min cross-head displacement. It is found that as the temperature is raised samples of lowest molecular weight tested exhibit a ductile (stable macroscopic necking) to brittle transition, and that there is a sharp drop in failure stress in the transition region for the aged samples. The ductile-brittle transition moves to higher temperature with increasing M, and indeed the highest molecular weight samples remain ductile at all temperatures under

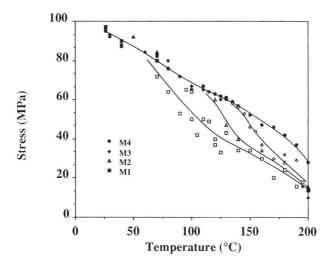

Fig. 5.5 Tensile tests for four different molecular weights of PES, aged at 180°C for 70 hours; M1, M2, M3, M4 have molecular weight averages Mw of 47,000, 52,000, 58,000 and 63,000 respectively; open symbols represent brittle fracture, filled symbols represent ductile necking.

these test conditions. At low temperatures, the initial deformation mode is diffuse shear banding, developing into relatively sharp necking. Where brittle failure does not intervene, the necks become progressively less well defined with increasing temperature, such that deformation at 200°C is homogeneous. On the other hand, the angled microshear banding preceding macroscopic neck formation is best defined at intermediate temperatures (Fig. 6).

Whilst such behaviour is consistent with the craze initiation behaviour observed in thin films, craze stability may also play a role in the high temperature ductile/brittle transition. Indeed, whereas fracture appears simultaneous with craze initiation in the low M_w samples, stable crazes can be observed in the high M_w samples at elevated temperatures, since shear yielding intervenes prior to craze breakdown. The improved stability of crazes in high M_w samples is consistent with the previous discussion in section 2.1.3.

Fracture Behaviour; The Low Temperature Ductile-Brittle Transition   The fracture behaviour of PES compact tension specimens has been discussed using the concept of 'mixed mode fracture' [46]. This has nothing to do with the loading mode, but refers to the observation that tensile fracture surfaces are characterised by an interior plane strain region which fails by crazing at room temperature, and an outer plane stress region which fails by shear, giving rise to characteristic shear lips.

Thus, G_{Ic} and, in particular, crack resistance curves will contain contributions from both crazing and shear, whose proportions will depend on sample dimensions and test conditions, including the width of the shear lips. The width of the shear lips may, for example, be diminished by aging below T_g, which raises the yield stress (thus providing a convenient test for this interpretation [46]). Whilst in what follows it is implicitly assumed that plastic zone sizes are very much smaller than the sample dimensions, it is important to realize the importance of size effects in regimes of behaviour where fracture mechanics applies.

True plane strain critical strain energy release rates in mode I loading should nevertheless represent the cohesion of crazes ahead of a crack, and thus show a molecular weight dependence as discussed in section 2.1.3. *Davies et al.* [47] have measured plane strain K_{Ic} values (equal to $(EG_{Ic})^{1/2}$) for unaged PES single edge notched three point bend tests for the same range of M_w used for the tensile tests in the previous section, at different temperatures between $-65°C$ and room temperature. In this regime K_{Ic} shows little temperature dependence, but an increase of between 25% and 30% as Mw is increased from 47000 to 63000. In conjunction with tensile yield stress measurements such data may be used to account for observations of a low temperature ductile-brittle transition using simple fracture mechanics. In the Griffith approach

$$\sigma_b = K_c/(Ca^{1/2}) \tag{6}$$

where C is a geometrical factor, σ_b is the breaking stress and a is some characteristic incipient flaw size, assumed to be constant. If one assumes that brittle failure in a tensile test takes place when sb is less than the tensile yield stress, the criterion for brittle behaviour is

$$\left(\frac{KI_c(T)}{\sigma_y(T)}\right)^2 \leq Ca. \qquad (7)$$

Thus the transition from ductile to brittle behaviour as the temperature is decreased may be explained by the increase in the yield stress σ_y. *Davies et al.* [47] were able to use this approach to account for falling weight impact test data which showed the ductile-brittle transition temperature to decrease with increasing M_w in PES (from $-25°C$ for $M_w = 47000$, to $-65°C$ for $M_w = 63000$). Whilst σ_y is approximately independent of M_w, the decrease in $K_{Ic}(T)$ with decreasing M_w lowers the value of $\sigma_y(T)$ for which equation 7 is satisfied, and thus the transition moves to higher temperature. In Izod tests, PES is found to be both relatively brittle and highly notch sensitive, to the extent that un-notched samples may not fracture under impact conditions. Values at room temperature range from 76 to 120 Jm^{-1}, again increasing with molecular weight [48].

2.1.5 Long Term Properties

In both fatigue and creep rupture tests, transitions from ductile to brittle failure are found, although there are important differences between the fatigue and creep rupture behaviour [47]. In room temperature fatigue there is a ductile-brittle transition with decreasing load/increasing number of cycles, which is relatively weakly dependent on M_w. This has been discussed in terms of fracture mechanics [47, 49]. It has been argued that the plastic zone size r_p at crack tips will decrease with the number of cycles, as reflected by the measured decrease in K_{Ic} during fatigue loading, and that embrittlement will occur when r_p becomes small with respect to the sample dimensions.

In creep loading, although no ductile-brittle transition was seen by *Davies et al.* [47] at room temperature in the range 0 to 10^6 s, a strongly molecular weight depen-

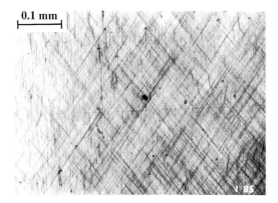

Fig. 5.6 Shear banding and crazing in PES, molecular weight average 63,000 deformed at 170°C.

dent and temperature dependent transition became apparent as the temperature was raised, moving to shorter times with decreasing M_w and increasing temperature. (Even so, at temperatures as high as 150°C, proprietary grades of PES will still perform as well as, if not better than many thermoplastics do at room temperature [50].)

This behaviour appears to be a further manifestation of the onset of disentanglement induced crazing. As may be inferred from Fig. 4, disentanglement crazing will be favoured by high temperature, low M_w and long times. Further, as implied by the results in section 2.1.4, at low molecular weights, the lack of craze stability means the onset of brittle behaviour will coincide closely with the onset of disentanglement crazing (regardless of the presence of intrinsic flaws). At higher molecular weights, improvement in failure resistance has also been linked to improved craze stability [51].

Fatigue failure appears more difficult to interpret in simple microscopic terms. Results for PES are generaly consistent with those for other high v_e amorphous polymers. However, even where the behaviour is a reflection of the competition between shear and crazing, the micromechanics of fatigue failure show a complex dependence on both internal and external variables such as temperature [52]. Important factors may include not only changes in crack propagation, craze formation and craze breakdown mechanisms [43, 44, 52, 53], but also ageing effects, particularly for long times and high temperature [43, 52, 53]. A better understanding of these processes would be of particular importance, given that the relatively poor fatigue performance of PES and other amorphous polymers represents one of their principal disadvantages with respect to semi-crystalline polymers such as PEEK.

2.1.6 PES as a Composite Matrix

There are at present relatively few examples of continuous fibre reinforced PES in aeronautical applications owing to concern over solvent resistance. Whether such reticence is still justified, in view of the decreasing use of aggressive solvents for environmental reasons, is debatable, but doubts such as those detailed above over the long term behaviour of the matrix may continue to limit the applications. Nevertheless, a number of authors have published studies of the properties of carbon/PES composites.

For example, *Lee* and *Trevett* used carbon/PES in a study of different compression tests and found that both film stacked and melt impregnated unidirectional PES composites had consistently low strengths, around 615 MPa. They attributed this to the low modulus of PES [54]. Other published results on PES composites have indicated good impact behaviour in comparison with other carbon reinforced thermoplastic laminates [55]. Fatigue results for PES composites with aligned short carbon fibres were less promising owing to poor fibre-matrix bond strength [56].

The delamination resistance of carbon/PES unidirectional composites with 60% by volume of fibres was studied in a round robin exercise by the European Group on Fracture [57]. G_{Ic} values rising steeply from around 700 J/m^2 at initiation to well over 2000 J/m^2 during propagation were measured. The high propagation values could be explained by the appearance of multiple cracking, but these values indicate

5. Structure and Properties of Matrices 157

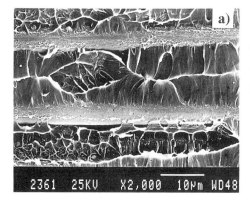
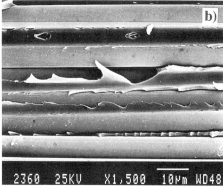

Fig. 5.7 Fracture surfaces of unidirectional carbon fibre (AS4) reinforced PES composites after (a) mode I and (b) mode II delamination.

nevertheless that this is a reasonably tough material. It is not as tough as PEEK but similar to many toughened epoxies and much tougher than the first generation carbon/epoxies currently flying in military aircraft [58]. More detailed studies were undertaken subsequently [59-61] in which mode II and mixed mode results were presented. Of particular interest was the comparison of fracture surfaces after mode I and mode II tests, Fig. 7, which revealed sharply contrasting fractures. In mode I the crack propagated in the matrix suggesting a reasonable fibre-matrix interface. Under mode II loading however, completely clean fibres were observed and mode II delamination resistance was quite low. In fact G_{IIc} values are quite similar to those for G_{Ic} for this material, so that the mode I delamination resistance may not be the critical property to be considered in design.

2.2 Polyphenylenesulfide (PPS)

2.2.1 PPS Matrix

PPS has a long history, with reports of its synthesis dating from the 19th century [62]. The development of a process to synthesize the polymer commercially is quite recent however, and is discussed in detail in a review by *Lopez* and *Wilkes* [63]. The polymer is notable for its good chemical resistance, good thermal resistance (DTUL of 260°C for fibre filled grades) excellent flame resistance, good mouldability and electrical properties, notably its arc resistance, whence wide application in the electrical/electronics industry. Melt temperatures for injection moulding are generally in the range 300 to 360°C As far as its mechanical properties are concerned these are dependent on the crystalline state of the material and it is this aspect which has attracted most attention in recent studies.

The kinetics of crystallization in PPS have been studied by a number of authors, [64-69], and Avrami equations have frequently been used to model the isothermal crystallization process from the melt using the expression:

$$C = 1 - e^{-Kt^n}$$

where C is the volume fraction of crystals at time t, K is the Avrami rate constant and n is the Avrami exponent. Values of n around 3 would indicate that three-dimensional spherulites form from homogeneous nucleation. Very different values of n have been published, as shown in Fig. 8.

The main reason for the difference in the values obtained is that different grades of PPS were tested, varying from films to powder, although the insensitivity of double logarithmic plots probably also hinders such investigations. Nevertheless Fig. 8 does underline the care necessary in inferring the crystallization behaviour of composites from matrix properties; values of n vary from 2 to 3 so crystal growth could be concluded to be either two or three dimensional.

In studies on the influence of molecular weight *Lopez* and *Wilkes* found lower values of n for a lower molecular weight grade of PPS and suggested that intermediate sheaf-like structures may be forming [66]. Both they and *Song et al.* [67] showed that crystallization rate increases with decreasing molecular weight, as would be expected. This is important when continuous fibre composites are considered, as low molecular weight grades are usually preferred in order to achieve good impregnation.

Among the few results linking structure and properties of PPS, *Brady* showed a surprising drop in tensile strength as degree of crystallinity was increased by annealing [70]. This was attributed to the rather low molecular weight and slightly crosslinked nature of the material tested. Chain extension and crosslinking can occur rapidly in PPS on heating above its melting temperature (285°C) in air [65], and this can be an important factor in the composite moulding processes. It may be necessary to heat prepreg in an inert atmosphere if optimum properties are to be obtained.

Toughness data are not available as a function of crystalline state, but some published results suggest that PPS is not a particularly tough polymer. *Karger-Kocsis* and *Friedrich* tested an unreinforced impact modified grade and found a value of K_{Ic} of 0.8 MPam$^{1/2}$, which corresponds to a G_{Ic} value of around 200 J/m^2 [71]. More recently, *Vanderschuren et al.* quoted a value of 650 J/m^2 obtained by an ASTM task group [72], although once again these grades may differ considerably from those employed in composites.

2.2.2 PPS Composites

The development of carbon fibre reinforced PPS composites has been hindered by considerable problems in achieving a good fibre-matrix interface bond. This has resulted in consistently poor mechanical properties in spite of good impregnation of fibre tows. For example, very low transverse tensile strengths around 30 MPa, and an interlaminar shear strength of 70 MPa have been reported [73]. These compare with values of 90 MPa and over 100 MPa respectively for PEEK composites with the same fibres [74]. Reliable 0° compression strength values are difficult to obtain, but these are also low, around 940 MPa compared with 1100 MPa for carbon/PEEK. Both these values are low compared with those measured on many epoxy composites, as discussed in chapter 6. In a comparison of the impact behaviour of composites of PPS and PEEK, both reinforced with AS4 fibres, the PPS composites exhibited compression after impact strengths of approximately half of those of the

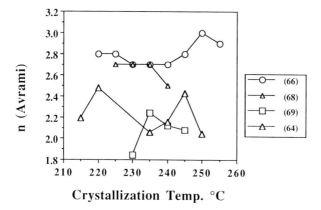

Fig. 5.8 Published values of Avrami exponent "n", for unfilled PPS, values replotted from references [64, 66, 68, 69].

PEEK composites and similar to those of an untoughened epoxy. This was due to significantly larger damage zones in the PPS and epoxy composites after impact [75].

Influence of Fibres on Crystallization of PPS Given the semicrystalline nature of the PPS matrix and the relatively poor fibre-matrix adhesion it is understandable that a considerable number of studies have focussed on the influence of different types of fibre on crystallization kinetics, crystalline morphology, and transcrystallinity. These studies have been complemented by a number of studies of the crystal structure in composites formed from prepreg.

Studies in which a few fibres were added to polymer melts and the interfaces examined during cooling have been described by *Waddon et al.* [76], *Camararo et al.* [77], *Khoury* [78] and *Lopez* and *Wilkes* [79]. These have clearly illustrated the importance of processing conditions, matrix molecular weight and fibre type on interface morphology. *Caramaro et al.* showed that for T300 fibres the development of a transcrystalline region is encouraged by a higher melt temperature (350°C instead of 300°C). *Khoury* examined seven types of carbon fibre, four PAN- and three pitch-based. The high modulus HMU and HMS4 fibres induced transcrystallization growth in the temperature range 270–280°C as did the pitch based fibres, whereas the lower moduli AS4 and IM7 did not. Interestingly, *Gaur et al.* used IM6 fibres in a study using a microbond technique to determine interfacial bond strength [80]. They were able to improve the IM6/PPS bond strength from 49.5 to 73.2 MPa by a plasma treatment. The study of *Lopez* and *Wilkes* on high strength low modulus fibres (XAU, AU4 and T300U) showed increasing nucleating efficiency with increasing fibre roughness [79], but as these authors underlined, many parameters are varied when the fibre type is varied.

These studies have clearly indicated that it is possible to nucleate crystals on, or very near to fibres, and to produce transcrystalline layers. However, they give little insight into whether such a layer will enhance fibre-matrix adhesion. Of more immediate interest to the user is the influence of processing conditions on the behaviour of composite components and this aspect has also received much attention.

Measurements of the crystalline state of the matrix in PPS composites have been performed using DSC (e.g. [69, 81]), FT-IR spectroscopy [82] and X-ray diffraction [83]. As far as higher volume fraction composites are concerned, it was first shown by *Jog* and *Nadkarni* [64] that the addition of fibres, glass in that case, reduced crystallization time and also resulted in a lower degree of crystallinity than for unreinforced PPS. *Desio* and *Rebenfeld* [68] confirmed this result for carbon fibre composites while for aramid fibres a minimum in degree of crystallinity was measured for a crystallization temperature of around 230–235°C. The simplest parameter to employ to correlate crystallization behaviour with moulding parameters is the average cooling rate. Published data for carbon/PEEK composites processed at temperatures of 380°C indicate that in order to suppress crystallization in these materials cooling rates of the order of 2000 K/min or more are necessary [84]. In contrast, for carbon/PPS processed at 330°C amorphous matrix composites can be obtained at cooling rates around 100 to 120 K/min [69, 81]. Annealing cycles may therefore be necessary if fast moulding cycles are required.

Crystal Structure and Mechanical Behaviour While many studies have concentrated on single fibre composites, very few results are available to show how the crystalline state of the matrix influences the composite behaviour. An exception is the work on interlaminar fracture toughness. High delamination resistance is frequently cited as a reason for adopting thermoplastic matrix composites and indeed the fracture behaviour of carbon/PEEK composites is exceptional (see chapter 6). Early mode I results for AS4/PPS indicated very high values of G_{Ic}, around 1400 J/m^2 [85]. Subsequent published values were rather lower, from 830 for AS4 fibres to 1020 J/m^2 for PPS with IM6 fibres [86]. *Davies et al.* tested AS4/PPS in the as-moulded state and then tested a second set of specimens which had been heat treated at 150°C for one hour [87]. The degree of crystallinity, measured by X-ray diffraction, of the as-moulded material was quite low (around 5%) but after heat treatment it had increased to around 30%. Further heat treatment for longer times and higher temperatures did not significantly increase this value further. Mode I and mode II results are shown in Table 2.

Mode I propagation values were similar to published values and dropped slightly (13%) after the heat treatment. However, mode I propagation values included a significant contribution from fibre bridging, as the crack followed the fibre-matrix interface (Fig. 9a). The initiation value may give a more sensitive indication of the state of the matrix and these values dropped by 23% after heat treatment, suggesting a significant embrittlement.

Mode II values were measured using the end notched flexure (ENF) specimen from mode I precracks, and these are also shown in Table 2. The initiation in this case occurs at the fibre-matrix interface, as was confirmed by in-situ tests in the scanning electron microscope, and is not sensitive to the crystalline state of the matrix. On the other hand the amount of damage which the composite can support before macroscopic propagation, the values corresponding to maximum load, are substantially reduced by the heat treatment. Once again fracture surfaces show clean fibres (Fig. 9b).

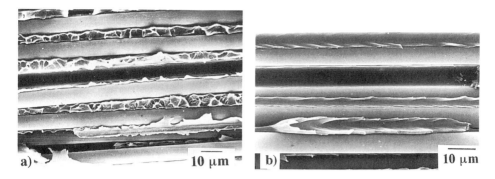

Fig. 5.9 Fracture surfaces of unidirectional carbon (AS4) fibre reinforced PPS after delamination under (a) mode I and (b) mode II loading.

Improved PPS Composites Development work on commercial PPS Composites has been performed principally by Phillips Petroleum, and several families of composite have been produced. After the first generation prepreg, based on PPS and introduced in 1984, two grades of prepreg based on polyarylene sulfides followed, trade-named PAS1 and PAS2. The T_g and melting temperatures of PAS1 were very similar to those of PEEK but transverse tensile strength was even lower than that of carbon/PPS, at around 25MPa, while interlaminar shear strength was improved at 90 MPa [88]. G_{Ic} was similar to that of the earlier PPS composites. For PAS2, a high temperature amorphous matrix containing 60% by volume of fibres and a tensile strength of 1380 MPa, a compression strength of around 900 MPa was quoted.

Then more recently, an improved PPS grade was launched [89]. This matrix is claimed to be five times as tough as the original grade and to have increased ability to adhere to fibres. Significantly improved transverse tensile strength was measured when both glass and carbon fibres were used as reinforcements, 69 MPa for the latter. However, tests on the carbon fibre reinforced grade do not show significantly higher mode I toughness values than those measured on the original grade. The principal difference between the two is that delamination propagation is no longer completely stable and regions of instability similar to those seen in early carbon/PEEK specimens are observed. Fig. 10 shows fracture surfaces from the stable and unstable regions. While the interface is tougher than the matrix during rapid propagation the stable propagation regions (Fig. 10a) show similar features to those in the specimens tested previously (Fig. 9a).

Table 5.2 Delamination Resistance of AS4/PPS Composites in J/m^2, Mean Values (Standard Deviations in Brackets). NL: Value at Non-Linearity on Load-Displacement Curve

	Mode I		Mode II	
Material	Init. (NL)	Propagation	Init. (NL)	Max. Load
As-moulded	610 (40)	894 (79)	461 (27)	1125 (46)
Annealed	470 (80)	774 (88)	477 (6)	948 (10)

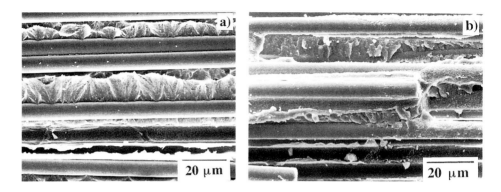

Fig. 5.10 Fracture surfaces of improved carbon fibre/PPS specimens after mode I delamination, showing regions from (a) stable and (b) unstable propagation.

2.3 Liquid Crystalline Polymers

The position of thermotropic liquid crystalline thermoplastics (TLCPs), that is, the thermoplastic LCPs, is somewhat ambiguous as regards their role in composite materials; TLCPs have been considered for use both as fibre reinforcements in conventional thermoplastic matrix composites [90], and as matrices for carbon fibre composites [91]. Indeed unreinforced TLCP injection mouldings and extrudates behave in a similar fashion to short fibre filled conventional thermoplastics, to such an extent that filler addition is often required in order to *limit* anisotropy in TLCPs.

A more detailed discussion of the mechanical properties of injection mouldings of the Vectra™ (Hoechst-Celanese) TLCP is given in chapter 8. Here it is proposed briefly to consider some potential alternative fabrication routes covering the use of TLCPs both as reinforcements and as matrices. For this purpose it suffices to recall that the molecular architecture of a TLCP may be to some extent idealized as that of a rigid rod, and that flow, and in particular extensional flow, results in high molecular alignment in the flow direction. Owing to the relatively low molecular mobility of TLCPs, the flow induced structures are readily quenched into the solid state giving rise to highly anisotropic structures.

2.3.1 TLCP Fibres in Conventional Continuous Fibre Composites

For the high degrees of alignment achievable for example during fibre spinning or extrusion, the moduli in the flow direction approach the axial moduli of the molecules themselves. Where the degree of alignment is high, then assuming a uniform stress distribution along the fibre, the axial fibre modulus E'_{33} is given approximately by

$$\frac{1}{E'_{33}} = \frac{1}{E_{33}} + \frac{1}{G}\langle \sin^2 \phi \rangle. \tag{8}$$

E_{33} and G are the molecular axial and shear moduli and ϕ is the angle between the molecular axis and the fibre axis locally [92, 93]. Typically E_{33} is two orders of magnitude greater than G, whence for melt spun fibres, where values of $\langle \sin^2 \phi \rangle$ of less than 0.01 can be achieved,[1] E'_{33} should exceed $\frac{2}{3}E_{33}$. Consequently, commercial Vectran™ fibres (Hoechst-Celanese), based on a 73% hydroxybenzoic acid (HBA) and 23% 2,6-hydroxynaphthoic acid (HNA) random copolymer, for which E_{33} has been estimated at 130 GPa at room temperature [93], have a modulus in the as-spun state of approximately 70 GPa.

The room temperature tensile strength of the as-spun Vectran fibres is approximately 1.3 GPa, but can be more than doubled by suitable heat treatment below the melting point. This increase has been argued to be a result of the continued solid state polymerization known to occur during such heat treatments, and there is little corresponding change in modulus [94, 95].

On a weight-for-weight basis the tensile strength of Vectran compares well with those of carbon, glass and Kevlar fibres. The specific modulus, however, is only about a third of that normally quoted for carbon fibres, and about half that which can be achieved in Kevlar, for which E_{33} has been estimated at 240 GPa [96], but this is compensated by a somewhat higher strain to fail of about 3%.

The main potential drawback of TLCP fibres with respect to Kevlar is the relatively weak interchain bonding, leading to poor transverse properties. This is partly a consequence of the deliberate suppression of crystallinity in TLCPs in order to facilitate melt processing. Kevlar fibres, which are not melt processable, are highly crystalline and characterized by strong intermolecular hydrogen bonding [97].

There is little in the literature to indicate the potential of Vectran or other TLCP based fibres as reinforcements in conventional continuous fibre composites. Brady and Porter [90] produced continuous-fibre unidirectional composites from combinations of Vectran fibres and: (i) a 58/42 HBA-HNA resin; (ii) PC, with fibre volume fractions of 0.40 and 0.45 respectively. They found that fibres properties degraded if processing was carried out in excess of the fibre melting point, presumably because of relaxation of the fibre orientation. More importantly the composite transverse properties were relatively low, regardless of the matrix. Scanning electron microscopy (SEM) of the transverse fracture surfaces indicated that this was due to fibre splitting. Thus surface treatment (ozonation), in order to improve the fibre matrix bonding, merely increased the extent of fibre splitting, and did not substantially improve the transverse properties. In the transverse direction of the PC based composites, for example, the tensile strength was approximately 13 MPa compared with 65 MPa for the matrix.

2.3.2 TLCP Fibre Reinforcement in 'in-situ' Composites

Whilst the poor transverse properties of TLCP fibres such as Vectran appear to constitute a considerable disadvantage, TLCPs remain of particular interest in they offer the possibility of forming 'in-situ' composites by the extrusion, or injection

---

[1] $S = \frac{1}{2}(3\langle \cos^2 \phi \rangle - 1)$
which is equal to 1 for perfect alignment and 0 for random orientation. $\langle \sin^2 \phi \rangle$ in equation (8) is obtained from $\langle \cos^2 \phi \rangle = 1 - \langle \sin^2 \phi \rangle$.

moulding of blends with conventional thermoplastics, thus avoiding problems associated with fibre wetting and complex geometries.

In general, blends of flexible and rigid or semi-rigid polymers show little miscibility, not least because geometrical constraints force the flexible polymers to adopt other than random coil conformations [98]. Nevertheless, TLCPs have a marked plasticizing effect on conventional polymer melts, addition of a few weight per-cent of HNA-HBA, for example, resulting in large viscosity reductions [99–101]. These and other wider applications of both LCP- and LC-flexible polymer blends have been the subject of several recent reviews [102–105].

Assuming that both phases are well separated, and may be melt processed simultaneously, suitable process conditions may be adopted to impart a high aspect ratio fibrillar morphology to the TLCP component. Injection moulding, extrusion and fibre spinning have all been used to obtain high aspect ratio fibrillar morphologies, examples of which are shown in Fig. 11. However, careful consideration not only of the flow conditions, but also of the rheology and proportions of the respective components is required. For example, the viscosity of the TLCP phase should be less than or approximately equal to that of the flexible polymer, particularly for shear flow, making the choice of processing temperature and thermal history of the TLCP component critical [106–108].

In order to improve molecular orientation in the TLCP fibrils and thus optimize the reinforcement it is generally necessary to draw or stretch the blended component. For example in a blend of PC with 20% of a low melting point TLCP HNA-HBA/10 % Terephthalic acid-10% Hydroquinone (Vectra RD-500), *Weiss et al.* obtained moduli increasing to approximately 7 GPa for a maximum draw ratio of 1000 which compares with values of less than 2 GPa for the starting extrudate [103, 109].

The modulus of the pure TLCP reaches about 26 GPa under similar conditions, whence it was shown that a simple rule of mixing applies up to TLCP contents of approximately 40% [110]. This is reasonable, given that TLCP fibril aspect ratios of the order of 100 can be achieved, permitting the application of continuous fibre composite theory [110, 111]. However, given the relatively modest stiffness of the fibres (cf. the melt spun fibres described in section (i)), the magnitude of the modulus reinforcement is more comparable to that in conventional short fibre filled composites.

The tensile strengths of *in-situ* composites depend strongly on the adhesion between the two phases, which will in turn depend on their compatibility. In general the aromatic advanced thermoplastics, PES and PEEK for example, are likely to show the best compatibility with aromatic TLCPs [112]. On the other hand, perhaps a more promising application of *in-situ* methods is the reinforcement of cheaper, low performance thermoplastics such as polyethylene terephthalate (PET) and polypropylene (PP) where the weakness of the TLCPs perpendicular to the orientation direction is likely to be less critical.

Baird [108] has used reactive extrusion coupled with hot drawing, blow moulding and solid forming routes to produce PP components reinforced with HNA-HBA with good fibre-matrix adhesion. The overall mechanical properties of these compare well with the mechanical properties of conventional short glass fibre reinforced PP,

 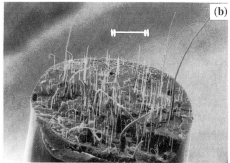

Fig. 5.11 In situ composites: (a) SEM of the core region of the fracture surface of injection moulded PS-25 wt% Vectra B950; (b) SEM of a spun fibre of PS-10 wt% Vectra A950 (scale bars correspond to 100 µm, photographs courtesy G. Crevecoeur).

but have the advantage of better surface finish, and lack of effective fibre degradation during post forming [108]. *Akkapeddi et al.* have also reported improvements in compatibility in *in-situ* composites containing a particulate filler such as talc, silica or carbon black [113].

2.3.3 Lamination of Oriented TLCP Sheets

The lamination of oriented (extruded) TLCP sheets would permit design approaches similar to those used in conventional continuous fibre composite technology. Indeed, the angular dependence of the mechanical properties of an oriented TLCP sheet may be modelled using expressions derived for uniaxial continuous-fibre prepregs [114]. The applications need not be limited to the exploitation of the structural properties however. The matching of the thermal expansivity to that of certain metals by suitable lamination, for example, would open up wider possibilities in the electronics industry [115], in which TLCPs have already made substantial inroads as a result of suitability for precision moulding, and their thermal and chemical resistance.

The main problem is one of bonding without loss of orientation. Two possible methods are as follows:

(i) Attempts have been made to produce unidirectional composites by compression of carbon fibres and oriented extruded HBA-HNA films, with the orientation of the films perpendicular to the fibre direction, using various processing conditions [91]. Pressing was carried out at 280°C (just above the main melting point) and 335°C. At the higher temperature, film orientation was substantially lost, but at the lower temperature approximately 80% of the initial orientation was retained during processing, at the expense of poor fibre wetting. This reflects the relatively long relaxation times in HNA/HBA just above the melting point. An exponential decay in orientation has been observed with a decay time constant of 36 minutes at 285°C [114].

(ii) *Avramova* and *Fakirov* have also used pressing to bond oriented films of the copolyester Polyethylene terephthalate (PET) 60% HBA (Eastman) [116]. Pressing was carried out for several hours at approximately 100 K below the softening point. In this case the bonding mechanism appeared to be chemical in origin, rather than resulting from fusion, and consequently there was no loss in orientation.

Conclusions

Three types of thermoplastic matrix have been considered in this chapter, amorphous, semicrystalline and thermotropic liquid crystalline polymers. This has highlighted the material-specific nature of much current research on thermoplastic composites, although important areas remain unexplored. The deformation mechanisms in semicrystalline composites, the influence of matrix structure on composite properties, the effects of heat treatments on properties of such materials, and long term behaviour of thermoplastic composites are examples. As a result few general guidelines are available to enable potential users to optimise material performance and applications remain limited. If these fundamental issues are not studied there will be little incentive for investment in the processing technology necesary to switch from the (continually improving) thermoset matrix materials. Finally, the exploitation of TLCPs in composite technology has been briefly considered. Whilst very much at the experimental stage, much effort has been directed toward the exploitation of the potential of these materials.

Acknowledgments

The authors are grateful to Philippe Béguelin (EPF Lausanne) and Walter Bradley (Texas A&H, College station), for useful suggestions and discussions, and to Guido Crevecoeur (Katholieke Universiteit Leuven) for the use of the micrographs in Fig. 11.

References

1. Döring, E., Kunststoffe **80**, 1149 (1990).
2. Leach, D.C., Chapter 2 in "Advanced Composites", Ed. Partridge, I.K., Elsevier Publishers (1990).
3. Carlsson, L.A., (Ed.), "Thermoplastic Composite Materials", Composite Material Series Vol. **7**, Elsevier Publishers (1991).
4. Hoggatt, J.T., 20th Nat. SAMPE Conf., May (1975), p 606.
5. Maximovich, M.G., 19th Nat. SAMPE Symp. (1974), p 262.
6. Maximovich, M.G., ASTM STP **617**, (1977), p 123.

7. Husman, G.E., Harness, J.T., Proc. 24th SAMPE Symp. (1979) p 21.
8. Hergenrother, P.M., Jensen, B.J., Havens, S.J., Proc. 29th Nat. SAMPE Symp., April (1984), p 1060.
9. British patent 1570000, June (1976).
10. Phillips, L.N., Murphy, D.J., RAE Technical Report 76140, October (1976).
11. Brownbill, D., Modern Plastics Intl., March (1985), p 30.
12. US Patent no. 4058581, 15th Nov. (1977).
13. European patent, 82300150.8, filed 12th Jan. 1982.
14. Brevet français no. 8405627, April (1984).
15. Hartness, T., Proc. European SAMPE Conf., Milan (1986), p 571.
16. Brevet français no. 8121545, 18th Nov. (1981)
17. European patent 82 401904.6, 18th Oct. (1982).
18. Clemans, S.R., Western, E.D., Handermann, A.C., "Hybrid yarns for high performance thermoplastic composites", BASF product literature (1987).
19. Dzral, L.T., SAMPE Jnl., Sept/Oct., (1983) p 7.
20. Swain, R.E., Reifsnider, K.L., Jayaraman, K., El-Zein, M., J. Comp. Matls. **3**, 13 (1990).
21. Favre J.-P., in Proc. "Interfacial Phenomena in Composite Materials '89", Ed. Jones, F.R., Butterworths, Sept. (1989) p 7.
22. Yee, A.F., in "Toughened Composites", ASTM STP 937.
23. Bascom, W.D., Jensen, R.M., Cordner, L.W., Proc. ICCM6, Elsevier, London (1987).
24. Luippold, D.A., Proc. 30th Nat. SAMPE Symp., March (1985) p 809.
25. Seitz, J.T., presented at the 5th Golden Jubilee of the Rheology Society, Boston (1979).
26. Plummer, C.J.G., Donald, A.M., J. Appl. Poly. Sci. **41**, 1197 (1990).
27. Donald, A.M., Kramer, E.J., Polymer **23**, 1183 (1982).
28. Donald, A.M., Kramer, E.J., Polymer **23**, 461 (1982).
29. Kramer, E.J., Adv. Poly. Sci. **52/53**, ch1, Ed. HH Kausch (1983).
30. Plummer, C.J.G., Donald, A.M., J. Poly. Sci. - Poly. Phys. Ed. **27**, 325 (1989).
31. Donald, A.M., J. Mat. Sci. **20**, 2634 (1985) .
32. DeGennes, P.G., J. Chem. Phys. **55**, 572 (1971).
33. Doi, M., Edwards, S.F., J. Chem. Soc. Far. Trans. **74**, 1789, 1802 (1978).
34. McLeish, T.C.B., Plummer, C.J.G., Donald, A.M., Polymer **30** 1651 (1989).
35. Kramer, E.J., Berger, L.L., Adv. Poly. Sci. **91/92**, ch1, Ed. Kausch, H.H. (1990).
36. Plummer, C.J.G., Donald, A.M., Macromolecules **23**, 3929 (1990).
37. Frost, H.J., Ashby, M.F., "Deformation Mechanism Maps", Pergamon Press, Oxford 1982.
38. Plummer, C.J.G., Donald, A.M., J. Mat. Sci. **26**, 1165 (1991).
39. Yang, A.C-M., Kramer, E.J., Kou, C.C., Phoenix, S.L., Macromolecules **19**, 2010 (1986).
40. Plummer, C.J.G., Donald, A.M., Polymer **32**, 409 (1991).
41. Brown, H.R., Macromolecules **24**, 2752 (1991).
42. Yang, A.C-M., Kramer, E.J., J. Mat. Sci. **21**, 3601 (1986).
43. Döll, W., Könczöl, L., Adv. Poly. Sci. **91/92**, ch 4, Ed. Kausch, H.H. (1990).
44. Schirrer, R., Adv. Poly. Sci. **91/92**, ch 5, Ed. Kausch, H.H. (1990).
45. Plummer, C.J.G., Donald, A.M., J. Appl. Poly. Sci. **41**, 1197 (1990).
46. Hine, P.J., Duckett, R.A., Ward, I.M., Polymer **22**, 1745 (1981).
47. Davies, M., Moore, D.R. Slater, B., "On the Fracture Behaviour of Polyethersulphone", Inst. of Phys. Conf. Proc. (1988).
48. "Properties & Introduction to Processing", Ref. no. VS2, ICI Advanced Materials.
49. Moore, D.R., Ch. 10, "Thermoplastic Composite Materials", Composite Material Series Vol. **7**, Elsevier Publishers, Ed. Carlsson, L.A., Elsevier (1991).
50. Haas, T.H., Ch. 26, "Handbook of Plastic Materials and Technology", Ed. Rubin, I.I. (1990).
51. Gothan, K.V., Turner, S., SPE Technical Papers **19**, 171 (1973).
52. Takemori, T.I., Adv. Poly. Sci. **91/92**, ch6, Ed. Kausch, H.H. (1990)

53. Clark, T.R., Hertzberg, R.W., Mohammadi, N., Proc. 8th International Conference on Deformation, Yield and Fracture of Polymers, Cambridge, April (1991), Paper 31.
54. Liu, L.B., Lewis, J.C., Yee, A.F., Gidley, D.W., Proc. 8th International Conference on Deformation, Yield and Fracture of Polymers, Cambridge, April (1991), Paper 29.
55. Lee, R.J., Trevette, A.S., Proc. ICCM6, Elsevier, London, p1.278 (1987)
56. Stori, A.A., Magnus, E., in "Composite Structures 2", Ed. Marshall, I., Elsevier, p 332 (1983).
57. Schulte, K., Friedrich, K., Horstenkamp, G., J. Mat. Sci. **21**, 3561 (1986).
58. Roulin, A.C., Davies, P., Proc ECF7, Budapest, EMAS Publishers, p 416 (1988).
59. Davies, P., Benzeggagh, M.L., ch 3 in "Application of Fracture Mechanics to Composite Materials", Ed. Friedrich, K., Elsevier (1989).
60. Partridge, I.K., Davies, P., Parker, D.S., Yee, A.F., in Proc. PRI Conf. on "Polymers for Composites", Solihull, Dec. (1987), p 5/1.
61. Hashemi, S., Kinloch, A.J., Williams, J.G., Comp. Sci. & Tech. **37**, 429 (1990).
62. Grenvesse, P., Bull Soc. Chim France **17**, 599 (1897)
63. Lopez, L.C., Wilkes, G.L., JMS-Rev. Macromol. Chem. Phys. **C29(1)**, 83 (1989).
64. Jog, J.P., Nadkarni, V.M., J. Appl. Polymer Sci. **30**, 997 (1985)
65. Lovinger, A.J., Davis, D.D., Padden, F.J., Polymer **26**, 1595 (1985).
66. Lopez, L.C., Wilkes, G.L., Polymer 29, 106 (1988).
67. Song, S.S., White, J.L., Cakmak, M., Poly. Eng & Sci. **30**, 944 (1990).
68. Desio, G.P., Rebenfeld, L., J. Appl. Polymer Sci. **39**, 825 (1990).
69. Maffezzoli, A., Kenny, J.M., Torre, L., Nicolais, L., Proc. SAMPE European meeting, Basel Switzerland, Ed. Hornfeld, H., p 307 (1990).
70. Brady, D.G., J. Appl. Polymer Sci. **20**, 2541 (1976)
71. Karger-Kocsis, J., Friedrich, K., J. Mat. Sci. **22**, 947 (1987).
72. Vanderschuren, J. et al., Proc. SAMPE European meeting, Basel Switzerland, Ed. Hornfeld, H., p 441 (1990).
73. O'Connor, J.E., Beever, W.H., Geibel, J.F., Proc. 31st SAMPE Symp. April (1986), p 1313.
74. ICI data sheets on APC2 composites.
75. Spamer, G.T., Brink, N.O., Proc. 33rd SAMPE Symp., March (1988), p 284.
76. Waddon, A.J., Hill, M.J., Keller, A., Blundell, D.J., J. Mat. Sci. **22**, 1773 (1987).
77. Caramaro, L., Chabert, B., Chauchard, J., Proc. J.N.C 7, Lyon, (1990), p 145.
78. Khoury, F., Proc. 48th SPE Conf., ANTEC **90**, p 1261, (1990).
79. Lopez, L., Wilkes, G.L., Polymer Preprints, ACS **30**, 2, Sept (1989), p 207.
80. Gaur, U., Desio, G., Miller, B., Proc. ANTEC89, p 1513 (1989).
81. Lindenmeyer, P.H., Gerken, N.T., Sheppard, C.H., Proc. 18th Int. SAMPE Tech. Conf., Oct. (1986), p 45.
82. Cole, K.C., Noël, D., Hechler, J.-J., J. Appl. Polymer Sci. **39**, 1887 (1990).
83. Johnson, T.W., Ryan, C.L., Proc. 31st SAMPE Symp., April (1986), p 1537.
84. Blundell, D.J., Osborn, B.N., SAMPE Quarterly, **17**, 1, Oct. (1985) p 1.
85. Martin, C.C., O'Connor, J.E., Lou, A.Y., Proc. 29th SAMPE Symp., April (1984) p 753.
86. O'Connor, J.E., Lou, A.Y., Beever, W.H., Proc. ICCM5, San Diego, p 963, (1985).
87. Davies, P., Benzeggagh, M.L., de Charentenay FX, Proc. 32nd SAMPE Conf., Anaheim, April (1987), p 134.
88. O'Connor J.E., Beever W.H., Geibel J.F., Proc.31st SAMPE Conf., April (1986), p 1313.
89. Hagenson, R.L., Bonazza, B.R., McKillop, B.E., Avtel product literature, 1990.
90. Brady, R.L., Porter, R.S., J. Composite Materials **3**, 252 (1990)
91. Marais, C., Dartyge, J.M., Broqua, N., Proc. Journées Nationales sur les Composites 7, Lyon, Nov. (1990), p 833.
92. Northolt, M.G., Polymer **21**, 1199 (1980).
93. Troughton, M.J., Davies G.R., Ward, I.M., Polymer **30**, 58 (1989).

94. Calundann, G.W., Jaffe, M., "The Robert A. Welch Foundation Conferences on Chemical Research", XXVI, Houston, Nov. (1982).
95. Yoon, H.N., Colloid & Polymer Science **268**, 230 (1990).
96. Northolt M.G., "Recent Advances in Liquid Crystalline Polymers" Ed. Chapoy, L.L., ch 20 (1985).
97. DeTeresa, S.J., Porter, R.S., Farris, R.J., J Mat. Sci. **23** 1886 (1988).
98. Flory, P.J., Macromolecules **11**, 1112 (1977).
99. Cogswell, F.N., Griffin, B.P., Rose, J.B., U.S. Patents 4,386,174 (1983), 4,433,083 and 4,438,236 (1984).
100. Siegmann, A., Dagan, A., Kenig, S., Polymer **26**, 1325 (1985).
101. James, S.G., Donald, A.M., MacDonald, W.A., Mol. Cryst. Liq. Cryst. **153**, 491 (1987).
102. Crevecoeur, G., "In situ composites", PhD Thesis, Katholieke Universiteit Leuven (1991).
103. Dutta, D., Fruttwala, H., Kohli, A., Weiss, R.A., Poly. Eng. & Sci. **30**, 1007 (1990).
104. Brostow, W., Polymer **31**, 979 (1990).
105. Baird, D.G., Ramanathan, R., in "Contempory Topics in Polymer Science Vol. **6**; Multiphase Macromolecular systems", Ed. Culberton, B.M., Plenum, New York (1989).
106. Acierno, D., Amendola, E., Carfagna, C., Nicolais, L., Nobile, R., Mol. Cryst. Liq. Cryst. **153**, 533 (1987).
107. Nobile, M.R., Amendola, E., Nicolais, L., Acierno, D., Carfagna, C., Poly. Eng. & Sci. **29**, 244 (1989).
108. Baird, D.G., Priv. Commun., Lausanne (1991).
109. Weiss, R.A., Kohli, A., Chung, N., Dutta, D., unpublished work (see (105)).
110. Kohli, A., Chung, N., Weiss, R.A., Poly. Eng. & Sci. **29**, 573 (1989).
111. Isayev, A.I., Modic, M.J., Polym. Compos. **8**, 158 (1987).
112. Baird, D.G., ACS Symposium Series No. 435 "Liquid Crystal Polymers", Eds. Weiss, R.A., Ober, C.K., American Chemical Society (1990)
113. Akkapeddi, M.K., DeBona, T., Li, H.L., Prevorsek, D.C., US Patent 4,611,025 (1986).
114. Ide, Y., Chung, T., J. Macromol. Sci.-Phys. **B23**, 497 (1984).
115. Weiss, R.A., Priv. Commun., Basel (1991).
116. Avramova, N., Fakirov, S., J. Appl. Poly. Sci. **42**, 979 (1991).

Part II
Evaluation of the Composites

Chapter 6

The Short-Term Properties of Carbon Fibre PEEK Composites

W.J. Cantwell and P. Davies

Abstract

In the first part of this chapter, the short-term properties of carbon fibre poly ether ether ketone (PEEK) are examined and compared to those of similar high performance composite materials. In particular, attention is given to characterizing the interlaminar fracture resistance and compressive properties of these materials since these are areas of particular interest and sometimes considerable concern.

Subsequently, the fracture properties of carbon fibre/PEEK are examined in more detail and the influence of varying certain processing parameters on the mechanical behaviour of this material is considered.

It is concluded that carbon fibre reinforced PEEK composites offer a range of superior mechanical properties, perhaps the most important being its excellent toughness under low velocity impact conditions. It is also suggested that some care has to be taken when processing this material since the higher level of crystallinity associated with slow cooling as well as the increased residual stresses resulting from fast cooling may influence adversely certain key properties.

1 Introduction

Carbon fibre reinforced composites offer many advantages over conventional metallic materials, the most frequently cited being their superior specific strength and stiffness properties as well as the possibility of manufacturing complex shapes in relatively short manufacturing cycles. The first generation of high performance composites were based on relatively brittle epoxy matrices exhibiting low strains to failure. As a result of this relatively brittle behaviour, conventional epoxy resins are incapable of absorbing and dissipating large energy amounts of energy without incurring significant damage and suffering large reductions in load-bearing properties. Indeed, low velocity impact tests on a standard carbon fibre/epoxy have shown that energies

---

W.J. Cantwell, Laboratoire de Polymères, Ecole Polytechnique Fédérale de Lausanne, CH-1015 Lausanne, Switzerland; P. Davies, IFREMER, Laboratoire de Matériaux, Centre de Brest, F-29280 Plouzané, France

as low as 4 Joules are capable of reducing the compressive strength of the material by over 50 per cent [1]. More recently, a number of high performance composites based on thermoplastic matrices have been developed, perhaps the best known amongst these being carbon fibre reinforced poly ether ether ketone (PEEK) manufactured by I.C.I. PEEK is an aromatic polymer with a glass transition temperature of 143°C and a melting temperature of 334°C [2]. It is a semi-crystalline polymer in which the maximum achievable degree of crystallinity is around 48 per cent [3], under normal processing conditions, however, the degree of crystallinity is usually lower, typically 30 per cent. At room temperature and low rates of strain PEEK is capable of undergoing extensive plastic deformation typically failing at strains approaching 100 per cent. This distinct difference in stress-strain behaviour is summarized in tensile loading curves presented in Fig. 1.

PEEK-based composites are frequently manufactured by a compression moulding technique in which the pre-preg is stacked in a picture frame mould and consolidated under pressure at 380°C. Other techniques can be used to manufacture components from this material, these include autoclave moulding, hot stamping and diaphragm forming. The processed material is generally of high quality containing a low level of voids and offering excellent surface finish characteristics.

In the first part of this chapter several of the fundamental short-term mechanical properties of carbon fibre PEEK will be evaluated and compared to those of similar high performance thermoplastic and thermosetting-based composites. Much of the work presented in the literature is based on tests undertaken on AS4 carbon fibre PEEK composites although more recently considerable data have been obtained for IM6 carbon fibre PEEK. The properties of the reinforcing fibres obviously influence the overall mechanical response of the composite and this needs to be borne in mind when comparing data from materials with different fibres types. It should also be stressed that the properties quoted in the following section may not necessarily be the optimum values but only reflect a particular processing condition and specimen geometry. However, they can be considered as being typical of the range of values reported in the literature. A more detailed assessment of the mechanical response of this material will follow in the latter half of this chapter.

2 Comparison of the Mechanical Properties of Carbon Fibre PEEK with Other High Performance Composites

2.1 Interlaminar Properties

Interlaminar fracture or delamination failure has for many years been seen as a major limitation to the more widespread application of high performance composites in both primary and secondary aircraft components. Low energy impact loading results in large areas of delamination which in turn reduce dramatically both the short and long-term properties of the composite most particularly in compression [1]. In the first generation of composites based on epoxy resins the mode I fracture

6. Short-Term Properties of Carbon Fibre PEEK Composites

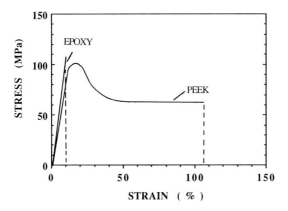

Fig. 6.1 Typical stress-strain curves for PEEK and a standard epoxy resin.

energies were frequently as low as 50 J/m^2 [4]. Such relatively poor interlaminar fracture toughness is undoubtedly related to the low strain to failure of the epoxy resin and therefore the reduced ability to absorb and dissipate energy in localized plastic deformation.

Composites based on thermoplastic matrices such as carbon fibre reinforced PEEK and PES generally offer significantly higher interlaminar fracture toughnesses than their epoxy-based counterparts as shown in Table 1.

From the data presented in this table it is clear that the three thermoplastic composites, i.e. those based on PEEK, PPS and PES exhibit superior mode I initiation fracture toughnesses to those of the two epoxy composites. It is also interesting to note that of these thermoplastic composites, the PEEK-based material offers the highest overall toughness. An examination of the fracture surfaces of these samples indicates that failure occurs largely within the polymeric matrix and that considerable plastic deformation and drawing occurs during the passage of the crack, Fig. 2.

Table 6.1 Comparison of G_{IC} Initiation Values for Various Epoxy and Thermoplastic Carbon Fibre Composites. Data Taken from Ref. [5]

Material	Specimen Thickness (mm)	G_{IC} Initiation (J/m^2)
Carbon fibre epoxy (T300 / 914)	5	185
Carbon fibre epoxy IM6 / 6376	5	204
Carbon fibre PEEK AS4 / APC2	5	1464
Carbon fibre PES AS4 fibre	3.8	743
Carbon fibre PPS AS4 fibre	5	610

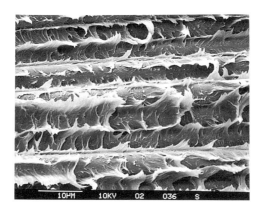

Fig. 6.2 Scanning electron micrograph of the mode I fracture surface taken from the zone of stable crack propagation in a carbon fibre PEEK specimen.

The figure also indicates that the fibre-matrix interfacial strength is good. A closer examination of the fracture surface also highlights the presence of a large number of broken fibres spread across the fracture surface. The presence of such broken fibres suggests that considerable fibre bridging has taken place during the failure process. Such bridging occurs as a result of fibre nesting during the processing cycle and is likely to stabilize the propagating crack and also increase the resulting fracture toughness of the composite after initiation.

A comparison of the mode I fracture energies of AS4, IM6 and IM7 carbon fibre composites reported in refs. 5 to 7 indicates that increasing the energy absorbing capacity of the fibre (i.e. the area under the stress-strain curve) results in an appreciable increase in the interlaminar fracture energy of the composite. This suggests, therefore, that although interlaminar fracture occurs supposedly between the individual plies of a multi-layer composite, the fibres still play a significant rôle in determining the resulting interlaminar fracture propagation energy.

From Table 1 it is clear that the PES-based composite does not perform as well as its PEEK counterpart, the lower value probably resulting from the reduced strain to failure of the amorphous PES thermoplastic. The PPS-based composite offers the lowest interlaminar fracture energy of the three thermoplastic composites. Post-failure analyses of the fracture surface of this material [5] have indicated the presence of a large number of fibres devoid of resin suggesting that the fibre/matrix interfacial bond strength in this material is poor. Phillips 66 the manufacturers of this material have recently developed a modified version of carbon fibre PPS in which it is claimed that the fibre/matrix bond has been greatly improved. If this is the case then one would expect a similar improvement in G_{IC}.

Although it is interesting to compare the mode I fracture toughnesses of the various thermoplastic and thermosetting matrices available, it has been shown that the mode II toughness of a long fibre composite is more likely to reflect the post-impact compression properties of the material, a parameter of fundamental importance in the aerospace industry. Indeed, *Masters* [8] has shown that a direct relationship

6. Short-Term Properties of Carbon Fibre PEEK Composites

Table 6.2 Comparison of G_{IIC} Non-Linearity and Maximum Force Values for Various Epoxy and Thermoplastic Carbon Fibre Composites. Data Taken from Ref. [5]

Material	$G_{IIC\ Non\ Linear}$ (J/m²)	$G_{IIC\ Max.\ Force}$ (J/m²)
Carbon Fibre Epoxy (T300 914)	518	518
Carbon Fibre Epoxy (IM6 6376)	578	703
Carbon Fibre PES (AS4 FIBRE)	759	1312
Carbon Fibre PPS (AS4 FIBRE)	460	1125
Carbon Fibre PEEK (AS4 FIBRE)	1109	3045

exists between the mode II interlaminar fracture toughness and the compression after impact strength of long fibre bismaleimide and epoxy composites.

Table 2 summarizes the mode II interlaminar fracture toughnesses as measured using the end notch flexure specimen of the materials presented in Table 1. Here again, the thermoplastics tend to out-perform the epoxy matrix composites although the non-linearity value for the PPS composite is again low due to the poor fibre/matrix interfacial strength. It is interesting to note that the relative performance of the thermoplastics to the thermosets is not as impressive when the mode II fracture properties are considered. The PEEK-based composite clearly offers the highest mode II fracture toughness values. An examination of the damage zone at the tip of the main crack in an AS4/PEEK composite, Fig. 3a indicated the presence of a large number of cracks oriented at approximately 45° to the main crack. This zone of extensive shear cracking extends some considerable distance in front of the crack tip and clearly re-distributes the stress field at the tip of the crack. A post-failure analysis

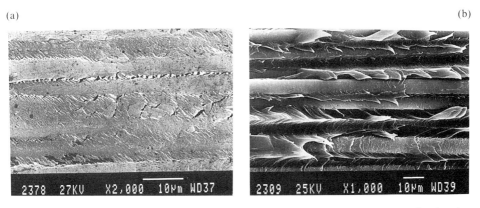

Fig. 6.3 Scanning electron micrographs of mode II AS4 carbon fibre/PEEK samples showing (a) microcracks in front of the crack tip and (b) the resulting fracture surface Ref. [5].

of the fracture surface shows extensive ductility with the matrix appearing to have been drawn considerably in the direction of shear, Fig. 3b.

2.2 PEEK In-Plane Properties

Thermoplastic-based composites appear to be inferior to their thermosetting-counterparts under conditions of axial compressive loading [9, 10]. Indeed, available experimental data suggests that the compressive strength of certain fibre reinforced thermosets can be up to three times higher than a similarly reinforced thermoplastic, Fig. 4. The fundamental reasons for such significant differences in mechanical response are not at present well understood. The shear modulii of the thermoplastic and thermosetting matrices are not significantly different rendering invalid arguments based on the differing amount of support the matrix applies to the fibres. *Johnston* and *Hergenrother* [9] suggested that the inferior compressive properties of thermoplastic composites are due to poorer fibre-matrix adhesion as well as greater levels of fibre waviness. Indeed, polished sections removed from carbon fibre/PEEK laminates manufactured following the recommended cycle have shown significant fibre waviness [11]. The effects of such fibre distortion is likely to be considerable since *Wisnom* [12] has shown that a fibre misalignment of less than one degree is sufficient to reduce the compressive strength of a long fibre reinforced composite by over 30 per cent.

Leeser and *Leach* [10] undertook a series of open hole and compression after impact (CAI) tests on a wide range of high performance composites and concluded that although the initial undamaged compression strength of the thermoplastic composites was inferior to those of reinforced epoxies, their open hole and CAI properties were equal to if not superior to those of their thermosetting counterparts. They conclude that since the open hole and CAI tests are more representative of the in-

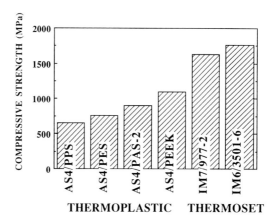

Fig. 6.4 Comparison of the compressive properties of a number of thermoplastic and thermosetting-based composite materials. Note the variations in fibre type (Data taken from refs. [9, 10 and 13]).

service requirements, these values are of greater fundamental importance in materials selection.

Composite components are frequently required to operate in hostile environments without suffering significant reductions in their load-carrying capability. *Kriz* [14] has shown that absorbed moisture influenced the residual stress state and reduced the load-bearing properties of a quasi-isotropic carbon fibre/epoxy laminate. It is likely that this problem will be further accentuated in rubber-toughened systems where the absorption of moisture and any corresponding reduction in high temperature properties is likely to be greater. *Buchman* and *Isayev* [15] examined the process of water absorption in a number of thermoplastic composites. They found that the solubility of water in carbon fibre/PEEK was approximately 0.2% and that in a carbon fibre reinforced PES system as high as 2.5%. A PEI-based composite, although absorbing intermediate levels of water was found to undergo severe changes leading to leaching out of the surface layer and the formation of craze-like structures. As a result of their semi-crystalline morphology, PEEK-based composites are also likely to offer superior resistance to acids than those manufactured from amorphous polymers.

2.3 PEEK Impact Properties

The data presented in Table 2 suggest, therefore, that the compression after impact properties of the carbon/PEEK laminate should also be good. Low velocity impact tests [16] on AS4/PEEK and T400H/6376 carbon fibre reinforced epoxy have indeed shown this to be the case. Some of their impact data is presented in graphical form in Fig. 5 where the residual compressive strength of the damaged laminates as a function of impact energy is shown. It should be noted that the T400H fibres offer a superior strain to failure to their AS4 counterparts (1.8% compared to 1.5% [16]).

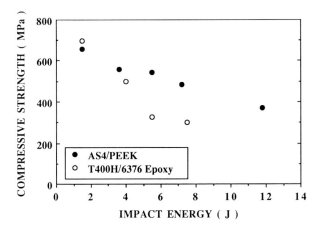

Fig. 6.5 The variation of compression strength with impact energy for AS4/PEEK and T400H/6376 epoxy 16 ply (0, ±45) laminates. Ref. [16].

At low energies where little or no damage is incurred in either system, the epoxy laminate appears to exhibit superior residual properties, probably as a result of the relatively poor compressive properties of the thermoplastic composite. At higher energies, however, the superior residual properties of the thermoplastic composite are clearly evident. Here, the damage zone in the thermoplastic is considerably smaller for a given energy and therefore the support offered to the load-bearing 0° fibres is greater.

Unfortunately, the excellent low velocity impact resistance of thermoplastic composites does not appear to be maintained at higher rates of loading [17]. Following tests on a number of long fibre composites, *Dan-Jumbo et al.* [17] concluded that the residual strengths of APC2 and Torlon (a poly amide imide composite) after high velocity impact loading were no better than a relatively brittle bismaleimide composite. At first this finding appears somewhat surprising. However, high rate fracture mechanics-type testing on pure PEEK samples has shown that the material is very rate-sensitive and suffers a significant loss in toughness at high rates of deformation [18]. This will be discussed in more detail in a subsequent section.

2.4 Summary

The relative merits of PEEK and epoxy-based composites are summarized in tabular form in Table 3. For each mechanical property the material offering the superior characteristics is indicated. Although the conclusions are general it is believed that they reflect the overall relative performance of the two types of material.

In tension, there appears to be no significant difference between epoxy-based and PEEK composites [19]. Here, the mechanical properties of the reinforcing fibres is likely to dominate the response of the material. In compression, however, the epoxy composites are superior most probably as a result of the lower amount of fibre waviness. The interlaminar properties of the thermoplastic composite are undoubtedly superior to those of epoxy composites and this is in turn reflected in an enhanced compression after impact performance. Carbon fibre/PEEK has been shown to be more notch sensitive than epoxy composites [19]. It is believed that this increased notch sensitivity occurs as a result of a smaller size of the stress redistributing damage zone at the tip of the crack.

3 A More Detailed Characterization of the Mechanical Properties of Carbon Fibre PEEK

In the previous section it has been shown that in most areas carbon fibre PEEK offers superior mechanical properties over a wide range of test conditions. The values quoted above correspond to individual tests undertaken on a specific specimen geometry following a particular processing condition. Indeed, the short-term properties of carbon fibre/PEEK have been shown to depend directly upon a wide number of parameters including processing conditions [5, 21, 22] as well as specimen geometry

6. Short-Term Properties of Carbon Fibre PEEK Composites

Table 6.3 A Relative Comparison of the Principal Mechanical Properties of Carbon Fibre Epoxies and Carbon Fibre/PEEK. X Signifies a Superior Performance. If the Property is Similar for Both Materials a Cross Has Been Placed in Both Columns

Property	CF / Epoxy	CF / PEEK
Tensile strength	X	X
Tensile modulus	X	X
Compressive strength	X	
G_{IC}		X
G_{IIC}		X
Compression after impact		X
Holed strength Tension	X	
Holed strength compression		X
Temperature/moisture stability		X

(Obtained using data reported in refs. [5, 13, 19, 20]

[6]. One good example of this is the significant variations in mode I interlaminar fracture toughness data reported in the literature. For example, *Russell* and *Street* [23] reported a value for G_{IC} of approximately 1500 J/m^2 whereas *Leach et al.* [24] measured a value approaching 3000 J/m^2 on the same APC2 material. Before components can be designed with full confidence, it is important to determine the influence of such parameters on the basic fracture properties of the material.

3.1 Interlaminar Properties

PEEK is a semi-crystalline polymer, it is likely that the mechanical properties of the composite will depend upon the crystalline morphology of the polymeric matrix. Early work on AS4 carbon fibre PEEK has indeed shown this to be the case [22, 25]. *Curtis et al.* [25] conducted mode I double cantilever beam tests on a series of AS4/PEEK panels manufactured using cooling rates ranging between 1 and 19 K/minute. This range of processing conditions yielded levels of crystallinity between 32 and 24 per cent respectively. The subsequent mode I crack propagation was found to be almost entirely unstable and it was necessary to determine $G_\text{I instability}$ and G_Iarrest values. The variation of $G_\text{I instability}$ and G_Iarrest with cooling rate is shown in Fig. 6. Examination of the data indicates an alarming drop in fracture toughness between cooling rates of 3 and 1 K/minute.

Similar results were also obtained by *Talbott et al.* [22]. It was tentatively suggested that such a loss in fracture toughness was the result of a weakening of the interspherulitic boundary [25] with the consequence that the crack passed around the spherulite rather than through it. These early results clearly indicated that care had to be taken when processing PEEK in order to avoid incurring high levels of crystallinity with the obvious consequences on interlaminar fracture toughness. More recent tests on IM6 carbon fibre reinforced PEEK did not produce such a distinct cooling rate dependency [26]. Here, the interlaminar crack always propagated in a stable manner

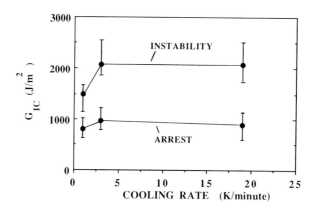

Fig. 6.6 The variation of $G_{I\,instability}$ and $G_{I\,arrest}$ values as a function of cooling rate for AS4/PEEK. Ref. [25].

with there being no evidence of the extensive stick-slip crack propagation observed in the AS4 composite. Since the IM6 carbon fibre reinforced composite is a more recent development of the AS4 material it is possible that the PEEK matrix has been modified, for example by altering the molecular weight or weight distribution. It should be noted, however, that IM6 fibres differ from AS4 fibres not only in the fact that the former have superior mechanical properties but also in that IM6 fibres are significantly smaller than their AS4 counterparts (5 µm as opposed to 7 µm). The increased fibre surface for the same volume fraction may influence the type and extent of crystal nucleation and in turn the fracture properties of the composite. However, no significant difference was observed when thin sections removed from similarly processed AS4 and IM6 fibre composites were examined in polarized light. It is also possible that the smaller IM6 fibres may permit greater fibre nesting leading to a higher level of fibre bridging during crack propagation. This is a distinct possibility and might explain the lack of cooling rate dependency and the increased crack stability observed in the IM6 composites. In order to examine this in more detail the centre plies of the double cantilever beam specimen were offset at several degrees to each other in a manner similar to that adopted by *Johnson* and *Mangalgiri* [27]. The effect of such an offset was to reduce the amount of fibre nesting likely to occur during interlaminar crack propagation and therefore accentuate effects resulting from variations in the matrix morphology. The results from these tests as well as data from a series of tests in which a 100 µm layer of unreinforced PEEK was placed at the mid-plane of the laminate are presented in Fig. 7.

It is interesting to note that the data in Fig. 7 follow the trends reported previously by *Curtis et al.* [25] following tests on AS4/PEEK, that is, a drop in fracture toughness at low cooling rates. Also, the mode of crack propagation in both types of laminate was again unstable with the crack advancing in a stick-slip manner. These results appear to suggest that the lack of cooling rate dependency associated with the IM6/PEEK composite is related to an increased level of fibre bridging, a phenomenon that is also likely to conceal the consequences of variations in matrix morphology.

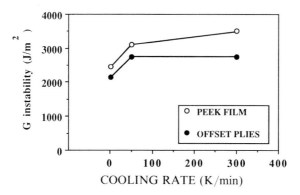

Fig. 6.7 The variation of $G_{I\ instability}$ with cooling rate for DCB specimens with their centremost plies offset by approximately 5° and for DCB specimens containing a 100 μm thick pure PEEK layer. Ref. [11]

The effect of varying test temperature on the interlaminar fracture properties of this material are summarized in Fig. 8. Here, the variation of G_{IC} with temperature of the AS4 and IM6 fibre composites are shown as a function of test temperature. Both composites show a systematic increase in G_{IC} with temperature over the range of conditions examined. At room temperature the IM6 fibre composites clearly offer a higher fracture toughness, this being a result of the higher energy absorbing capacity of the IM6 fibres. However, this difference appears to diminish with increasing temperature with the G_{IC} values for the two materials being almost identical at 120°C. This suggests that the energy associated with breaking the bridging fibres is swamped by the increased energy dissipation due to plastic flow of the matrix.

At lower temperatures it appears that PEEK is somewhat less ductile. *Barlow et al.* [28] have shown that PEEK fails in a brittle manner at −85°C with there being little plastic deformation in evidence on the fracture surface of specimens loaded in

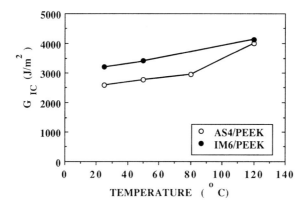

Fig. 6.8 The variation of the mode I propagation fracture toughness with temperature for AS4 and IM6 composites. The AS4 data are taken from Ref. [5].

tension. This is also reflected in the composite [5] where the interlaminar fracture toughness G_{IC} at $-30\,°C$ has been shown to be some twenty-five per cent lower than its room temperature value.

Great care has to be taken when undertaking interlaminar fracture tests in order to minimize geometrical effects. Work by *Davies et al.* [6] has shown that both the measured values of mode I and mode II interlaminar fracture toughnesses of APC2 can be influenced by the geometry of the test specimen. Two of the most important geometric parameters appear to be the thickness of the insert and the overall thickness of the test sample. The thickness of the insert or starter film appears to have a significant effect upon the initiation fracture toughness as defined by the deviation from linearity in the load-displacement curve. For example, increasing the thickness of the insert from 25 to 50 µm may result in a fifty per cent increase in $G_{Ic\ initiation}$ [29]. It is believed that that dramatic increase in initiation toughness results from increased blunting and plastic deformation at the tip of the insert [29]. Similar variations in mode I propagation toughness with specimen thickness have been observed by *Davies et al.* [26]. Here, using data published by a number of workers it was shown that the mode I propagation toughness increased from approximately 1500 J/m^2 for a three millimetre thick sample to approximately 3000 J/m^2 for a 6.5 mm specimen. A close examination of the specimens during and after testing indicated that secondary cracking above and below the principal crack was extensive in the thicker samples. It was also clear that the fracture surfaces of these thicker samples was considerably rougher than those of the 3 mm thick specimens. The reasons for these phenomena are not clear at present and more work needs to be undertaken in order to clarify this.

3.2 In-Plane Properties

Although a number of research studies have concentrated on the effects of matrix morphology on the interlaminar fracture properties of APC2, few workers have attempted to evaluate the effect of varying processing parameters on the in-plane properties of the material. *Curtis et al.* [25] found that the tensile properties of an eight ply ($\pm 45°$) AS4/PEEK laminate were strongly dependent upon the cooling conditions employed after consolidation. Similar conclusions were drawn following tensile tests on IM6/PEEK [21]. Fig. 9 shows the variation of failure strain with temperature for a series of slow and fast cooled ($\pm 45°$) laminates. A distinct difference between the two types of laminate is apparent with the fast cooled panels failing at significantly greater strains.

It was believed that such distinct differences in fracture behaviour were not simply a result of variations in matrix morphology alone. In order to cast further light on these differences, a number of specimens were sectioned, polished and examined under an optical microscope. These sections indicated that the level of damage developed during the tensile deformation process was considerably greater in the faster cooled samples. This is clearly seen in the polished sections presented in Fig. 10.

Here, damage in the slow cooled sample is clearly less severe than that in the fast cooled specimen. In the latter, large voids and extensive delamination developed

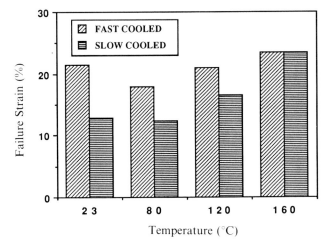

Fig. 6.9 Tensile failure strain vs temperature for slow (1K/minute) and fast cooled (50K/minute) ±45° IM6/PEEK samples [21].

Fig. 6.10a Polished section of a slow cooled (1K/minute) eight ply (±45°) tensile specimen.

Fig. 6.10b Polished section of a fast cooled (50K/minute) eight ply (±45°) tensile specimen.

during the loading process. It has been suggested that the occurrence of such extensive damage is linked to the higher residual stresses resulting from rapid cooling [21]. Upon loading, cracks develop within the material and the subsequent opening of these cracks during the test contributes to the greater overall strains measured at the point of rupture. The presence of sizeable residual strains within APC2 has been noted by a number of workers [30–33]. *Young et al.* [33] used Raman spectroscopy to measure the residual strain field in HMS4 carbon fibre/PEEK pre-preg samples and found that the average residual compressive strains in the surface layer were of the order of 0.28%. Subsequent compression testing on more typical quasi-isotropic (0°, 90°, ±45°) laminates suggested that the cooling rate effect was not as significant as that apparent in Figs. 9 and 10 [30].

3.3 Impact Properties

Carbon fibre reinforced PEEK is renowned for its excellent impact resistance and subsequent residual load-bearing capability. Several workers have observed a top surface dent as well as localized fibre failure following low velocity impact loading on APC2. A typical example of the impacted face of a quasi-isotropic laminate is shown in Fig. 11. The presence of such top surface damage can be a positive advantage since both the location as well as the severity of the damage can be readily determined.

Again, a distinct processing-dependent response has been observed in the impact properties of APC2 [30, 34]. *Davies et al.* [30] showed that the overall delaminated area in a multi-directional IM6 carbon fibre/PEEK laminate is greater under conditions of slow cooling after consolidation, Fig. 12. Similar observations have been made by *Leicy* and *Hogg* [35] who by removing and polishing sections from impacted panels observed that damage in quenched APC2 took the form of localized matrix cracking over a region immediate to the impact location.

These results agree qualitatively with the findings of the interlaminar fracture studies reported above, that is that the fracture toughness is lower in slow cooled panels. An optical examination of the impacted specimens indicated that the amount of top surface fibre damage was greater in the fast cooled laminates. It is believed that this increased fibre damage was a result of the greater residual stresses in the fibres in the 50 K/minute cooled laminates. Therefore, although rapid cooling results in lower levels of delamination and therefore superior compression properties for a given incident energy, the greater degree of fibre damage may result in more significant losses in tensile strength.

Unfortunately, the excellent damage tolerance of APC2 observed under conditions of low velocity impact does not appear to be maintained at higher velocities [16, 17]. *Morton* and *Godwin* [16] found that high velocity impact loading created large areas of rear surface delamination and spalling at relatively low incident energies. *Dan-Jumbo et al.* [17] went further and concluded that the residual load-bearing properties of APC2 following high velocity impact were no better than a relatively brittle bismaleimide composite. In order to explain this abrupt transition it is necessary to consider the rate-dependent properties of the PEEK matrix itself. *Béguelin* and

6. Short-Term Properties of Carbon Fibre PEEK Composites   187

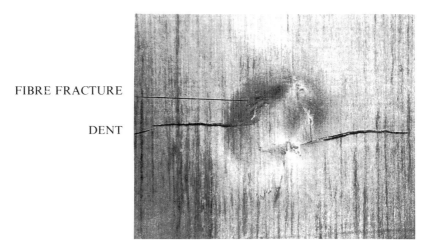

Fig. 6.11 Top surface damage in a sixteen ply $((0°, 90°, \pm 45°)_2)_s$ laminate subjected to low velocity impact loading.

Kausch conducted fracture mechanics-type tests over a wide range of strain rates on a number of thermoplastic polymers [36]. Their results indicated that PEEK offers excellent toughness characteristics at low rates of strain with values of K_{IC} exceeding 6 MPa m$^{1/2}$. At higher rates, however, the toughness of the polymer decreases rapidly. Indeed, at crack tip opening rates approaching 10^5 MPa m$^{1/2}$ s^{-1} the fracture toughness of the polymer dropped to approximately 2 MPa m$^{1/2}$. Under these conditions it appears that all of the polymers exhibit similar fracture toughness values with the K_{IC} of reputedly tough PEEK and POM having dropped to values similar to that of

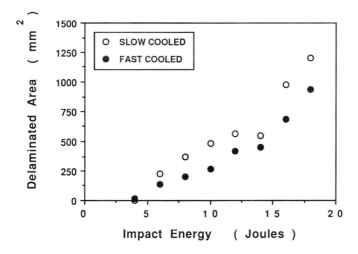

Fig. 6.12 Variation of delaminated area with impact energy for a fast (50 K/minute) and slow (0.6 K/minute) cooled $((0°, 90°, \pm 45°)_2)_s$ laminate.

notoriously brittle PMMA. A subsequent post-failure analysis of the fracture surface indicated that the amount of stress whitening at the tip of the pre-crack in the POM and PEEK samples diminished with increasing rate suggesting that the amount of plastic deformation and therefore the plastic zone size had also decreased considerably.

Significant reductions in the interlaminar fracture toughness of carbon fibre/PEEK composites with increasing strain rate have been observed by a number of workers [18, 37–39]. *Smiley* [39] showed that the mode I fracture toughness remained roughly constant over a range of low and intermediate crack tip opening rates but dropped dramatically at high rates, Fig. 13. *Béguelin* and *Barbezat* [18] found that the stable mode of crack propagation observed in IM6/PEEK at low crosshead speeds developed into a highly unstable mode at high crosshead speeds.

At first view, it appears that the behaviour of the PEEK composite is significatly different to that of the unreinforced polymer. *Friedrich et al.* [40] explained these differences in terms of the size of the damage zone at the crack tip and the inter fibre spacing. They suggested that reductions in the mode I fracture toughness occur once the size of this damage zone is smaller than that of the interfibre spacing. Therefore, over a wide range of rates the size of this zone is considerably greater than that of the fibre spacing and the toughness remains approximately constant. Once a critical rate is reached at which the damage zone approaches this inter-fibre distance, the amount of bridging developed during crack advance will drop and the rate sensitivity of the matrix will become more apparent. As indicated above, the double cantilever beam specimen does suffer the disadvantage that significant levels of fibre bridging may occur the crack propagation. The presence of even a small number of bridging fibres might, however, conceal the true viscoelastic response of the base polymer. High rate interlaminar fracture tests on DCB specimens with their centre plies offset at several degrees to each other have highlighted a significantly greater rate sensitivity than that apparent in Fig. 13 [11]. These results are summarized in Fig. 14 where the average mode I G_{IC} values for two crack tip strain rates are presented. Here, the strain rate at the crack tip, $\dot{\varepsilon}$, was determined using the equation used by Smiley [39]:

$$\dot{\varepsilon} = 3h\dot{\delta}/(4\,a^2)$$

where $\dot{\delta}$ is the crosshead rate
          h half the beam thickness
          a the crack length

From Fig. 14 it is clear that rate effects are considerably greater in the specimens containing offset centre plies. Indeed, over the modest range of rates considered here the mode I interlaminar fracture toughness of the coupons with offset centre plies decreased by some 35 percent compared to 8 percent in the standard material. It is also interesting to note that the mode of crack propagation in the offset DCB was inherently unstable with the crack propagation in a stick-slip manner similar to that observed by *Béguelin* and *Barbezat* [18]. The influence of even limited amounts of fibre bridging is therefore clearly evident. These findings are important since engineering composite structures are based on plies oriented in a number of directions in which the oppurtunity for fibre nesting is limited.

6. Short-Term Properties of Carbon Fibre PEEK Composites   189

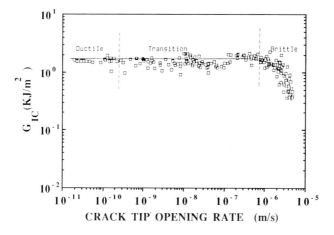

Fig. 6.13 Mode I propagation toughness vs. rate for AS4 carbon fibre PEEK. Ref. [39].

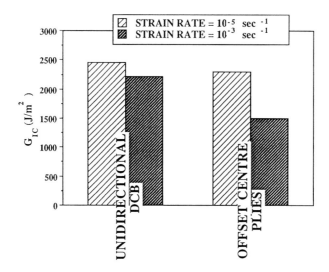

Fig. 6.14 The effect of loading rate on the mode I interlaminar fracture toughness of IM6/PEEK.

3.4 Summary

As a result of the superior mechanical properties of the PEEK matrix, carbon fibre reinforced PEEK offers outstanding mechanical properties over a wide range of test conditions. The strain to failure of the pure poly ether ether ketone polymer is many times that of a conventional epoxy matrix and this is in turn manifests itself in a very high interlaminar fracture toughness under static conditions. The high energy

Table 6.4 Summary of the Effect of Slow and Fast Cooling on the Mechanical Properties of AS4 Carbon Fibre/PEEK. The Cross Indicates the Condition Yielding Superior Properties. Based on Data Taken From Refs [25, 34 and 41]

Property	Slow Cooled	Fast Cooled
Tensile strength ±45 laminate		X[1]
Tensile modulus ±45 laminate	X	X
Compression strength	X	
G_{IC}		X
G_{IIC}		X
Compression after impact		X

[1] Tests on IM6/PEEK [21] have shown that fast cooling results in larger residual stresses and greater volume damage. The fact that this does not result in a lower tensile strength probably relates to the notch insensitivity of (±45°) laminates.

absorbing capacity of the matrix is reflected by a very good resistance to localised low velocity impact loading. This superior impact resistance is unfortunately not apparent at high velocities where limited testing has shown that the load-bearing properties of this material may be no better than a brittle thermosetting composite. Some sensitivity to certain processing conditions is apparent. The cooling rate dependency of AS4/PEEK is summarized in Table 4. In brief, faster cooling appears to result in an overall superior mechanical response. The exception to this is the compression strength of the material which is probably related to the degree of manufacturing-induced fibre misalignment and residual stresses

References

1. Cantwell, W.J., Morton, J. and Curtis, P.T., in 'Structural Impact and Crashworthiness' Vol. 2, Ed. J. Morton, Elsevier Appl. Sci. Publ. 1984, 521.
2. Blundell, D.J. and Osborn, B.N., Polymer, **24**, 1983, 953.
3. Jones, D.P., Leach, D.C. and Moore, D.R., Polymer, **26**, 1985, 1385.
4. de Charentenay, F.X., Harry, J.M., Prel, Y.J. and Benzeggagh, M.L., in 'Effect of Defects in Composite Materials', ASTM STP836, 1984, 84.
5. Davies, P., PhD thesis, Université de Technologie de Compiegne, 1987.
6. Davies, P., Cantwell, W., Richard, H., Moulin, C. and Kausch, H.H., Proc. ECCM3, Eds. A.R. Bunsell, P. Lamicq and A. Massiah, Elsevier Applied Science 1989, 747.
7. ICI data sheet 3b for IM7 carbon fibre/PEEK.
8. Masters, J.E., Proc. of the 6th ICCM and 2nd ECCM Conf., Eds. F.L. Matthews, N.C.R. Buskell, J.M. Hodgkinson and J. Morton, Elsevier Applied Sci., 1987, Vol **3**, 3.96.
9. Johnston, N.J. and Hergenrother, P.M., Proc. 32nd SAMPE Symp., 1987, 1400.
10. Leeser, D., and Leach, D., Proc. 34th Int. SAMPE Symp. 1989, 1464.
11. Cantwell, W.J., Unpublished results.
12. Wisnom, M.R., Composites, **21**, 1990, 403.
13. Leeser, D. and Banister, B., in 'Advanced Materials: Cost Effectiveness, Quality Control, Health and Environment', Eds. A. Kwakernaak and L. van Arkel, SAMPE/Elsevier Sci. Publ. 1991, 97.

14. Kriz, R.D., in 'Effect of Defects in Composite Materials', ASTM STP836, 1984, 250.
15. Buchman, A. and Isayev, A.I., SAMPE Journal, **27**, 1991, 30.
16. Morton, J. and Godwin, E.W., Composite Structures, **13**, 1989, 1.
17. Dan-Jumbo, E., Leewood, A.R. and Sun, C.T., in Composite Materials: Fatigue and Fracture, 2nd volume, ASTM STP 1012 P.A. Lagace Ed., 1989, 356.
18. Béguelin, P. and Barbezat, M., Proc. of the 5th Journée Nationale DYMAT Bordeaux, 1989.
19. Dorey, G., Bishop, S.M. and Curtis, P.T., Comp. Sci. and Tech., **23**, 1985, 221.
20. Hartness, J.T., SAMPE Journal **20**, 1984, 29.
21. Cantwell, W.J., Davies, P. and Kausch, H.H., Comp. Structures, **14**, 1990, 151.
22. Talbott, M.F., Springer, G.S. and Berglund, L.A., J. Comp. Mat., **21**, 1987, 1056.
23. Russell, A.J. and Street, K.N., in 'Toughened Composites', ASTM STP 937, Ed. N. Johnston, 1987, 275.
24. Leach, D.C., Curtis, D.C. and Tamblin, D.R., in 'Toughened Composites', ASTM STP 937, Ed. N. Johnston, 1987, 358.
25. Curtis, P.T., Davies, P., Partridge, I.K. and Sainty, J.-P., Proc. of the 6th ICCM and 2nd ECCM Conf., Eds. F.L. Matthews, N.C.R. Buskell, J.M. Hodgkinson and J. Morton, Elsevier Applied Sci., 1987, Vol **4**, 4.401.
26. Davies, P., Cantwell, W., Moulin, C. and Kausch, H.H., Comp. Sci. and Tech,. **36**, 1989, 153.
27. Johnson, W.S. and Mangalgiri, P.D., NASA Technical Memorandum 87716, 1986.
28. Barlow, C.Y., Peacock, J.A. and Windle, A.H., Composites, **21**, 1990, 383.
29. Davies, P., Cantwell, W.J. and Kausch, H.H., J. Mats. Sci. Letters, **9**, 1990, 1349.
30. Davies, P., Cantwell, W.J., Jar, P.-Y., Richard, H., Neville, D.J. and Kausch, H.H., to be published in 'Fatigue and Fracture' Ed. T.K. O'Brien, ASTM STP 1110.
31. Galiotis, C., Melanitis, N., Batchelder, D.N., Robinson, I.M. and Peacock, J.A., Composites, **19**, 1988, 321.
32. Jeronimidis, G. and Parkyn, A.T., J. Comp. Mats., **22**, 1988, 401.
33. Young, R.J., Day, R.J., Zakikhani, M. and Robinson, I.M., Composites Sci. and Tech., **34**, 1989, 243.
34. Vautey, P., PhD thesis, Université de Technologie de Compiegne, 1989.
35. Leicy, D. and Hogg, P.J., Proc. ECCM3, Eds. A.R. Bunsell, P. Lamicq and A. Massiah, Elsevier Applied Science 1989, 809.
36. Béguelin, P. and Kausch, H.H., Proc. of the 8th Int. Conf. on Deformation Yield and Fracture of Polymers, The Plastics and Rubber Inst. 1991, paper 22.
37. Mall, S., Law, G.E. and Katouzian, M., Proc. of the SEM Spring Conf. on Experimental Mechanics, 1986, 412.
38. Gillespie, Jr. J.W., Carlsson, L.A. and Smiley, A.J., Comp. Sci. and Tech., **28**, 1987, 1.
39. Smiley, A.J., MSc thesis, Univ. of Delaware 1985.
40. Friedrich, K., Carlsson, L.A., Smiley, A.J., Walter, R., and Gillespie Jr. J.W., J.Mater. Sci., **24**, 1989, 3387.
41. Curtis, D.C., Davies, M., Moore, D.R. and Slater, B., to be published in 'Fatigue and Fracture' Ed. T.K. O'Brien, ASTM STP 1110.

Chapter 7

Long-Term Mechanical Properties of Aromatic Thermoplastic Continuous Fibre Composites: Creep and Fatigue

D.R. Moore

Abstract

The creep and fatigue properties of a number of aromatic thermoplastic continuous carbon fibre composites are presented and discussed.

Tensile creep measurements are presented for four different types of aromatic thermoplastic matrix in the form of [±] laminate composites i.e. the laminate configuration is especially selected in order to highlight the role of the matrix. Two semi-crystalline and two amorphous matrix materials are included where Tg's range from 145°C to 265°C. Creep measurements are discussed for the temperature range 23°C to 180°C. Particular attention is given to the various ways in which creep data can be presented and guidance on objective presentations for material comparisons are given.

Fatigue measurements are presented in the form of stress-log n curves for zero-tension and zero-compression loading. Observations cover a temperature range from −55°C to 120°C where different fibre types, different matrix materials and different lay-up configurations for the laminates are detailed. Some discussion of the various failure mechanisms accompanies the interpretation of the results. The work represents a review of some eight years activity in the author's laboratory on the long term testing of aromatic polymer composites. The chapter includes 13 references and 29 figures.

1 Introduction

Aromatic thermoplastic continuous fibre composites are used in aerospace applications where load bearing characteristics are crucial. The nature of the loading characteristics and the environment of the structure will vary from case to case. However a combination of static and dynamic loading will need to be accommodated in temperature regimes which may vary from those experienced at high altitude (sub

---

D.R. Moore, ICI Materials, Box 90 Wilton Centre F315, Middlesbrough, Cleveland, TS6 8JE, England

zero temperatures) to those that give rise to high temperatures (e.g. high speed flight or a sun-baked runway). This presents a requirement for knowledge of the response of these composite materials to these conditions of loading and temperature. A description of long term properties such as creep and fatigue behaviour can satisfy some of these requirements and these properties will be examined in this chapter.

In describing these long term properties there is also a need to accommodate the influence of some other factors. These will incorporate the many ways that the composite materials can change. For example, changes in the choice of matrix material or fibre type; changes in the laminate by selection of different lay-up configurations, the number of plies or from different processing effects. There will also be a need to contemplate the method used for acquiring a particular property; that is, the way that specimen geometry could play a part or the influence of some of the test variables.

There are two particular aims in describing long term properties for aromatic polymer composites in this chapter. First, to discuss the creep function where the deformation response to static stress will be described. In this context, the influence of time under load and temperature will be examined for a modulus which we will show can be defined in a number of ways. Second, to describe some fatigue functions in the form of stress versus logarithm of the number of cycles to fracture where the strength response to dynamic loading is accommodated.

There have been a number of individual and collective approaches to these general issues and indeed this subject matter could be tackled in various laudable ways. The approach that we will adopt relates to experimental studies in the author's laboratory which have been conducted over the last 8 years. During this period, we have acquired a relatively large database of long term properties of aromatic thermoplastic composites where much of the work is unpublished. Naturally, in restricting attention to specific results from our own laboratory and only referencing the work of others, in a relatively limited way, we are not assuming that the activities of others have not contributed significantly to this subject. In fact, there have been some notable major contributions [1, 2], however a detailed review of other work is outside the scope of this chapter.

The chapter is in three main parts. We start with a description of the material subjects of the work, then describe their creep behaviour and follow with an account of fatigue strength.

2 Materials

The material subjects of this chapter are a range of continuous carbon fibre reinforced aromatic thermoplastic composites. Each composite system has been prepared by the Thermoplastics Composites Group of ICI Materials by a pultrusion methodology into pre-impregnated (prepreg) sheet of about 125 microns thickness and then consolidated into laminated sheet by a compression moulding method which has been described in the manufacturer's literature. Some of the composite systems are commerically available under the trade name APC (e.g. APC-2/AS4), although many of the systems described here relate to special experimental grades of aromatic polymers that have been included in the work because they enable certain important changes to be made in the composites, in a manner that will be detailed later. In all cases, the carbon fibre content is 61% by volume. There are a number of material variables that have been incorporated into this study and these include different matrix materials, different carbon fibre types, different lay-up configurations, different numbers of plies in a laminate and different cooling histories from the melt subsequent to compression moulding. In general terms, a common cooling rate after compression moulding is used (as recommended in the manufacturer's literature) and this standard cooling rate can be assumed unless otherwise stated. The principal material variations can now be described. There are four different aromatic thermoplastic polymers that have been used. Most of the work to be reported will relate to polyether-ether-ketone (PEEK), a semi-crystalline polymers with a DMA Tg of 145°C. In order to investigate changes in the morphological and molecular configurations of the thermoplastic matrix, a number of experimental polymers have also been used. Each is an aromatic thermoplastic polymer, although details of the chemistry is outside the scope of this article. For example, we include another semi-crystalline polymer designated ITX with a Tg of 175°C. There are also two amorphous polymer systems; one with a Tg of 225°C, which we designate ITA, and another with a Tg of 260°C, which we designate HTA. The abbreviations imply that "A" stands for amorphous, "X" stands for semi-crystalline, "IT" stands for intermediate temperature and "HT" for high temperature.

The carbon fibre systems are all manufactured by Hercules and three different types are used in this work; AS4, IM6 and IM8. IM6 fibres have a higher tensile stiffness and strength than AS4 fibres. IM8 fibres have higher tensile stiffness and about the same tensile strength as IM6 fibres. Experimental composites are those systems that incorporate a matrix based on ITA, ITX, or HTA.

Our description of a composite will indicate the fibre type followed by the matrix type. For example, AS4/PEEK, IM8/ITA, IM8/ITX etc. The total number of fibre/matrix possibilities can therefore be seen to be relatively large and in fact some seven different composites will be included based on different material variables.

The laminates themselves also provide further variants of the composites and three different types are included:

$$(-45, 0, +45, 90)_{ms}$$

$$(\pm 45)_{ns}$$

$$(0)_l$$

where m, n, l are integers and enable the number of plies to be defined.

The cooling rate from the melt after compression moulding provides the final option of change for the laminate. A standard cooling rate is generally used and this is about 50°C per minute. On occasions a slow cooling rate is used in order to increase the level of crystallinity of the polymer matrix. When this is employed the cooling rate is 1°C per minute for cross-ply laminates and 0.5°C per minute for unidirectional laminate and these are designated "slow" cooled laminates. There is one undirectional laminate that has been cooled faster than all other laminates (80°C per minute) and this is designated a "fast" cooled laminate.

These composite systems have variously been used for the creep and fatigue studies that will now be described.

3 Creep

3.1 The Creep Experiment

Creep is a description of the way that strain changes with time under load subsequent to application of a constant stress. A creep function can be expressed in quite a number of ways some of which are objective and useful and others of which are misleading and perhaps uninformed. Creep studies on a range of materials have shown that there are some simple key parameters that need to be identified and controlled for clear expression of the creep function; these include the magnitude of applied stress (σ), the time under load (t), the test temperature (T) and the range of the creep strain (e.g. less than 0.5%, upto 1%, more than 1% etc.). Therefore, the viscoelastic function or the creep strain-time response can be expressed as:

$$\varepsilon = f(t, T, \sigma, \ldots)$$

The nature of the material will also have a significant influence on this function, that is, whether the material is linear elastic, isotropic, non-linear viscoelastic and so on. The type of material that we will be dealing with in contemplating the creep behaviour of aromatic polymer composites can be classed as non-linear anisotropic and visco-elastic. Simply stated this means that whatever feature one defines about the material then almost certainly this feature will have an influence on its creep function!

In this particular creep study, laminates have been selected in order to highlight the viscoelasticity of the thermoplastic matrix and therefore a [±45] lay-up has been used. Consequently, many of the additional complications mentioned above are removed since we will be limiting discussion to specific specimens taken from these [±45] laminates. Naturally, it should not be overlooked that a "worst-possible scenario approach" has been adopted and that many practical laminates will have creep functions that accommodate little or even no time dependence (e.g. [0] laminates). In this context, the creep experiments should be designed in order to describe the function expressed above. In our laboratories [3] we have developed a methodology for limiting the number of tests that need to be conducted by combining experimental data with a computerised interpolation procedure. In fact our experimental programme for a given material at a specific temperature can be restricted to a stress-strain measurement and a strain-time measurement. These data are then introduced into a computer system which then provides a version of the full stress-strain function at that temperature. The detail of this procedure have already been published [3], but the nature of continuous fibre reinforced composites provides an additional experimental complication. Historically, our procedures have been linked to experimental data derived from some special purpose-built lever loading creep machines where a specimen has a width of some 5.5 mm [4]. The nature of the continuous fibre reinforcement places some special restrictions on specimen size. For example, it is likely that a uniform stress across the width of a creep specimen in an angle-ply laminate will be disrupted by the presence of a single fibre traversing the width of a specimen. This creates an "edge-effect" where a proportion of the width does not have a uniform stress; by "rule of thumb" this proportion of the width approximates to twice the specimen thickness. Therefore particular attention must be given to the specimen design, especially in relation to the width to thickness ratio since the larger this ratio the most uniform the stress field. Our parallel-sided creep specimens have a width to thickness ratio of $6\frac{1}{4}$ which we believe is adequate for generating uniformity of the stress. The actual specimen dimensions are width 12.5 mm, thickness 2 mm, overall length 175 mm, and extensometer gauge length 50 mm.

3.2 The Experimental Programme and Data Manipulation

Two tensile tests are conducted on parallel-sided specimens in order to define the creep function. All specimens are machined from $(+45, -45)_{4s}$ laminates such that applied stress is directed at 45° to the fibre direction. Measurements are made with a servo-hydraulic Instron 8033 machine which is equipped with an environmental chamber for elevated temperature work. All force, time and deformation data are input to a software package housed on a Hewlett-Packard 9816 machine where data collection and initial analysis occurs. These data are subsequently passed to a Microvax computer for further analysis, presentation and storage with our laboratory's "MEDOC" software package.

The first test involves a stress-strain measurement at 1 mm/min at a specific temperature in the range 23°C to 180°C. These data enable the selection of a creep stress which aims to achieve a tensile strain of about 0.5% at short loading times. The force commensurate with this stress is then applied to a new specimen at a rate of application of 2 kN/sec. Depending on the test temperature, this loading rate achieves the desired stress in a time period of 0.3 to 0.8 secs; once this stress has been reached it is maintained constant for the remainder of the test whilst strain-time data are collected. It is important that the monitoring of creep strain relates to the viscoelastic response from an applied constant stress. In particular we wish to avoid the transients that occur on application of the stress. As a general guide, it has been shown that creep strains are viscoelastic responses to constant stress after $10 \times$ the loading time [4]. Therefore, we make no attempt to interpret experimental strains prior to 3 to 8 secs.

The results of these two tests in combination with our "MEDOC" software package enables a full strain-stress-time function to be determined by an interpolation procedure [3]. It is important to recognise that no attempt is made to exceed the strain-stress-time boundary established by the measurements. Nevertheless, it is extremely helpful to manipulate these data in order to set-up various ways of presenting the creep functions. There are three principal approaches that we use for making creep data comparisons:

(i) Secant modulus versus strain at a particular temperature.

$$E_s(T, \Delta\varepsilon) = \frac{\sigma}{\varepsilon}$$

Where the secant modulus E_s is the ratio of stress to strain at the specific strain rate $\Delta\varepsilon$ and at the temperature T.

(ii) Modulus at a common stress versus time under load.

$$E_\sigma = \frac{\sigma_s}{\varepsilon(t, T)}$$

Where the common stress modulus E_σ is generated from the interpolation procedure. A common stress (σ_s) of 15 MPa has been used.

(iii) Modulus at common levels of strain.

$$E_\varepsilon = \frac{\sigma(t, T)}{\varepsilon_s}$$

Where the common strain modulus E_ε is derived by a computer cross plotting of the interpolated creep data.

The common strain (ε_s) is 0.2%.

An additional form of presenting the data is as direct experimental creep modulus. In this case we revert to a secant modulus (E_s) since the constant applied stress (σ_s) will be changing from one test to the next:

$$E_s = \frac{\sigma_s}{\varepsilon(t, T)}$$

In defining and employing these various moduli functions, we hope to demonstrate the value in control of the deformation factors of stress and strain in presenting objective comparisons of the creep functions.

3.3 The Modulus-Strain Function

The modulus-strain function is derived directly from the stress-strain data at each of the test temperatures. The modulus is a secant value as described earlier (E_s). Figs. 1 to 4 show these functions for the four composites used in the creep study (IM8/PEEK, IM8/ITX, IM8/ITA and IM8/HTA) at four test temperatures, namely 23°C, 120°C, 150°C and 177°C. The intention now will be to discuss the behaviour as a function of temperature for each of these materials and then to consider comparisons between materials.

The modulus-strain function for IM8/PEEK is shown in Fig. 1. It is expected that considerable non-linearity will occur in the tensile stress-strain characteristic for these [±45] laminates, moreover, that this non-linearity will persist to quite large deformations (e.g. in excess of 20%). In contemplating the modulus property it becomes a little meaningless to think in terms of a modulus at strain magnitudes above a few percent. Beyond this region of deformation it is more meaningful to

7. Properties of Aromatic Thermoplastic Continuous Fibre Composites

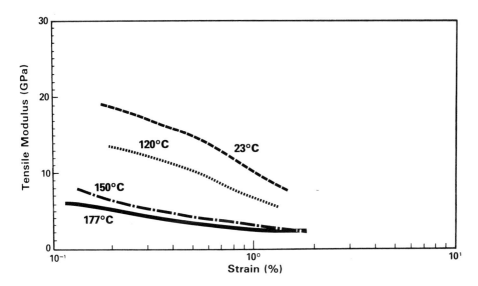

Fig. 7.1 Tensile modulus vs strain for [±45]$_{4s}$ laminates of IM8/PEEK at 1 mm/min.

describe the behaviour in terms of a failure strength. Consequently, our discussion of modulus will only relate to deformations below 3%. The modulus-strain function for IM8/PEEK is seen in Fig. 1 to be both strain dependent and to change its character of strain dependence with temperature. As expected, the magnitude of modulus at a specific strain is considerably temperature dependent.

The DMA Tg of PEEK is 145°C. This will be time dependent and therefore a slightly smaller value can be expected for the slower rate test that has been used for generating the modulus-strain function. This difference in Tg is not considered to be important in the context of our discussion of stress-strain data but will become significant when accommodating creep behaviour. With reference to the curves in Fig. 1 it can be seen that two of the curves relate to the composite being below its Tg (23°C and 120°C) whilst two of the curves relate to temperatures above the Tg (150°C and 177°C). As expected, the modulus will be relatively high for temperatures below the Tg and much lower above the Tg. This is observed but in addition it is noticeable that the strain dependence of modulus also seems to change with test temperatures.

Fig. 2 shows the modulus-strain function for IM8/ITX which has a DMA Tg of 175°C. There would appear to be a more strain dependence of modulus at test temperatures below Tg (23°C, 120°C, and 150°C) than that at 177°C.

The behaviour of IM8/ITA is shown in Fig. 3 and that for IM8/HTA in Fig. 4. Both of these amorphous matrix composites have been tested at temperatures below their Tg's and it is noticeable that there are smaller changes of modulus with temperature and a similar strain dependence throughout the testing regime. In addition, the material with the highest Tg (i.e. IM8/HTA) has the least sensitivity to temperature in the range of our measurements.

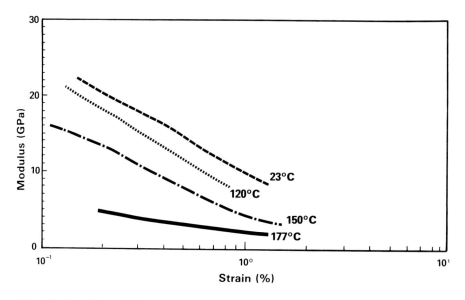

Fig. 7.2 Tensile modulus vs strain for [±45]$_{4s}$ laminates of IM8/ITX at 1 mm/min.

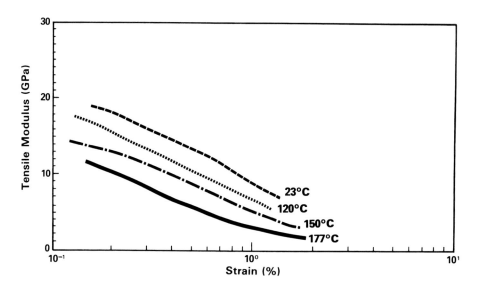

Fig. 7.3 Tensile modulus vs strain for [±45]$_{4s}$ laminates of IM8/ITA at 1 mm/min.

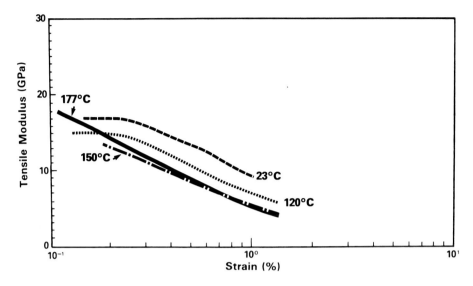

Fig. 7.4 Tensile modulus vs strain for [±45]$_{4s}$ laminates of IM8/HTA at 1 mm/min.

A modulus comparison between materials can be achieved from the modulus-strain functions. First, it is reasonable to select a temperature below the Tg for each of the materials e.g. 23°C. Then it is important to consider modulus at a common level of deformation e.g. 0.2% strain. An objective comparison can then be made where it is apparent that the moduli for the semi-crystalline matrix composites is similar and both are higher than that for the amorphous matrix composites. It would be possible to reverse this order if moduli at different and arbitrary levels of deformation are selected. Therefore objective comparisons are really quite important. To add to these observations, we have also measured the modulus-strain function for AS4/PEEK for comparison with IM8/PEEK. For the selected [±45] laminates no differences in behaviour were observed.

Clearly the location of the Tg relative to the test conditions is crucial in making modulus comparisons. This can be highlighted by comparison of the modulus-strain functions at 150°C for all four composites as shown in Fig. 5.

IM8/PEEK is above its Tg at this temperature while the other three composites are below their Tg. With reference to Fig. 5 it can be seen that the modulus-strain functions for the higher Tg materials are similar but quite different to that for IM8/PEEK.

3.4 Creep Modulus at Common Stress

Modulus-time plots provide a view of the creep function. In this section we will examine modulus at a common stress level and as defined in an earlier section two forms of modulus are possible:

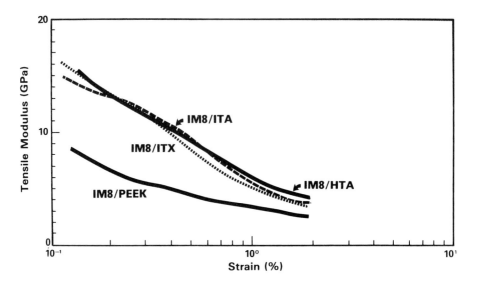

Fig. 7.5 Tensile modulus vs strain for $[\pm 45]_{4s}$ laminates of aromatic polymers composites at 150°C at 1 mm/min.

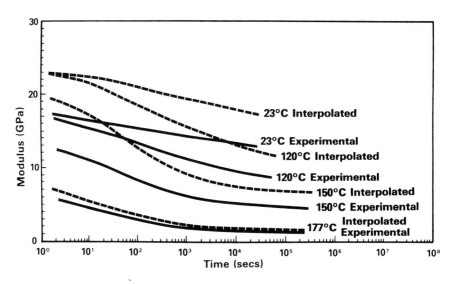

Fig. 7.6 Comparison between common stress modulus vs time by experiment and interpolation for $[\pm 45]_{4s}$ laminates of IM8/ITX.

7. Properties of Aromatic Thermoplastic Continuous Fibre Composites

$$E_\sigma = \frac{\sigma_s}{\varepsilon(t,\,T)}$$

or

$$E_s = \frac{\sigma_s}{\varepsilon(t,\,T)}$$

E_s is the direct experimental modulus whilst E_σ is an interpolated modulus which provides an opportunity for us to select a common stress level. The direct experimental modulus cannot easily be at a common stress level because use of a wide range of temperatures in the experimental programme makes it difficult to accommodate a wide range of deformations in the material if a single common stress is selected. Therefore the first issue that we will examine is whether experimental modulus or interpolated modulus provides a different view of creep performance. Fig. 6 shows both types of moduli plotted against log time under load for IM8/ITX at four different test temperatures. It is quite apparent that at the three test temperatures below the Tg (175°C) the view of creep performance and magnitude of modulus is radically different for the two types of presentation.

The experimental modulus relates to stress levels of 74 MPa, 45 MPa, 35 MPa and 17 MPa for the temperatures 23°C, 120°C, 150°C and 177°C respectively; whilst the interpolated common stress modulus relates to the common stress level of 15 MPa. Modulus depends both on time and deformation level (whether stress or strain), therefore merely expressing the experimental modulus based on inevitably different levels of deformation must be misleading. It is therefore vital to normalise the data in the mannerm made possible by the analysis available in our software package. This approach is therefore adopted for discussion of the creep results.

Examination of the creep function at various temperatures requires the position of the matrix material's Tg in time and temperature to be accommodated in accounting for the creep functions. A criterion for whether the Tg of the matrix material is influencing the creep performance can be simply established via the relative magnitude of modulus. For composites of [±45] laminates it is apparent that a common stress modulus of about 20 GPa is commensurate with a material in a time/temperature regime before/below its Tg; whilst a modulus of about 5 Gpa or less would indicate performance after/above the Tg. This criterion will be adopted in commenting on the creep functions.

Fig. 7 shows the common stress interpolated modulus-time function for IM8/PEEK, Fig. 8 for IM8/ITX, Fig. 9 for IM8/ITA and Fig. 10 for IM8/HTA at the four test temperatures. The creep curves for the IM8/PEEK composites enables us to illustrate the influence of the Tg of the matrix polymer on the creep function. The curve at 23°C portrays the viscoelastic response in a temperature/time regime which is always below the temperature or before the time that constitutes the molecular activities that are commensurate with the Tg of the matrix material. The curves at 150°C and 177°C indicate viscoelasticity in a regime which is above (temperature) and after (time) the molecular activities that constitutes the Tg. The data at 120°C starts in a time/temperature regime which is at a temperature below and at a time before the Tg. By the end of the test, the time/temperature regime would seem to be

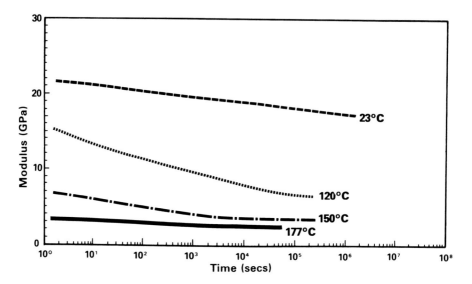

Fig. 7.7 Interpolated common stress modulus vs time for $[\pm 45]_{4s}$ laminates of IM8/PEEK (stress = 15 MPa).

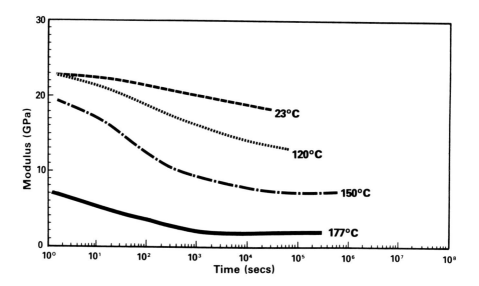

Fig. 7.8 Interpolated common stress modulus vs time for $[\pm 45]_{4s}$ laminates of IM8/ITX (stress = 15 MPa).

7. Properties of Aromatic Thermoplastic Continuous Fibre Composites

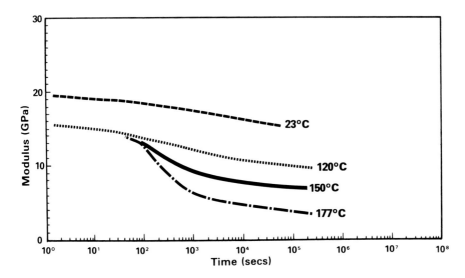

Fig. 7.9 Interpolated common stress modulus vs time for [±45]$_{4s}$ laminates of IM8/ITA (stress = 15 MPa).

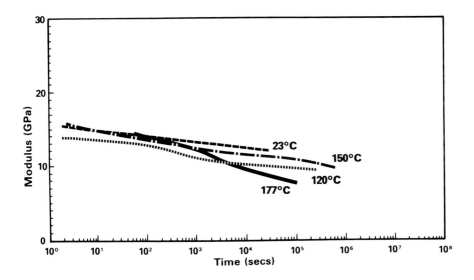

Fig. 7.10 Interpolated common stress modulus vs time for [±145]$_{4s}$ laminates of IM8/HTA (stress = 15 MPa).

commensurate with the material exhibiting behaviour below (temperature) its Tg, but moving towards (time) its Tg, at least judging by the speed at which modulus falls with time.

Further application of this criterion for the creep function for IM8/ITX in Fig. 8 indicates that at 23°C and 120°C that the Tg is not influencing the creep behaviour, although the time dependence of modulus increases as temperature increases. At 177°C the Tg is having a dominant effect on the creep function; in fact above the Tg the time dependence seems to reduce significantly. At the intermediate temperature of 150°C it would appear that the testing regime in more or less at a time just prior to and at a temperature just below the transition.

The creep behaviour of IM8/ITA (shown in Fig. 9) can be described in an identical manner to that just given for IM8/ITX even though the amorphous matrix material has a Tg some 50°C higher than that for the semi-crystalline material. Therefore a comparison of the creep functions of these composites illustrates the stiffness benefits that arise from the crystallinity of the matrix.

The creep curves for the IM8/HTA in Fig. 10 show that the time/temperature regime for measurement is both before in time and below in temperature the Tg of the matrix. Again there are suggestions of an increasing time dependence with increasing temperature.

A further interest in these results comes from comparative material performance in creep. The data so far presented enables a wide choice of comparisons, but there are two specific comparisons at 120°C and 150°C that will be discussed now. In order to make these comparisons we will again use the interpolated data at the common stress of 15 MPa, but on this occasion express the results directly as creep curves i.e. strain-time functions.

Fig. 11 shows these strain-time functions at 120°C where three of the materials are quite similar in creep but the IM8/PEEK although similar at short times, begins to exhibit much greater time dependence at longer times. This is due to the longer times under load at 120°C being nearer to the Tg for this matrix material.

The creep functions at 150°C are examined next with the results for the four composites shown in Fig. 12. IM8/PEEK is at a temperature above its Tg and shows a distinctly different creep characteristic than the other three materials. The composites based on ITA and ITX are similar. Again the higher Tg of ITA is being offset against the cryustallinity of the ITX. The matrix material with the highest Tg (HTA) exhibiting the least time dependence in creep because the composite is at a time/temperature regime which is furthest away from the Tg.

3.5 Creep Modulus at Common Strain

Interpolation of the creep data provides an ability to create many creep curves at other stress levels. When these are cross plotted at common strain levels then a description of common strain modulus can emerge as defined earlier:

$$E_\varepsilon = \frac{\sigma(t, T)}{\varepsilon_s}$$

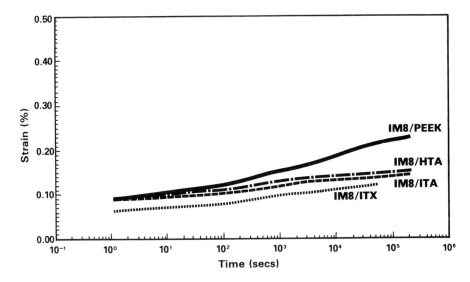

Fig. 7.11 Strain vs time creep curves after interpolation for [±45]$_{4s}$ laminates at a common stress of 15 MPa at 120°C.

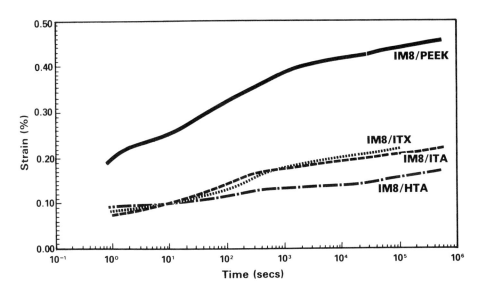

Fig. 7.12 Strain vs time creep curves after interpolation for [±45]$_{4s}$ laminates at a common stress of 15 MPa at 150°C.

This modulus should not be vonfused with a relaxation modulus because the experimental data for deriving the modulus relate to a common applied stress (i.e. retardation modulus).

The creep functions of modulus (at 0.2% strain) versus log time under load at the various temperature are shown in Fig. 13 for IM8/PEEK, Fig. 14 for IM8/ITX, Fig. 15 for IM8/ITA and Fig. 16 for IM8/HTA. What is noticeable is that the detailed magnitudes of modulus in this set of figures is slightly different than those in Figs. 7 to 10. This is because the definition of modulus is different and in particular that a stress of 15 MPa as used in defining the common stress modulus is not consistently equivalent to a strain of 0.2% strain as used in the definition of the common strain modulus. This again reinforces the imprtance of clarity of definition in the derivation of modulus. In other words, modulus is at least a function of time, temperature and deformation level (whether expressed as stress or strain).

The outline trends expressed in in the results in Figs. 13 to 16 are similar to those already discussed in the section on common stress creep functions and therefore we do not need to repeat the commentary here.

3.6 Concluding Comments

The results presented for the creep behaviour of these four aromatic thermoplastic composites (PEEK, ITX, ITA and HTA) in the form of [±45] laminates shows clearly that modulus depends on time under load, temperature, stress/strain level. The characteristics of these dependencies have been described where it is apparent that modulus will reduce for each of these parameters increasing in magnitude.

It has been possible to illustrate that matrix morphology is also important. In particular, there are stiffness benefits from crystallinity in the matrix material. However, the Tg of the matrix material would seem to have an over-riding importance. At temperatures just below the Tg then time under load features as a significant factor thus confirming the time/temperature nature of matters that relate to the Tg.

In general there are few concerns about creep of these materials provided that a brief for their likely service needs can be off-set against a knowledge of their viscoelasticity. Establishing a context for these matters should also be achieved in as much as the selection of the laminate system in this study has significantly exaggerated the creep function. Changing the lay-up configuration of the fibres within the laminate would dramatically reduce the contribution of creep.

4 Fatigue

4.1 The Fatigue Experiment

There are two main categories of fatigue experiment. One is a stress versus log number of cycles to failure or fracture; the other a crack growth type approach where a Paris Law type of presentation is used. Neither is all embracing and inevitably both approaches are necessary for anything that might attempt a full understanding of fatigue behaviour of composites.

Only one of these approaches is used in this work, namely the stress-log n presentation. This is not selected because it is scientifically better since both approaches have their limitations, but because it is simpler and provides a starting point for accommodating many of the material features which our work is attempting to describe.

It is possible in a simple stress-log n approach to select a particular laminate where one or more of the possible failure mechanisms can be occurring.

The failure mechanisms that occur in fatigue of continuous carbon fibre composite components depend on both the applied stress field and the lay-up, but include interlaminar fracture (delamination between the ply regions), intralaminar fracture (matrix cracking that can occur within the structure of the laminate not just within an inter ply region), debonding and translaminar fracture (fibre fracture). It is possible for these mechanisms to occur independently as well as in an inter-connected manner but no generalities exist. Three specific combinations of stress field and lay-up geometry will be used in this particular study. Limiting to this choice is somewhat arbitrary and other combinations require additional work. Tensile fatigue of $(\pm 45)_{ns}$ laminates can be shown to include interlaminar and intralaminar fracture [5]. Tensile fatigue of $(-45, 0, +45, 90)_{as}$ laminates starts with intralaminar fracture in the 90 degree plies followed by matrix cracking in the 45 degree plies and then delamination crack growth between the ply layers [6]. Compressive fatigue of unidirectional laminates involves matrix cracking (in the form of shear induced cracks), debonding between fibre and matrix and fibre failure (probably in the form of buckling of fibres [7–11]. Therefore fatigue experiments for these three types of laminate can furnish

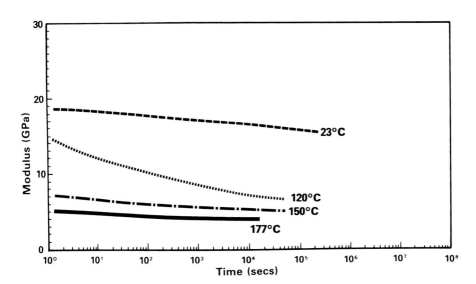

Fig. 7.13 Interpolated common strain modulus vs time for $[\pm 45]_{4s}$ laminates of IM8/PEEK (strain = 0.2%).

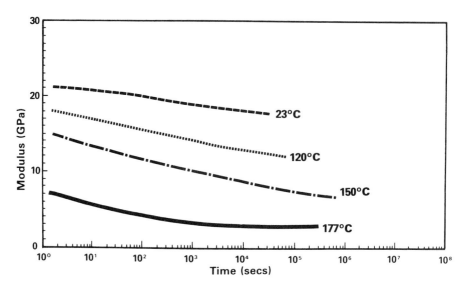

Fig. 7.14 Interpolated common strain modulus vs time for $[\pm45]_{4s}$ laminates of IM8/ITX (strain = 0.2%).

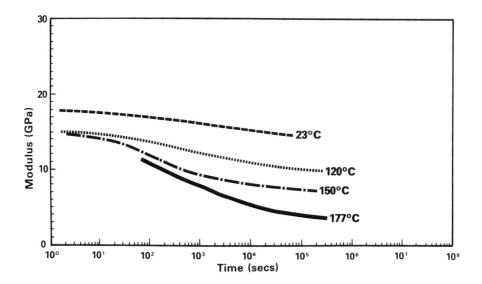

Fig. 7.15 Interpolated common strain modulus vs time for $[\pm45]_{4s}$ laminates of IM8/ITA (strain = 0.2%).

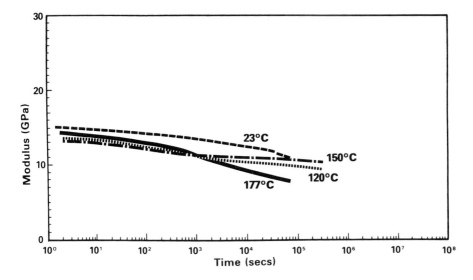

Fig. 7.16 Interpolated common strain modulus vs time for $[\pm 45]_{4s}$ laminates of IM8/HTA (strain = 0.2%).

information on a combination of the basic failure mechanisms mentioned earlier. In addition, by varying the test temperature or the cooling rate from the melt in the preparation of the laminate or even the type of matrix material or fibre reinforcement, then considerably more understanding on the role of these mechanisms can ensue.

The description of the various mechanisms cited in these literature references do not always relate to aromatic polymer composites. Nevertheless, it is assumed that the individual mechanisms are dominated by the lay-up and stress field, albeit that the balance and mix will be related to the fibre type, matrix type or interface. Consequently, it will be possible to approach failure mechanisms by adopting this knowledge.

The potential use of angle ply laminates for aerospace structure brings with it an interest in the fatigue strength of these laminates i.e. $(\pm 45)_{ns}$, $(-45, 0, +45, 90)_{ns}$. The option to have data at ambient temperatures, particularly in contemplating the flight of an aircraft, would require fatigue properties in a temperature range of $-5°C$ to $23°C$. In addition, fatigue strength plots for laminates prepared at different cooling rates from the melt will add practical significance. Compressive fatigue of a thermoplastic composite might focus interest at both $23°C$ and at some elevated temperature where stiffness of the resin will be less.

4.2 Test Procedures

Fatigue tests were concluded in a 'zero'-tension load controlled mode or a 'zero'-compression load controlled mode. Two types of fatigue machine were used; either an Instron servo-hydraulic machine or a pneumatic fatigue machine built in our laboratory [12]. The fatigue tests involved stress-log n measurements on unnotched specimens, in the temperature range −55°C to 120°C. Square load waveforms were generally employed in order to enable tests to be conducted on any of the fatigue sites. Inter-test variability was negligible as discussed in a previous publication [12]. In the majority of cases, temperature rise during fatigue was measured in order to establish whether test conditions gave rise to any autogenous heating. This was achieved through contact probes attached to the test specimens. No forced cooling of specimens during testing was employed.

Specimen preparation and design proved to be a critical aspect of the fatigue measurements. The tensile fatigue tests employed both end-tabbed coupon specimens and width-waisted specimens; these are shown in Fig. 17. Discussion of the specimen design and performance is presented elsewhere [12]. However, either specimen could be used to generate fatigue data provided that data stemming from failures in the vicinity of the clamped regions were ignored.

Specimen and test design for compressive fatigue was developed in our laboratory in order to conduct this phase of the work. A number of specimen designs were tried where the aim was to achieve a buckling failure in the central region without damage

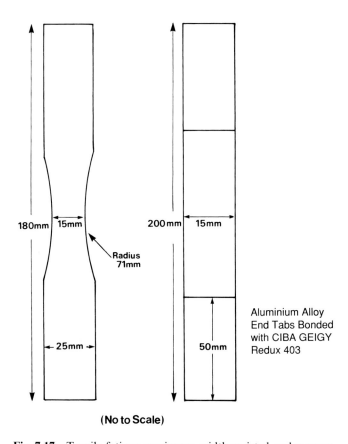

Fig. 7.17 Tensile fatigue specimens, width-waisted and coupon.

7. Properties of Aromatic Thermoplastic Continuous Fibre Composites   213

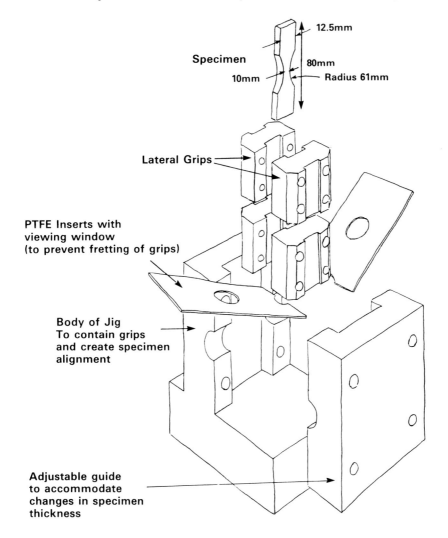

Fig. 7.18 Specimen and jig configuration for compressive fatigue.

elsewhere i.e. no end crushing and no delamination away from the central region. For example, attempts to use the specimen geometry of an ASTM compression test (ASTM D-695) with its thicker regions at the end of the specimen proved successful for short term fatigue, but at long times under load it merely gave rise to a delamination fracture in the region where there was a change in specimen thickness. This rendered it ineffective for obtaining compressive fatigue data. A successful specimen design emerged in the form of a short width-waisted specimen which was mounted into a special jig for gripping and supporting the specimen. The specimen and the jig are shown in Fig. 18. The role of the jig was to support the specimen during the fatigue test, ensuring that the centre region of the specimen was as unrestrained as possible. The success of the specimen is noted by the type of failure obtained in compression, namely failure at the waisted region with collapse of the fibre reinforcement angled across the width of the waist. Static tests on this waisted specimen in comparison with the compressive strength obtained with the ASTM D-695 approach at 23°C gave agreement to within about 10%, albeit that our waisted specimen regularly gave lower strength values.

In the preparation of all specimens it was necessary to use a diamond cutting disc for the preparation of parallel strips and then to use a silicon carbide grinding wheel in the preparation of the waisted regions of the specimen. These techniques ensured minimal damage to the edges of the specimens.

4.3 Tensile Fatigue

4.3.1 The Influence of Some Test Factors

In a previous publication [12] we have discussed the influence of a range of test parameters on the fatigue behaviour of AS4/PEEK, angle ply laminates. Some of these data are a reference for this work and therefore we will revisit the influence of waveform frequency in order to present and discuss other results now. Fig. 19 shows fatigue strength versus log number of cycles to failure for $(-45, 0, +45, 90)_{2s}$ laminates of AS4/PEEK at two frequencies (0.5 Hz and 5 Hz). The applied stress being aligned with the 0° fibres. Autogenous heating was observed at 5 Hz with a temperature rise of upto 25°C [12], but at 0.5 Hz the temperature rise was always less than 3°C.

It is known that the failure mechanisms that occur in a quasi-isotropic laminate (as discussed in earlier) include matrix cracking in the 90° plies, matrix cracking in the 45° plies and delamination between the ply layers. The sequence of these events is not known in terms of the fatigue tests although the data included in the fatigue curves of Fig. 19 are assumed to reflect all of these mechanisms. However, the fact that frequency of load waveform influences the fatigue function and that a temperature rise occurs in the tests conducted at the higher frequency implies that at least one of these mechanisms is temperature dependent.

A similar analysis can be conducted for the (±45) laminates with 16 plies. Fig. 20 shows the fatigue curves at the same two waveform frequencies, where the applied stress is at 45 degrees to the fibre direction. On this occasion the temperature rise at 0.5 Hz is typically 30°C and at 5 Hz is 140°C. As can be seen from Fig. 20 at 10,000 cycles the change in fatigue strength is around 80%. There are known to be two failure mechanisms that occur in the tensile fatigue of this laminate, namely delamination (interlaminar) crack growth between the plies and matrix cracking between the 45° fibres (intralaminar). Reference to the fatigue curves in Fig. 20 indicates that at least one of these mechanisms is temperature dependent. In addition, there is an indication that at long times under load that the fatigue strength will be similar for the two test frequencies. It has also been observed and reported that the temperature rise during fatigue is stress dependent [12], so it is possible that at the lower stress levels associated with these long term data that the temperature dependence is not so large if at

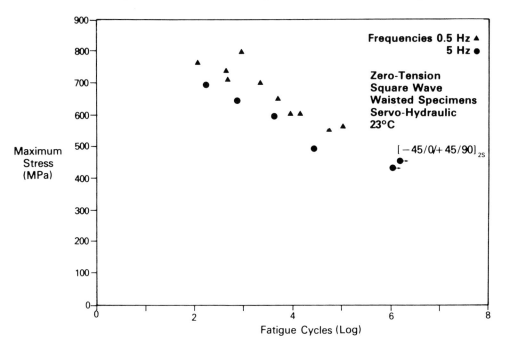

Fig. 7.19 The influence of test frequency for waisted specimens (AS4/PEEK).

all. This might be an explanation as to why the fatigue curves at those two waveform frequencies converge at the long times.

It can be noted that for both types of laminate system discussed so far that there are two common failure mechanisms. These are delamination and matrix cracking in the 45° fibres.

These observations of temperature rise are related to our choice of waveform frequency and waveform type (square). For example, it is known that a square wave can include some high frequency harmonics and that such higher waveform frequencies can lead to autogenous heating. We wished to avoid these temperature rises in the specimen during our fatigue programme, however we persisted with the use of a square waveform, albeit at a low frequency of 0.5 Hz, because this provided us with more scope in selecting a test machine which was necessary for some of the longer term tests which lasted for more than a year.

4.3.2 The Influence of Temperature

Zero tension fatigue has been conducted on $(-45, 0, +45, 90)_{2s}$ laminates at $-55°C$ in order to study the influence of temperature on practically relevant laminates but also to investigate any change in mechanisms. Fig. 21 shows the results from these tests by plotting normalised fatigue strength against log n. The normalised strength is the ratio of fatigue strength to static strength, thus enabling inter-laminate vari-

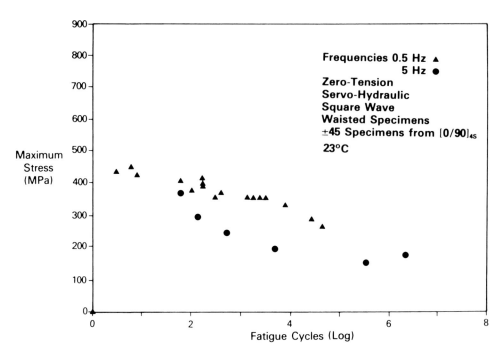

Fig. 7.20 Influence of frequency on fatigue strength of ±45 specimens (AS4/PEEK).

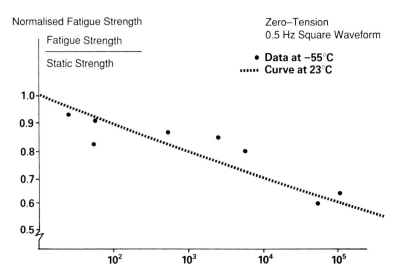

Fig. 7.21 Influence of temperature on the fatigue of $[\pm 45, 0, +45, 90]_{2S}$ laminates of AS4/PEEK.

ability to be removed from the data. It is now completely clear from the data in Fig. 21 that the trends with time under load at the two temperatures are the same. (This is not to say that the fatigue strengths are the same!) Consequently, the mechanisms that dictate fatigue behaviour at 23°C (as previously discussed) are likely to be the same as those mechanisms that dictate fatigue performance at −55°C. However, the contribution of at least one of these mechanisms will have changed in order to produce absolute differences in the fatigue curves at the two temperatures. At present it is not apparent as to which one or more of these mechanisms is exhibiting this temperature dependence out of the three processes of matrix cracking in the 90 plies, matrix cracking in the 45 plies or delamination between the plies.

4.3.3 The Influence of Fibre Type

The measurement of fatigue strength of AS4/PEEK provides some indication of the role of fibre type on the behaviour of $(-45, 0, +45, 90)_{2s}$ laminates. Fig. 22 shows stress-log n curves for these 16 ply quasi-isotropic laminates, where it can be seen that the stiffer and stronger fibres (IM8) exhibits the higher fatigue strength. This would of course be expected because the proportion of the 0° fibres will have a major influence on the strength characteristic. Moreover, this influence can be seen to be independent of time under load suggesting that the fibre properties are always over-riding other mechanistic effects.

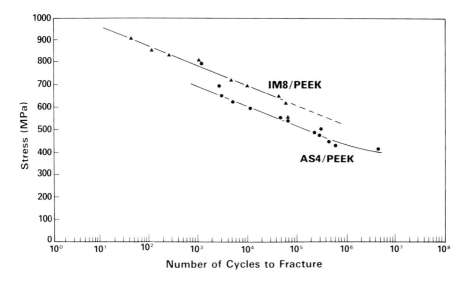

Fig. 7.22 The influence of fibre type on fatigue strength of $[-45, 0, +45, 90]_{2s}$ laminates at 23°C [0.5 Hz frequency, square waveform, zero-tension, width-waisted specimens].

4.3.4 The Influence of Matrix

The influence of thermoplastic matrix on the fatigue behaviour of these aromatic polymer composites has been studied for composite systems based on IM8 carbon fibres. Four composites have been used, namely IM8/PEEK and IM8/ITX (two semi-crystalline matrix composites) and IM8/ITA and IM8/HTA (two amorphous matrix composites). It should be remembered that the latter three of these composites are experimental materials and therefore their design is sub-optimum. In addition, two different angle ply laminates have been employed.

Measurements of fatigue strength on $(+45, -45)_{4s}$ laminates will be controlled by matrix fracture between the fibres and delamination between the plies. Both of these processes are strongly influenced by the matrix material. Fig. 23 shows the stress-log n curves for two of these materials, namely IM8/PEEK and IM8/HTA. It is apparent that the shape of these curves is quite different. The curve for IM8/PEEK shows a smooth reduction in strength with increasing number of cycles; that for IM8/HTA shows a gradually increasing reduction in strength then a flattening out of the curve. This latter shape is characteristic of a ductile-brittle transition in fracture; a feature often observed in the fatigue behaviour of amorphous polymers [2]. Semi-crystalline polymers, on the other hand, seldom exhibit a ductile to brittle feature in fatigue [2].

The $(+45, -45)_{4s}$ laminates will have a fatigue response dominated by the fracture behaviour of the matrix material, whether this is manifest as a intralaminar fracture or an interlaminar fracture. Therefore, the observation of a ductile-brittle transition for the IM8/HTA composite need be of no surprise.

A fuller view of the fatigue strength for these laminates of all four composites is shown in Fig. 24. Both amorphous matrix composites indicate a strong suggestion of a ductile to brittle fracture transition in that their fatigue strengths at long times under load is lower than that compared with the semi-crystalline matrix materials. There are other factors-taking part in the fatigue behaviour also. For example, the short term fatigue strength also favours the semi-crystalline matrix composites; at number of cycles to failure such as 100, the fatigue strengths of these composites follows the order:

$$IM8/PEEK > IM8/ITX > IM8/ITA > IM8/HTA$$

At these short times under load and particularly at high levels of applied stress it would be expected that the matrices would all exhibit yielding behaviour. Therefore, other factors have also to be considered and delamination toughness would be a likely explicable factor. We have some limited interlaminar static toughness results for unidirectional laminates that shows that the delamination toughness for IM8/PEEK is noticeably higher than that for IM8/HTA. Such observations provide only a partial account but support the view of order of fatigue strength for these composites together with supportive hints as to why this order is as expected.

We have also observed the fatigue behaviour of $(-45, 0, +45, 90)_{2s}$ laminates. All four composites were tested and results are shown in Fig. 25. The alignment of $0°$ fibres with applied stress reduces significantly the influence of matrix material on the fatigue strength. However, it does not eliminate the influence of the matrix altogether.

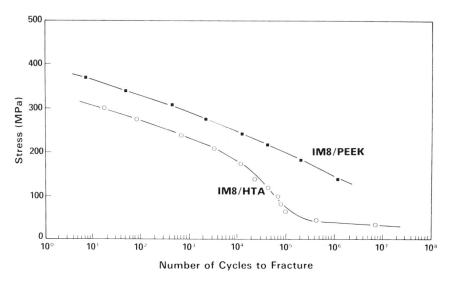

Fig. 7.23 The influence of matrix on fatigue strength for [±45]₄ₛ laminates at 23°C [0.5 Hz frequency, square waveform, zero-tension, width-waisted specimens].

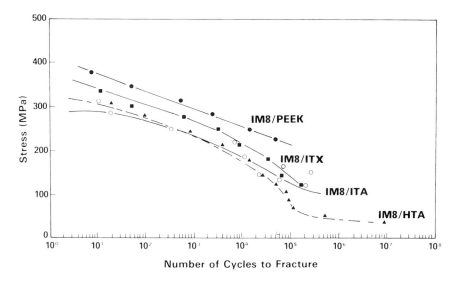

Fig. 7.24 The influence of different matrices on fatigue strength for 4 composites ... [±45]₄ₛ laminates [0.5 Hz frequency, square waveform, zero-tension, width-waisted specimens 23°C].

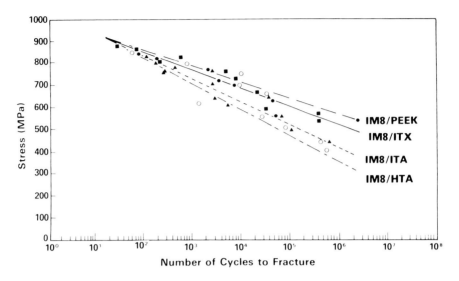

Fig. 7.25 The influence of matrix on fatigue strength for $[45, 0, +45, 90]_{2s}$ laminates at 23°C [0.5 Hz frequency, square waveform, zero-tension, width-waisted specimens].

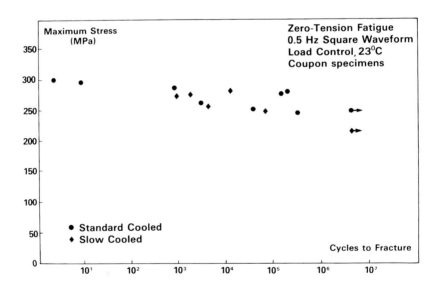

Fig. 7.26 Comparison of fatigue curves in tension for $[\pm 45]_{2s}$ laminates of IM6/PEEK at different cooling rates.

At short number of cycles the fatigue strength of these materials is similar, but at longer times under load it is clear that the two semi-crystalline matrix composites show higher fatigue strength than the two amorphous matrix composites. We could see from the results on the $(+45, -45)_{4s}$ laminates in Fig. 24 that a ductile to brittle fracture transition in the amorphous materials had established itself by just over 0.1 million cycles. The fatigue data on the $(-45, 0, +45, 90)_{2s}$ laminates also shows divergence of strength beyond these number of cycles when comparing semi-crystalline and amorphous matrix composites. Perhaps further evidence of a ductile to brittle transition. However, there are many other complications in interpreting the fatigue results on these quasi-isotropic laminates and therefore it would be desirable to seek additional supportive evidence.

A much more subtle investigation of the influence of matrix crystallinity can be investigated for IM6/PEEK laminates in the form of $(\pm 45)_{2s}$ laminates (8 ply) where different cooling rates from the melt subsequent to compression moulding were used. Standard and slow cooled panels resulted in matrix crytallinities of 26% and 40% respectively [13]. Fig. 26 shows the fatigue data obtained at 0.5 Hz square waveforms on parallel sided specimens at 23°C. The different cooling rates can be seen to have no influence on either the fatigue strength or the fatigue failure mechanisms for these ±45 laminates. It is expected that the PEEK matrix material will exhibit ductility and therefore such subtle differences in the morphology of the matrix would appear to be hidden by other features that influence fatigue strength.

Although a clear understanding of the mechanistic features that occur in these various angle ply laminates will certainly require more information, it is abundantly clear that the matrix has an important role in the overall fatigue characteristics.

4.4 Compressive Fatigue

4.4.1 The Influence of Temperature

Compressive fatigue was conducted on AS4/PEEK in the form of undirectional laminates containing 40 plies. There are primarily three failure mechanisms that are expected in zero-compression tests which include matrix cracking by a shear process between fibres, fibre failure through some buckling mechanisms and a matrix cracking mechanism between fibre and matrix.

The 'fast' cooled sheets of AS4/PEEK were used in order to compile some compressive fatigue data and to examine the influence of temperature. Fig. 27 shows the fatigue data at 23°C and 120°C obtained in load control at 0.5 Hz square waveforms. The data show a similar dependence of fatigue strength with log number of cycles at the two temperatures, although as expected, the curve at 120°C indicates a smaller fatigue strength. It is suggested that the lower stiffness of the matrix at 120°C encourages a higher contribution in absolute terms from the fibre buckling mechanism. In which case, it is implicit that the relative contributions from the matrix cracking mechanism is unchanged in this temperature regime. This is quite an encouraging view of the compressive fatigue performance of AS4/PEEK. Such a view would be

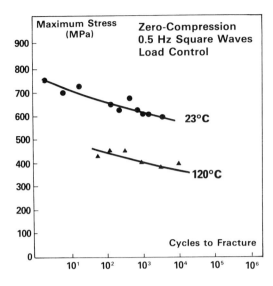

Fig. 7.27 Zero-compression fatigue for $[0]_{40}$ laminates of 'fast' cooled AS4/PEEK at 23°C and 120°C.

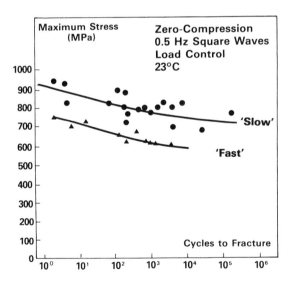

Fig. 7.28 Comparison of compressive fatigue for $[0]_{40}$ laminates of AS4/PEEK at 23°C for different cooling rates.

further validated if longer term fatigue data were available at the higher test temperature, but these are at present unavailable mainly due to the experimental difficulties in conducting longer term compressive fatigue tests at these high temperatures.

4.4.2 The Influence of Cooling Rate

Zero-compressive fatigue obtained under the same test conditions were also conducted on the 'fast' and 'slow' cooled AS4/PEEK unidirectional laminates at both 23°C and 120°C. Fig. 28 shows the compressive fatigue curves at 23°C for 'slow' and 'fast' cooled laminates, whilst Fig. 29 shows similar data but at 120°C. Curves have been fitted by eye to these data and within the scatter of the data the curves could be considered to be parallel. At both temperatures, the slow cooled material exhibits the higher fatigue strength. One possible explanation of this lies in the higher matrix stiffness resulting from the slower cooling from the melt leading to greater resistance to fibre buckling during compression.

The ratio of strength ('slow' to 'fast' cooled) can be calculated on the basis of the curves being parallel; this ratio then being the same at all cycles to failure. This strength ratio is 1:2 at both temperatures which then implies that the mechanistic processes that produces the reduced compressive fatigue strength for the 'fast' cooled laminate is merely changing the contribution of one of the failure mechanisms. As just mentioned, this is likely to be that due to fibre buckling.

4.5 Concluding Comments

The main purpose in selecting and presenting these fatigue results was twofold. First to present some long term strength data on these aromatic polymer composites, in as simple a manner as possible. (It can be added of course that many schools of thought on the fatigue of composites would not confirm stress-log n fatigue to portray the simple picture.) Second to provide some evidence on the material features that can influence fatigue strength. In so doing, we have attempted to open a view on the mechanistic aspects that dominate fatigue fracture, but as can be seen considerable other interpretive evidence is also required.

The simple view that we try to present on fatigue behaviour hides some important points. Fatigue is a relatively complicated business in as much that there are many experimental factors to accommodate but also many combining mechanistic contributions to the fracture process. Certain of these complications make the simple act of collecting fatigue data a tricky task on more occasions than not. Therefore we believe that an early objective in acquiring and understanding long term behaviour of composites must accommodate some sound observations.

In the fatigue results that we have presented there are perhaps a unique collection of long term fracture data i.e. beyond months of data collection and on occasions beyond one year. If composites are to find multiple use in aerospace applications where durability is of the essence, then clearly some confidence in performance as well as an understanding of material characteristics will need to be established. This has been one of the motivating factors behind our experimental study.

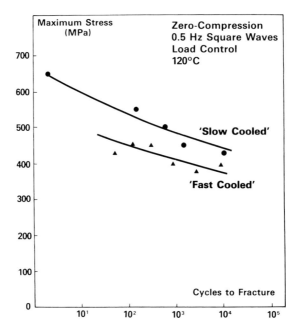

Fig. 7.29 Comparison of compressive fatigue for $[0]_{40}$ laminates of AS4/PEEK at 120°C for different cooling rates.

5 Other Thoughts on Long-Term Properties

There are perhaps four questions that can now be addressed on this discussion of long term properties of aromatic polymer composites.

What is required? Long term properties will be needed in order to provide a description of deformation and fracture characteristics of the composites in a manner that provides knowledge and confidence in the use of the materials. In particular, to provide design know-how and the avoidance of failure in service. How might this be achieved? There will be a need to present deformation properties and strength properties. A composite structure will comprise a certain laminate configuration for which a knowledge of the pseudo-elastic constants in combination with laminate theory can serve to predict performance. However, time and temperature dependence will also be required to complement a pseudo-elastic approach. This can then lead to prediction of the deformations of a structure. In parallel with this, there is a need to contemplate how near such deformations are to a failure point, therefore a time dependent strength function naturally complements the viscoelastic constant. To this end, fatigue results provide an important contribution.

What other factors should be incorporated? There are perhaps three other variables that should be taken into account; geometry, processing and materials. Geometric factors can describe the influence of lay-up configuration and the number of plies in the laminate (i.e. size effects). Processing variables will influence such things as morphology, structure within the matrix and residual stress distributions. Material factors accommodate the type of matrix and fibre characteristics (type, concentration) as well as the interface properties between fibre and matrix.

What contribution has this work made? We believe that we have made a start! For both creep and fatigue properties the functions that we have presented are by no means comprehensive. However, there are some observations on most of the key factors mentioned above as well as a presentation of properties in a context that we hope will make the long term characteristics useful in the context of what is required.

Acknowledgments

The experimental work is the outcome of many years of activity in the Mechanical Properties group in the Wilton Materials Research Centre of ICI and the contributions from Don Curtis, Mark Davies, Barbara Slater and Nabil Zahlan are acknowledged. The author also acknowledges the support and encouragement from David Leach in the Thermoplastic Composites group of ICI.

References

1. Reifsnider, K. L.: "Fatigue of Composite Materials", Vol. 4, Composite Materials Series, Elsevier (1991).
2. Carlsson, L. A.: "Thermoplastic Composite Materials", Vol. 7, Composite Materials Series, Elsevier (1991).
3. Barrie, I. T., Moore, D. R. and Turner, S.: Plastics & Rubber Proc. & Appl., *3*, 293–299 (1983).
4. Turner, S.: "Creep in Thermoplastics", British Plastics, July 1964.
5. Ogin, S. L., Smith, P. A. and Beaumont, P. W. R.: Comp. Sci. Tech. *22*, 23–31, (1985).
6. Beaumont, P. W. R.: "The Fatigue Damage Mechanics of Composite Laminates", CUEDF/C/MATS/TR 139, (1977).
7. Reifsnider, K. L. and Jamison, R. D.: Int. J. Fatigue, p. 187, Oct. 1982.
8. Jamison, R. D., Schulze, K., Reifsnider, K. L. and Stinchomb: ASTM STP, (1983).
9. Curtis, P. T. and Moore: B. B.: RAE Technical Report 82031 (1982).
10. Sturgeon, J. B., Rhodes, F. S. and Moore, B. B. RAE Technical Report 78031 (1978).
11. Barnard, P. M., Butler, R. J. and Curtis, P. T.: 3rd Int. Conference on Comp. Structures, Paisley, Scotland (1985).
12. Curtis, D. C., Moore, D. R., Slater, B. and Zahlan, N.: Composites *19*, 6, p. 338, 1989.
13. Davies P.: Department Des Materiaux, Ecole Polytechnique Federale de Lausanne, Lausanne, Switzerland, Private Communication (1989).

Chapter 8
Structure and Mechanical Properties of Filled and Unfilled Thermotropic Liquid Crystalline Polymers

Ch.J.G. Plummer

Abstract

The basic structural and mechanical properties of injection mouldings of various grades of Vectra™ thermotropic liquid crystalline polymer have been investigated with a view to evaluating their potential as engineering materials. Particular attention has been given to the effects of moulding conditions and of particle addition, both of which are of considerable practical concern for these materials.

1 Introduction

Structural thermotropic liquid crystalline polymers (TLCPs) are a relatively new class of materials, whose commercial development had its beginnings in the early 1970's [1–3], since when most leading polymer producers have at some stage invested in their development [4–6]. TLCPs differ radically from conventional polymers both in terms of their fundamental properties and of the approach which must be taken to exploit fully their potential advantages as engineering thermoplastics. These include:

- high orientability during flow and extended chain conformations, leading to excellent mechanical properties in the flow direction of moulded items;
- low thermal expansivity, advantageous in coatings for optical fibres, circuit boards etc., where large differences in thermal expansivity between organic and inorganic components may result in delamination;
- low melt shrinkage/die swell, highly advantageous for precision moulding;
- low heat of solidification, permitting fast cycle times for injection moulding;
- low viscosity and hence low injection pressures and minimal flash (savings on finishing costs, possibility of miniaturization);
- good chemical, flame and thermal resistance.

Christopher J.G. Plummer, Laboratoire de Polymères, Ecole Polytechnique Fédérale de Lausanne, DMX-D, 1015 Lausanne, Switzerland.

Flow induced orientation may also lead to problems in injection mouldings, particularly where these involve complex forms. Local orientation transverse to the loading direction such as is commonly associated with knit lines is both difficult to avoid, and a source of mechanical weakness [7]. Similarly, in spite of the low thermal expansivity and melt shrinkage, mouldings in which the orientation distribution is highly non-uniform may be subject to substantial internal stress variations, promoting buckling, particularly of asymmetric parts. For these reasons, temperature and geometry play a crucial role in the processing of TLCPs and one should not be mislead into thinking that TLCPs may simply be substituted for conventional polymers in a pre-existing processing set-up. Indeed, some of the solutions advanced for the problem of knit lines [8, 9], would require highly specialized technology, and substantial initial investment.

Given the relatively high price of TLCPs (upwards of 16 $/Kg in 1988), such considerations have tended to mitigate against their widespread use for structural applications, even though this was considered at the outset as the one area in which they showed great potential. Far more commercially attractive has been the use of TLCPs in precision mouldings where the mechanical properties are not at a premium, and in applications requiring high temperature resistance. On the strength of these, TLCPs had gained a share of about 3% of the world high-temperature polymer market in 1988, that is, an estimated 90 million kilograms [10], with the annual growth rate predicted to be between 20 and 50% [5]. In fact, of this, only around 12% was used for precision moulding. Considerably more production has gone into specialized TLCPs for applications such as 'dual-ovenable' cookware, which has to meet stringent temperature requirements and at the same time be microwave transparent. This accounts for about 80% of production of Amoco's Xydar^{TM} range, which includes grades whose heat deflection temperature exceeds 350°C [6].

Suppliers' data for injection moulded bars nevertheless suggest TLCPs to remain an attractive alternative to conventional thermoplastics in structural applications. A clear advantage in Young's modulus is seen; 10 GPa for unfilled TLCPs rising to over 30 GPa for certain of the short fibre reinforced grades, which compares with values of the order of 2 GPa and 10 GPa for unfilled and short fibre reinforced conventional thermoplastics respectively. Similarly, the room temperature strengths of injection moulded unfilled TLCPs may be up to double those of advanced thermoplastic resins such as PES (polyethersulphone) and PEEK (poly-ether-ether-ketone).

However, as with short fibre reinforced composites, there is often a very tenuous link between design data for TLCPs based on standard specimens and the properties of a given non-standard moulding. This inevitably necessitates supplementary research when considering such materials for a specific application [11, 12]. Here it is proposed to discuss some aspects of the basic structural and mechanical properties of TLCP injection mouldings, in the light of our own investigations of the Vectra^{TM} (Hoechst-Celanese) 'A' series of filled and unfilled TLCPs.

1.1 Basic Concepts

LCPs are a hugely diverse category of polymers, whose common structural characteristic is their incorporation of rigid mesogenic units capable of forming a liquid crystalline (LC) phase either in solution (*lyotropic* LCPs) or in the melt (TLCPs). In linear or 'main-chain' structural LCPs, the existence of an LC phase may be accounted for in terms of the statistical mechanical behaviour of a system of high aspect ratio rigid rods (the idealization of a rigid or a semi-rigid linear polymer, where in the latter case, the persistence length is the effective rigid-rod length) [13–17].

The basic requirements of rigidity and high aspect ratio may be satisfied in a vast range of polymer types and architectures, a diversity which is reflected by that of the types of molecular order present in the LC phase [18]. In structural TLCPs, however, the dominant form of long range order above the melting point is molecular orientation along a preferred axis in space **n** called the director. The degree of molecular alignment with **n** is described by the *order parameter*, given by the Hermans Orientation Function

$$S = \tfrac{1}{2}(3\langle \cos^2 \Theta \rangle - 1), \qquad (1)$$

where Θ is the angle between a given molecule and **n**, and such that $S=1$ for perfect alignment and $S=0$ for random orientation. This is called the *nematic* state.

It has long been recognized that an extended chain conformation in a fully aligned polymer should result in a material whose tensile properties depend directly on the bond length and bond angle force constants. Such a situation is difficult to achieve in conventional flexible chain polymers owing to their tendency to adopt a random coil conformation. However, in rigid polymers, chain conformations are in general highly extended, and moreover, in a TLCP melt there will be spontaneous molecular alignment. Since such alignment can generally be maintained during solidification, owing to the low molecular mobility associated with rigid polymers, one might thus hope to obtain solid state tensile properties in the direction of alignment which approach their theoretical limits.

In practice, there are two main problems with TLCPs. The first of these is their intractability. In general, rigid linear polymers tend to be highly crystalline in the conventional sense, making it difficult to obtain a stable nematic melt at temperatures below the decomposition temperature. There are various approaches to overcoming this problem, for example the use of flexible spacers in the chain backbones [2, 19–21], asymmetric ring substitution [22, 23] and the insertion of kinks or bends into the chain backbone [3, 4, 24, 25].

This latter strategy has been adopted in the case of the Vectra 'A' series of thermotropic copolyesters, based on random copolyesters of hydroxybenzoic acid (HBA) and 6-hydroxy-2-naphthoic acid (HNA) units, whose structure is illustrated in Fig. 1. The insertion of a certain proportion of the kinked HNA units into the chain backbone introduces a 'parallel offset' which serves to break up chain linearity and limit the degree of crystallinity [3, 4]. Thus whilst homopolymers of both HBA and HNA have crystalline-nematic transitions well in excess of 400°C [26–28] (indeed, for practical purposes, neither can be considered stable with respect to decomposition), in

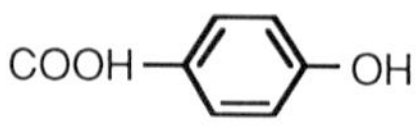

(a) Hydroxybenzoic acid

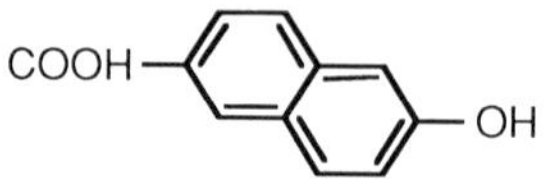

(b) Hydroxynaphthoic acid

Fig. 8.1   The constituent monomers of HBA-HNA.

random copolymers containing between 20 and 60% HNA, the crystalline-nematic transition is reduced to below 300°C [3, 4].

For the injection moulding grades of Vectra we have investigated we believe the composition to be 73% HBA-27% HNA, with a main melting point of approximately 280°C. This compares with Xydar, developed for high temperature resistance rather than processability or mechanical properties, in which the melting point is some 100°C higher [29, 30]. Lower melting point TLCPs also exist, such as Vectra RD-500 (melting point of 230°C), which are particularly suited to technologies based on blending with conventional polymers, in which the matching of processing temperatures is an important factor (see chapter 5).

The second problem is that the spontaneous alignment associated with the nematic state is not necessarily reflected by macroscopic anisotropy since, as is generally characteristic of nematic LCPs, there is tendency for orientational correlation with a single value of **n** to be limited to micron-sized *domains*. In the quiescent state, **n** in adjacent domains appears uncorrelated, so that the quantity given by equation 1, taken over much larger spatial ranges than the domain size, tends to zero. (Under these circumstances, since **n** is itself varying in space, S in equation 1 describes the macroscopic orientation, rather than the order parameter.)

The domain structure is illustrated in Fig. 2 for a thin film of 73/27 HBA-HNA (Vectra A950) cast from a solution in pentaflourophenol at 40°C onto a glass slide and dried; the micrograph is an optical micrograph taken under crossed polarizers, with the contrast thus representing changes in the local director orientation. This is essentially the frozen structure of the nematic texture of the LC melt, frequently referred to as a *Schlieren* texture. Windle and coworkers have also coined the term *tight structure* [31], and argued, on the basis of their observations of the textures of acid etched films under scanning electron microscopy (SEM), that the optical contrast is representative of an essentially continuous sinuous variation in the local director plus the presence of singularities known as *disclinations* [32].

Given that interchain interactions in TLCPs are generally weak, such structures would not account for the remarkable mechanical properties and anisotropy found in injection mouldings. As will be discussed in the following sections, it is the flow

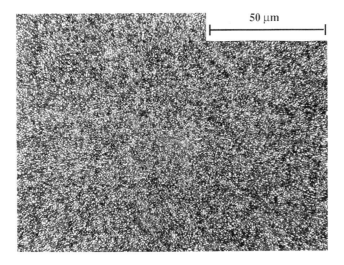

Fig. 8.2 The domain structure of a thin film of 73/27 HBA-HNA.

induced elongation and alignment of the domains which permits the development of useful mechanical properties in these materials.

2 Injection Mouldings

Injection moulding is among the most complex of commonly used polymer processing routes, and as such is extremely difficult to analyze in terms of basic rheological data. The polymer is melted by means of an electrically heated screw extruder, and then injected into a mould cavity with relatively cold walls using sufficient pressure to ensure complete filling of the mould. The mould may subsequently be heated for a short period to minimize residual stresses, after which the part is removed, and the cycle repeated.

The flow behaviour of orientable melts (short fibre thermoplastics, TLCPs) during injection moulding have been described by *Kenig* [33] and *Garg* and *Kenig* in terms of four main processes [34];

(i) diverging flow close to the inlet gates;
(ii) converging flow at the exit gates;
(iii) fountain flow at the free surface of the advancing melt front;
(iv) shear flow behind the melt front.

The first two are also of relevance to any abrupt change in mould cross-section as for example in the shoulder regions of a tensile test bar. In general [33], (i) will promote orientation perpendicular to the main melt fill direction, whereas (ii) will

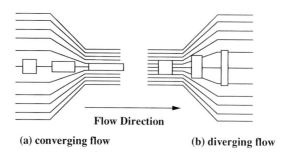

(a) converging flow   (b) diverging flow

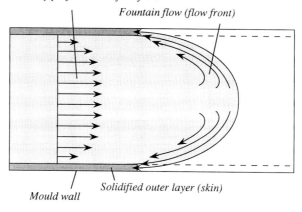

(c) flow fields in the sample barrel

Fig. 8.3 Flow patterns in injection moulding of TLCPs.

tend to have the opposite effect (Fig. 3). Thus in a dumbbell shaped injection moulded bar, with the inlet gate situated at one of the sample ends, transverse orientation developed at the inlet is compensated as the melt is subsequently squeezed through the constriction in the shoulder region, to an extent which will depend on the precise sample geometry (the possibility of 'jetting' directly into the sample barrel before complete filling of the sample end, may be reduced by placing the inlet to one side).

In general during mould filling there will be a gradient in flow velocity v_i across the mould width. Assuming that the advancing melt front moves with a velocity $\langle v_i \rangle$, elements of the melt entering the melt cavity at later times, and moving at $v_i > \langle v_i \rangle$ will tend to catch up with the melt front and thus experience *fountain flow*, illustrated in Fig. 3c. During fountain flow the melt is in extensional flow. Extensional flow is known to result in high degrees of orientation in TLCPs,[1] even at low flow rates

---

[1] Indeed, since the molecular relaxation times are long in TLCPs, simple phenomenological models can be justified in situations where extensional flows dominate, such as extrusion [33, 34].

[35, 36]. This orientation is frozen into a *skin layer* as the polymer comes into contact with the cool mould walls, giving rise to the skin-core structure characteristic of a wide variety of injection mouldings. The regions of the melt which do not reach the flow front are in shear (Poiseuille) flow during mould-filling. At the shear rates generally encountered in injection moulding, the domains have a tendency to remain stable and 'tumble' rather than elongate in the flow direction, leading to relatively weak orientation [37–40]. Further, owing to the characteristic yielding behaviour of TLCP melts, the velocity profile becomes very flat towards the centre of the mould, a phenomenon known as *plug flow*. Thus the central 'core' region of the moulding may be characterized either by a lack of orientation, or alternatively by orientation acquired earlier in its flow history, depending on the mould geometry and the relaxation rate (and hence also the temperature).

Considerable effort has gone into detailed simulation of injection moulding using mathematical and computational modelling [41], mostly aimed at short fibre filled thermoplastics, but also potentially applicable to the problem of orientation development in TLCPs (indeed the Doi model for LC dynamics [17] has also been adapted to fibre suspensions [42]). However, theoretical models for LCPs are not as yet able to provide working constitutive equations for detailed melt flow simulation. Hence, current simulations based on finite element analysis (FEA), whilst of great value to the technologist [43], are at best only able to give a general indication of the spatial distribution of orientation direction for complex geometries. Moreover, in practice, the structure of even the simplest TLCP mouldings as revealed by TEM of ultrasonically disintegrated fibres [44] and microtomed sections [45], polarized light microscopy of thin sections, SEM of acid and plasma etched surfaces, and fractography, may display many complex levels of hierarchy, extending far beyond that of the basic skin-core structure [34, 35, 43, 46–56].

2.1 Morphologies of Injection Mouldings

Both unfilled and filled samples of the Vectra 'A' series of 73/27 HBA-HNA TLCPs display a coarse multi-layered structure, as sketched in Fig. 4. The details of this structure depend to a great extent on geometry, moulding conditions and filler content, but it is broadly representative of the range of TLCP types investigated in the literature [34, 35, 43, 46–56]. The 4×10 mm cross-section dumbbell-shaped tensile test bars to be described in what follows, were injection moulded using a volume fill rate of approximately 10 cm^3 s^{-1}, and mean melt and mould temperatures of 280°C and 80°C respectively (following the suppliers' recommendations). They are typical in so far as they contain the full range of structural types we have encountered in other moulding geometries.

2.1.1 Unfilled Mouldings

The structure of unfilled mouldings is conveniently described in terms of five main regions (Fig. 4):

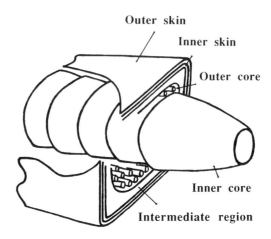

Fig. 8.4 Structure of an injection moulding of unfilled TLCP.

(i) The 'outer skin' layer, which extends 40 to 50 μm into sample surface, and is highly birefringent indicating high orientation. Polished sections transverse to the flow direction and etched in concentrated sulphuric acid show a characteristic layered structure as in Fig. 5a. The inner surfaces of outer skin layers separated from the rest of the sample are relatively smooth as shown in Fig. 6a.

(ii) The 'inner skin' layer, approximately 200 μm in total thickness and, like the outer skin, distinguishable from the sample interior by its high birefringence. In contrast with the outer skin, however, etching of polished transverse sections reveals little texture under SEM. Optical microscopy of thin sections, however, reveals a striking banded texture (Fig. 7a), and longitudinal freeze fracture surfaces (Fig. 6b) are suggestive of a coherent periodic variation in the molecular trajectory along the flow direction. This periodicity, and the associated banding, appear to characterize transient relaxation of a wide range of LCPs subject to high degrees of shear induced orientation [57–64]. Its precise origins nevertheless remain unclear.

(iii) The 'intermediate' layer, weakly birefringent compared with the skin layers, showing a predominantly tight optical texture (Fig. 7b) and, as with the inner skin, no coarse structure under SEM.

(iv) The 'outer core', for which SEM of etched transverse sections and longitudinal freeze fracture surfaces reveals a striking fibrillar structure (Fig. 5b, Fig. 5c and Fig. 6c). The fibrils have a diameter of approximately 5 μm (consistent with the observations of *Tharpar* and *Bevis* [46] and *Sawyer* and *Jaffe* [47]), each fibril appearing to consist of an oriented outer cylindrical region, and an inner region characterized by structures more akin to the unoriented tight texture. However, some continuity of structure in the inner part of these fibrils

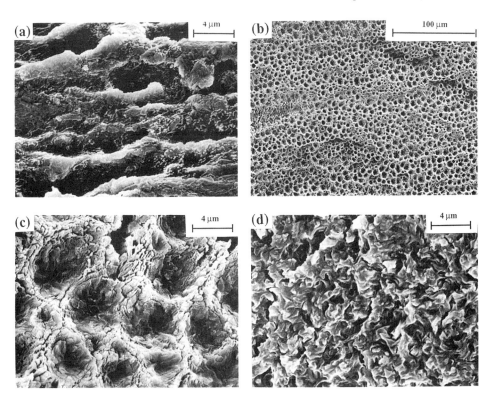

Fig. 8.5 SEM of etched (10 minutes in concentrated sulphuric acid) transverse sections of unfilled HBA-HNA mouldings (A950): (i) outer skin; (ii, iii) outer core; (iv) inner core.

may be inferred from their tendency to break up into elongated sub-fibrils with a diameter of approximately 1 μm (*Sawyer* and *Jaffe* have also identified further levels of fibrillar hierarchy down to 50 nm [47]).

(v) The 'inner core'. The bulk structure of the central part of the inner core generally shows features consistent with the tight texture, both optically and under SEM (Fig. 5d). However, superimposed onto the core structure at irregular intervals are approximately conic 'flow surfaces' concentric with the core centre, and which point along the sample axis. During fracture cracks often run along these flow surfaces giving locally smooth fracture surfaces as with delamination between the outer and inner skin. Towards the centre of the inner core however, the flow surfaces appear attenuated, the fracture surfaces being predominantly fibrillar (Fig. 6d). Particularly dense and coherent parabolic structures have been observed in the inner core of edge gated injection moulded plaque samples, being clearly visible as parabolic lines in etched longitudinal sections, and optical sections (Fig. 7c and Fig. 7d).

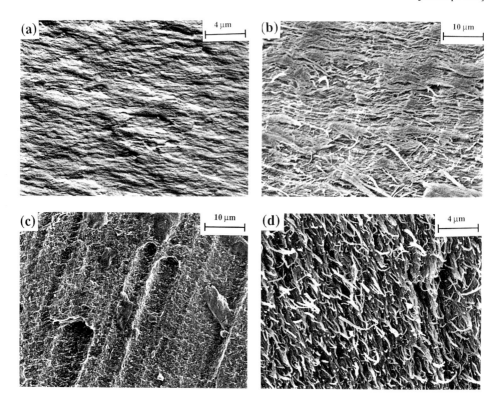

Fig. 8.6 SEM of longitudinal freeze fracture surfaces of unfilled HBA-HNA mouldings (A950); (i) interface between outer and inner skin; (ii) interface between inner skin and intermediate region; (iii) outer core region; (iv) inner core region (fracture along flow surface).

2.1.2 Filled Mouldings

In the mineral (wollastonite) filled Vectra grades, the filler particles themselves have aspect ratios not exceeding 10. Some random orientation exists in the central core, but in the bulk of the sample barrel the particles are aligned in the melt fill direction.

The outermost skin layer, while still essentially layered in structure, is much less coherent than for the unfilled mouldings, with many cavities and local distortions of the layered structure, particularly in association with filler particles, and a certain degree of continuity with the inner skin layer. Again, the inner skin layer is distinguished from the intermediate region by its high birefringence, and shows banding perpendicular to the plane of the layers.

The structure in the outer core often contrasts markedly with that of unfilled mouldings; with increasing filler content the fibrillar structure of the unfilled mould-

8. Structure and Mechanical Properties of Polymers 237

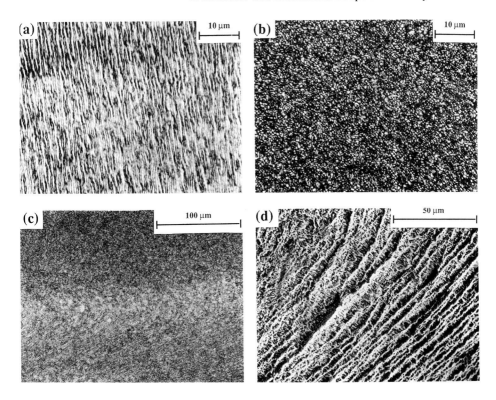

Fig. 8.7 Optical investigation of longitudinal thin polished sections of unfilled HBA-HNA mouldings (A950) under crossed polarizers: (i) inner skin; (ii) intermediate layer; (iii) inner core of 2 mm thick plaque; (iv) SEM of etched section corresponding to (iii).

ings is progressively broken up as shown in Figs. 8a to 8c. Thus the bulk structure of 40 wt% wollastonite filled HBA-HNA consists mainly of discontinuous layers tangential to the core centre (Fig. 8c). Indeed, there is no clear boundary between the different regions in the etched transverse sections and the main distinction between the inner core and the outer regions is the lack of orientation of filler particles in the core. The conic flow lines observed in the core of unfilled samples are either absent or weakly defined.

In short glass fibre filled injection mouldings, little coarse matrix structure can be seen. The fibres themselves are highly oriented in the flow direction in intermediate and outer regions, although considerable disruption of the surface structures occurs owing to 'swirling' of the fibres (Fig. 8d). The core of dumbbell-shaped samples generally shows random fibre orientation. In edge-gated plaques, however, there is often a high degree of transverse fibre orientation in the core (in the plane of the plaque).

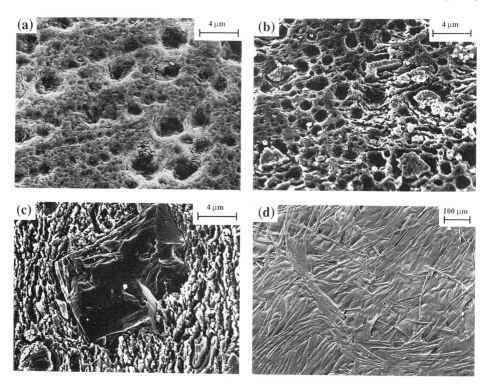

Fig. 8.8 The effect of filler addition: (i–iii) Etched transverse sections from outer core region for HBA-HNA, HBA-HNA/15 wt% wollastonite and HBA-HNA/40 wt% wollastonite respectively; (iv) surface (unetched) of HBA-HNA/30 wt% short glass fibre.

3 Mechanical Properties

3.1 Basic Tensile Properties

3.1.1 Effects of Filler on Properties in the Flow Direction of Injection Mouldings

Whilst it is generally accepted that filler addition to TLCPs has less effect than in the case of conventional thermoplastics, it is not entirely clear from the literature whether fillers degrade or enhance the tensile strength of TLCPs. This is compounded by the fact that sample geometries differ considerably, and that moulding conditions are often inadequately specified, making systematic comparisons of available data difficult. Results for unfilled and particle filled mouldings are given in Fig. 9. This shows the temperature dependence of the static mechanical properties for two types of injection moulded dumbbell-shaped test bars. The 3 × 6 mm cross-section bars

8. Structure and Mechanical Properties of Polymers 239

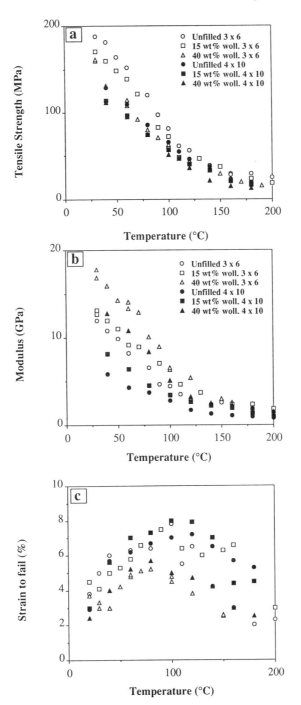

Fig. 8.9 Wollastonite particulate filler content and mechanical properties in test bars of HBA-HNA for two different cross-sections (crosshead speed 10 mm/minute).

were moulded in-house using a mould temperature of 90°C and a mean barrel temperature of 280°C, and a fill-rate of 1 cm^3 s^{-1}. The 4 × 10 mm cross-section bars were as described in 2.1.

In these samples it is found that in the flow direction, the tensile strength decreases and the Young's Modulus increases with filler content for fixed moulding conditions. The loss in strength appears greatest however, where the moulding conditions are such that the strength of the unfilled mouldings is relatively high (approaching 200 MPa). Where the strength of the unfilled mouldings is lower (less than 140 MPa), the filler has less effect on the strength, although modulus reinforcement is still evident.

30% glass fibre addition to the 3 × 6 mm samples resulted in a decrease in the room temperature tensile strength in the flow direction from approximately 200 to 170 MPa, consistent with the results of *Voss* and *Friedrich* for 3.2 mm thick plaques [65], but not with the suppliers' data for 4 mm thick samples, which indicate an increase from 156 to 188 MPa [12], or with our own results for the 2 mm thick plaques to be described in the next section, for which there was an increase from 145 to 150 MPa.

The results in Fig. 9 also illustrate the steep drop in modulus and tensile strength with T, correlated with an increase in the strain to fail, which peaks at approximately 100°C. This temperature appears to be associated with a second order transition, showing certain features of a glass transition [66] and which we refer to as the glass to nematic/crystalline transition, T_{gN} (there remains considerable doubt as to the possibility of a true glass transition in rigid polymers [30]). Thus, whilst at room T a 3 × 6 mm injection moulded bar of HBA-HNA/15 wt% wollastonite, for example, has a tensile strength of approximately 170 MPa, which is nearly double that of an unfilled conventional high performance thermoplastic, by 100°C it has dropped to values comparable with those for unfilled PES and PEEK, whose strengths show a weaker T dependence. Indeed, reinforced injection moulding grades of these latter may show superior performance above 100°C (PEEK/30% short glass fibre having a tensile strength of the order of 100 MPa at 150°C).

3.1.2 Anisotropy

The effect of orientation on mechanical properties in injection mouldings is illustrated in Figs. 10 and 11 for edge gated 80 × 80 mm plaques of unfilled HBA-HNA (Vectra A950), HBA-HNA/wollastonite mineral particulate filler and HBA-HNA/30 wt% glass fibre (supplied by Hoechst-Celanese Ltd.). The details of the moulding of such plaques are discussed in reference [52]; here we assume uniform properties across the sample width, consistent with microscopical observation of uniform microstructures in the plane of the plaques). The data given are the static tensile strength and the Young's modulus measured at 2 mm/minute for bars cut at different angles to the flow direction.

These results illustrate the remarkable degree of anisotropy of TLCP injection mouldings. Indeed, ratios of the longitudinal and transverse moduli of between 5 and 8 have been reported for similar plaques [52, 67], these depending crucially on the moulding conditions. Anisotropy is generally highest for small thicknesses, and

8. Structure and Mechanical Properties of Polymers 241

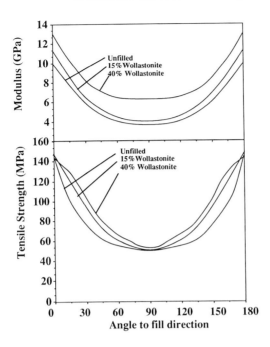

Fig. 8.10 Anisotropy of tensile properties in unfilled and mineral filled 2 mm thick plaques (crosshead speed 2 mm/minute).

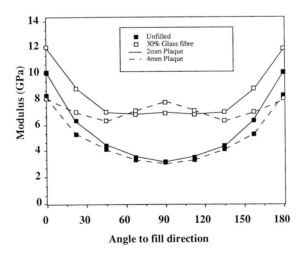

Fig. 8.11 Anisotropy of tensile properties in 2 mm and 4 mm thick unfilled and glass fibre filled plaques (crosshead speed 2 mm/minute).

slow fill rates, and tends to be reduced by filler addition. This is in contrast to conventional thermoplastics for which significant anisotropy is only observed in short fibre filled grades (although the anisotropy ratios do not generally exceed 2).

For 4 mm thick plaques of HBA-HNA/30 wt% short glass fibre, the tensile moduli perpendicular and parallel to the flow direction are very similar, with a minimum at intermediate angles. It is significant that in these particular samples, a high degree of transverse orientation of the glass fibres is observed (unusually, not in the centre of the core, where the orientation is in the flow direction, but in layers of approximately 0.7 mm in thickness and approximately 0.2 mm in from the sample surface). In the 2 mm plaques, on the other hand, for which data are also given in Fig. 11, transverse orientation is restricted to a narrow central region of approximately 0.3 mm in thickness. Further, tests on layers not containing transverse fibre orientation, indicated little difference in anisotropy to unfilled samples.

It has been argued that where molecular alignment already results in effective fibre reinforcement, addition of fibres aligned with the molecules is unlikely to have a strong influence, since the molecular and fibre modulus are of the same order of magnitude [67] (see also section 4). However, for transverse alignment, there is likely to be a substantial improvement in the transverse modulus, which seems a likely explanation for the large decrease in anisotropy for the HBA-HNA/30 wt% short glass fibre filled plaques. (The sample is effectively behaving as a cross-ply laminate, and may thus retain a high degree of anisotropy at a local level.)

Filler addition may also disrupt molecular orientation, as might be inferred from the microstructural observations of the previous section, although there appears to be no direct evidence for this in the literature. We shall discuss this possibility in more detail in 3.2. However, comparison of Figs. 10 and 11 indicates that the effect of particulate filler on the mechanical anisotropy is weak compared to that of the glass fibres.

3.1.3 Mechanical Properties and Wall Thickness

As suggested by Fig. 11, the increased anisotropy on wall thickness reduction is generally associated with improved mechanical properties in the flow direction [12]. This reflects the general decrease in the core-skin ratio for small cross-sections, the relatively high skin orientation being reflected by enhanced mechanical properties in the flow direction. It suggests that one might improve the properties by using for example, I-beam type cross-sections as opposed to solid sections, which would also save on materials costs and weight.

This has been investigated by using detachable inserts in the pre-existing moulds, as illustrated in Fig. 12 for 4×10 mm the dumbbell specimens. (Since the resulting mouldings were not necessarily symmetric, the filled grades (here 15 wt% wollastonite) were used in order to minimize warping.) Intrinsic values for the room temperature Young's modulus and tensile strength for comparable moulding conditions for a variety of cross-sections, are plotted against the effective wall thickness of the mould. Substantial increases in modulus are seen as the wall thickness decreases, such that the effective modulus (normalized with respect to the cross-sectional area of the standard 4×10 mm sample) is raised by up to 50% even though the cross-

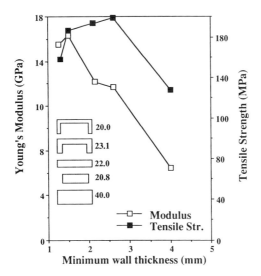

Fig. 8.12 Effect of different sample cross-sections on the tensile properties of HNA-HBA/ 15 wt% wollastonite (crosshead speed 2 mm/minute). The cross-sections and cross-sectional areas in mm² are given in increasing order of minimum wall thickness.

section contains 50% less material. The strength does not increase indefinitely, since at small wall thicknesses, large changes in effective mould cross-section close to the sample ends resulted in weak points at the ends opposite the gate, where there are locally strong diverging flows (see Fig. 3). This serves as an illustration of the importance of careful mould design in these materials.

3.2 Property Profiles within Samples

Given the highly inhomogeneous nature of TLCP injection mouldings, any attempt to rationalize fully their behaviour must involve investigation of their local properties.

This may be done in a number of ways, for example:

(i) by removing successive layers from a moulding, and measuring modulus each time, from which a modulus-distance profile may be calculated [34];
(ii) by measuring the properties of the extracted layers themselves [68];
(iii) by measuring orientation distributions within the mouldings [69–71].

3.2.1 Layer Removal

The first method has been used to produce the profiles in Fig. 13 for 3 × 6 mm cross-section filled and unfilled mouldings. The stress-strain curve was recorded up to 0.1% extension (in an attempt to avoid permanent deformation). A layer of chosen thickness was then cut from the sample surface using a rotary cutter, and the stress-

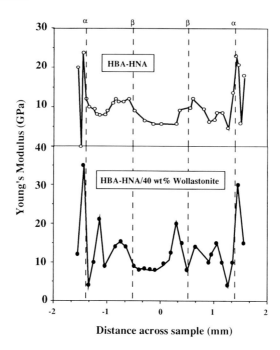

Fig. 8.13 Modulus profiles for unfilled and filled 3 × 6 mm cross-section injection mouldings (crosshead speed 1 mm/minute). a and b are the approximate positions of the skin-intermediate region, and the intermediate region-core boundaries respectively.

strain curve re-measured. The method of analysis used in the literature [34] assumes a simple rule of mixtures, that is, that a given layer contributes to the overall modulus in proportion to its thickness. One can then estimate the modulus of a given layer according to

$$E_l = \frac{E_s a_s - E_{s-1} a_{s-1}}{a_l} \tag{2}$$

where E_l, E_s and E_{s-1} are the moduli of the layer, the whole sample and the sample without the layer respectively and a_l, a_s and a_{s-1} are the corresponding cross-sections.

Given that the error inherent in equation 2 may exceed 30%, the E profiles obtained in this way are essentially qualitative, but reflect well the results of other workers [34, 68]. In the unfilled samples, E is relatively large for a thin layer at the sample surface, but drops sharply further into the sample. At approximately 200 mm from the surface there is a second large peak and finally a further, weaker peak at approximately 1 mm from the sample surface.

The positions of these peaks are found to correspond well to those of the inner and outer skin layers and the intermediate layer-core boundary. According to *Garg* and *Kenig* [34], the inner peaks arise from locally high shear rates where differing melt streams meet under non-isothermal conditions. Thus the high orientation in the inner skin is argued to result from shear, rather than directly from fountain flow. This seems reasonable in view of the marked contrast in the textures of the outer and inner skin in etched cross-sections, and the fact that the banded optical texture observed in the inner skin is generally associated with shear flow (see 2.1).

For the same geometry, the 40 wt% wollastonite filled mouldings, showed a qualitatively similar modulus profile to the unfilled material. The outer skin peak was either absent, or not detectable by this method, and the inner regions showed additional peaks, as well as the slight peak at the boundary of the core seen in the unfilled case.

3.2.2 Measurements on Individual Layers

The second method, can be used to measure the strength, and extension to fail in different regions as well as the modulus. Its resolution is somewhat limited owing to handling problems with samples whose thickness is less than 300 µm, and thus may not reflect features such as the large fluctuations in E in the skin region suggested by Fig. 13. Typical results for the 3 × 6 mm mouldings are illustrated in Fig. 14. The skin regions in unfilled samples are characterized by high strengths and low extensions to fail relative to the intermediate layers, but the extensions to fail are similar to those of the whole mouldings, suggesting that skin fracture determines the overall strength of these latter. This implies further that the contribution to overall strength

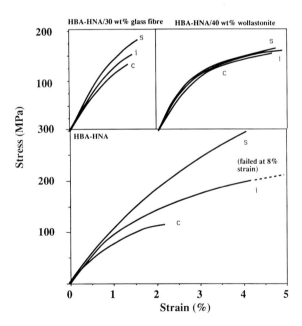

Fig. 8.14 Stress strain curves for individual layers in 3 × 6 mm injection mouldings (crosshead speed 10 mm/minute): s, surface layer 6 × 0.2 mm; i, intermediate layer, 5 × 0.4 mm centred at 0.5 mm into sample; c, 3 × 0.4 mm from centre of sample.

from the inner layers is dependent on their stiffness at the failure strain, rather than their strength.

The frequent observation of smooth conic fracture surfaces in the core suggest these to be points of particular structural weakness, and moreover the effect of such features appears exaggerated in the 3 × 6 mm where the stiffer outer layers have been removed. The strain to fail is often reduced below that of the whole sample, whereas there is no evidence for premature core failure during tests on the latter (the outer layers presumably concentrate strain energy away from defects in the core).

For both the particle filled, and glass fibre filled grades, addition of filler leads to a progressive decrease in the differences between the strengths of the different layers. Layers from 3 × 6 mm mouldings of the 40 wt% wollastonite filled grade in particular not only show little difference in tensile strength, but similar high strain behaviour. This immediately suggests an explanation for the conflicting data regarding the effect of filler addition. In the unfilled mouldings the load is concentrated in the skin, whereas for 40 wt% filler content, it is distributed throughout the sample cross-section at rupture. Thus where the proportion of skin is low, for example, in a moulding with a relatively large total cross-section, the strength of the unfilled mouldings may well fall below that of the filled mouldings. Indeed, for unfilled 4 × 10 mm mouldings with relatively low tensile strengths, the inner regions often continue to bear loads up to and beyond the load at which the skin fails.

3.2.3 Orientation Distributions

The use of wide angle X-ray (WAXS) methods is well documented in the literature, and involves measurement of the height of the main scattering peak as a function of the sample orientation direction, which in turn gives information on the angular distribution of the molecular orientation [70]. This is not the only method. Infra-red dichroism measurements have been used [71], although here the information is less quantitative owing to possible differences between the orientation of the principal axes of the optical indicatrix and the chain direction.

In general WAXS measurements of the Hermans orientation function, S, as defined by equation 1, such as shown for layer samples in Fig. 15 taken from the 4 × 10 mm mouldings, indicate an increase in orientation in the outer layers of the samples. Fig. 15 also suggests a tendency for wollastonite addition to lower the difference in orientation between the skin and the core, but significantly this is due to an increase in core orientation rather than a substantial decrease in orientation in the outer regions. Further, whereas a substantial drop in S is observed close the shoulder region opposite the gate in the unfilled HBA-HNA samples, this is less marked on filler addition. Thus at 15 wt% wollastonite content, the overall orientation appears to have increased with respect to the unfilled mouldings. Data are not available for a direct comparison with HBA-HNA/40 wt% wollastonite filled 4 × 10 mm mouldings, but measurements from the 3 × 6 mm samples suggests orientations ranging between approximately 0.5 in the core to 0.6 in the skin, with little variation along the sample length.

Results are given in Fig. 16a, showing the correlation between the modulus and S. S is in this case calculated from WAXS data for layers of between 0.3 and 0.5 mm

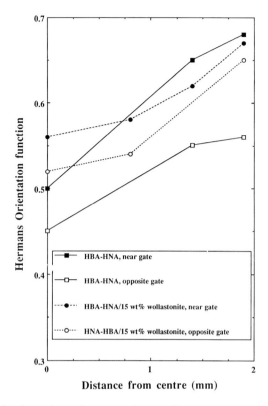

Fig. 8.15 Orientation in regions close to and opposite gate measured at different points in sample cross-section for the 4 × 10 mm mouldings.

in thickness cut from the unfilled and 40 wt% wollastonite filled 3 × 6 mm mouldings. Clearly wollastonite addition increases the modulus for a given value of S.

A similar correlation between tensile strength and S might be anticipated. However, it should be remembered that tensile strength is not an intensive property, so that if one assumes *a priori* that the strength decreases with S, it should be correlated with the minimum orientation values, rather than bulk or maximum values. Indeed the fact that the samples tend to break towards the end furthest from the gate, suggests this is a valid approach. Values for the maximum and minimum measured S are plotted against strength in Fig. 16b. It may be seen that use of the maximum values would imply an increase in strength on wollastonite addition for a given S; use of the minimum values however, suggests a decrease in tensile strength on wollastonite addition.

The considerable scatter, and in particular the apparently anomalous results for low S (corresponding to the core region) in the unfilled 3 × 6 mm samples, probably reflects the fact that the spatial resolution of these measurements (of the order of 1 cm^2) is insufficient to identify orientational variations on the scale of features such as the flow surfaces described in 2.1.1. As suggested in 3.2.1, these may result in premature failure of the core when the outer layers are removed.

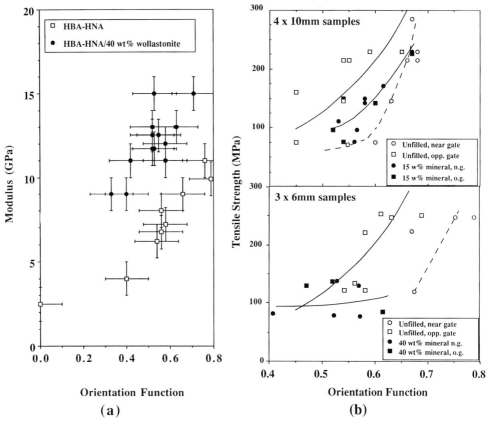

Fig. 8.16 (a) The relationship between modulus and orientation for layers cut from 3 × 6 mm samples; (b) The relationship between tensile strength and orientation for various samples (crosshead speed 2 mm/minute).

3.3 The Effect of Changing Moulding Conditions

Whilst for both TLCPs and conventional short fibre filled thermoplastics, flow geometry is the overwhelming factor affecting orientation, other potentially important factors include:

- melt temperature
- mould temperature
- injection speed

In both conventional short fibre filled mouldings and TLCPs, there is some indication that overall alignment may be increased at low flow rates [29, 52, 72–74]. In TLCPs, where molecular alignment is to a great extent controlled by heat transfer at the mould walls, low mould and melt temperatures, as well as low flow rates, result in thicker skin layers and improved properties [34]. Here we have used relatively low mould temperatures as recommended by the supplier, although it should be pointed out that higher mould temperatures may have advantages from the point of view of surface finish, colour uniformity and resistance to blistering during heat treatment.

The effect of fill rate has been investigated on the properties of the 4 × 10 mm cross-section tensile specimens, with the mean melt temperature and mould temperature maintained at 280°C and 80°C respectively (as previously).

On the basis of observations using a variety of techniques (SEM of etched specimens and of fracture surfaces; reflected light and transmitted light optical microscopy), three different types of morphology were distinguished, depending on the fill-rate, as sketched in Table 1. The relationship of these to injection rates and tensile strengths is shown in Fig. 17 for unfilled 73/27 HBA-HNA. Broadly, 'type 2' mor-

Table 8.1 Sample Morphologies in Injection Mouldings of Vectra

	Type 1	Type 2	Type 3
Region	% of cross section		
Outer and inner skin	10–20	5–10	20–30
Intermediate	50–60	20–30	50–60
Outer core	10–20	50–60	10–20
Inner core	10–20	10–25	2–8

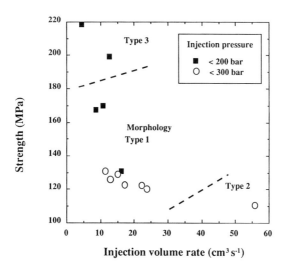

Fig. 8.17 The influence of moulding conditions on tensile strengths in HBA-HNA.

phologies, found at high injection rates (in excess of 30 cm³ s⁻¹) are distinguished by a relatively large outer core region and tensile strengths and Young's moduli of the order of 110 MPa and 6 GPa respectively for unfilled HBA-HNA, and 120 MPa and 12 GPa for the 40 wt% wollastonite filled grade. In 'type 1' morphologies, corresponding to moderate injection rates (10 to 30 cm³ s⁻¹), the extent of the intermediate layer and skin is increased at the expense of the outer core, and there are modest increases in tensile strength and moduli for both unfilled and filled grades. Finally, for unfilled HBA-HNA at lower injection rates, type 3 morphologies dominate, with a smaller, 'layered' core, and a somewhat higher proportion of skin, leading to tensile strengths and moduli as high as 220 MPa and 12 GPa.

Although the proportion of skin is also increased, it is not possible to distinguish clear 'type 3' morphologies at low injection speeds for 40 wt% filler (for which there is, in any case, a much less clear demarcation of the different structural regions). The corresponding increase in tensile strength is weaker, with maximum values not exceeding 140 MPa. The modulus on the other hand, may reach 21 GPa for fill rates of 5 cm³ s⁻¹.

Since failure of whole unfilled samples is observed to be initiated by failure of the skin layers, it is assumed that to a first approximation, the overall tensile strength σ_{tot} is equal to

$$\sum_l \sigma_l(\varepsilon_f)a_l \qquad (3)$$

where $s_l(\varepsilon_f)$ is the stress in each layer l at the failure strain, and a_l is the proportion of each layer in the sample cross-section. On the basis of stress-strain measurements, it has been found that differences in properties between the fast and slow unfilled injection mouldings are dominated by the increase in the proportion of the skin, but that there is a significant contribution from stiffening of the inner regions at high strains [75].

There is also evidence in the literature that simple mixing rules based on the proportion of skin may account for the variations in properties with fill rate (through the changes in the skin proportion) [52]. In view of the role of changes in the inner layers, this may be an over-simplification in the present case. Nevertheless, as Fig. 18 shows, it is possible to establish a correlation between mechanical properties and skin thickness, with data for samples produced under various moulding conditions [76].

It has also been found that stopping mould filling just before the melt front reaches the far end of the sample, leads to increases in tensile strength even though there are no variations in skin thickness. These increases are due to increases in the stiffness of the inner layers (such that they are bearing more load at the point of skin failure), possibly as a result of elimination of backflow effects [73].

3.4 Tensile Tests as a Function of Heat Treatment Time

Large improvements in room temperature properties are not reported for injection moulded samples on annealing below the melting point [4], even though such treat-

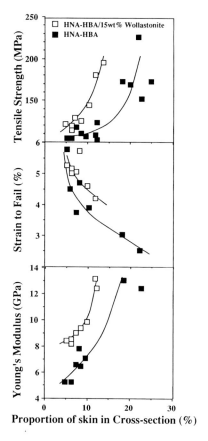

Fig. 8.18 Correlation between skin proportion and tensile properties in HNA-HBA and HNA-HBA/15 wt% wollastonite.

ments result in molecular weight increases, and consequently large improvements in the tensile strength of highly oriented fibres [4, 77, 78]. Nevertheless, one also expects the properties above T_{gN} to be dependent on the increases in crystallinity/crystalline order generally reported for annealing treatments [79–84]. This may be inferred for example, from the fact that in Ultrax TLCPs (BASF), which exist in both crystallizable and non-crystallizable forms, the non-crystalline forms have negligible tensile strength above T_{gN}.

Indeed, solid state annealing is recommended by the supplier [12] to improve the heat deflection temperature in injection mouldings. The standard heat treatment for Vectra A950 and materials based on Vectra A950 is a two stage treatment either in air or in nitrogen, during which the sample is heated to 220°C over two hours, then to 240°C over one hour, where it is held for two hours. This is then followed by a two hour treatment at 250°C [12], after which the area of the main melting peak at approximately 280°C is found by differential scanning calorimetry (DSC) to be of the order of 10 Jg^{-1}, which compares with 2 Jg^{-1} for the as-received material.

This particular choice of heat treatment appears to minimize sample damage both from evolution of volatiles (which may cause severe blistering of samples subject to abrupt heating) and thermal stresses, resulting in internal cracking between the skin and the inner layers and also in the core. With more prolonged treatment the sample becomes progressively darker in colour, particularly in air. This 'burning' effect also occurs during synthesis, resulting in the light brown colour of the as-received product (although its precise origins do not appear to be well understood).

Heat treatment resulted in little improvement in the room temperature tensile strength, and indeed decreases in strength of up to 10% were found for 3.1 × 6 mm test bars of the 15 and 40 wt% wollastonite filled HBA-HNA. At higher temperatures however, particularly above T_{gN}, where there was an increase in the tensile strength of between 10 to 20%. Thus the room temperature behaviour may be related to the development of extrinsic flaws, which become critical above a certain stress level but have a minimal effect on the tensile stress at higher T, where the matrix is softer. Holding the test bars beyond 1 hour at 250°C resulted in little further evolution of the tensile strength in any of the grades tested (indeed, much longer heat treatment times result in a further decrease in strength, as reported in [4]). There is nevertheless a progressive, albeit small increase in the low strain modulus as shown in Fig. 19 for HBA-HNA/40 wt% wollastonite for example.

The most marked effect of heat treatment, however, was a lowering by more than 50% of the strain to fail, which reflected a substantial increase in the steepness of the high strain region of the stress-strain curves. This occurred after between 15 and 30 minutes at the final heat treatment temperature, and appeared to stabilize for longer times.

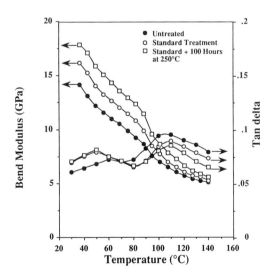

Fig. 8.19 The effect of heat treatment on the three-point bend modulus at 100 Rad/s for 3 × 6 mm mouldings of HBA-HNA/40 wt% wollastonite (measured using the Rheometrics RSA II dynamic mechanical analyzer).

3.5 Long Term Behaviour

Given the relatively high melting point, and heat deflection temperatures of upwards of 168°C, Vectra HBA-HNA materials show relatively good creep properties [12]. Our own tests have to date been limited to relatively short term investigations of the 3 × 6 mm injection mouldings in tension, using the Zwick tensile test apparatus. 200 second compliance-stress curves were measured at low strains in order to determine the range of linearity, whence the applied load was fixed at 50 N (2.8 MPa). Owing to the scatter in modulus within a given batch of samples, a single sample was used for each set of measurements over the chosen temperature range, being subsequently relaxed *in situ* for 5 times the duration of the previous test (to avoid disturbing the sample or the extensometer), and then conditioned for 5 hours at the new test temperature.

Horizontal shifts have been used to superimpose the data as shown in Fig. 20, along with the empirical shift factors. (Temperature scaling of the compliance would seem unjustified for rigid polymers, and the density changes are negligible.) Further, from DMA data for different frequencies in the same temperature range, it appears that only the α relaxation process at approximately 100°C, corresponding to the glass to nematic transition, is likely to influence these long term tests (Fig. 21). The lower temperature β peak which is clearly evident at 100 rads^{-1} moves to temperatures below room temperature for timescales associated with the creep tests. Finally, it is not believed that significant crystallization takes place in this temperature range for HBA-HNA, this being one factor which tends to mitigate against thermorheologically simple behaviour in semicrystalline materials in the neighbourhood of T_g.

Regardless of whether the above superposition is in any sense 'valid', it provides a useful way of representing the data. If one takes the glass to crystalline/nematic transition temperature at approximately 100°C (see 3.1.1) to be analogous to T_g in a conventional polymer, the overall behaviour is qualitatively similar to that of semicrystalline thermoplastics, for example PEEK [85]. Also given in Fig. 20 are data for HBA-HNA/40 wt% wollastonite and HBA-HNA/30 wt% glass fibre, shifted using the same shift factors as for the unfilled mouldings. The general form of the data is similar in all three cases, but whereas there is a substantial loss of reinforcement in the particle filled material with increasing time or temperature, the glass fibre filled material shows considerably improved creep properties. Indeed, the reinforcing effect of the glass fibres appears to increase with time and temperature.

Creep tests have also been carried out for different stresses for HBA-HNA and for HBA-HNA/15 and 40 wt% wollastonite. The corresponding room temperature creep rupture curves shown in Fig. 22. The similarity in slope suggests that the behaviour is essentially matrix dominated. Data have been given by *Chivers* and *Moore* in [67] for similar TLCPs, including glass fibre filled grades, showing much the same behaviour. These authors also considered the effect of sample orientation on creep rupture, and found a considerable decrease in strength perpendicular to the flow direction, but a similar time dependence to that parallel to the flow direction.

There is some evidence that the creep rupture data and static data for the tensile strength and failure strain of the 3 × 6 mm samples may be combined to give a

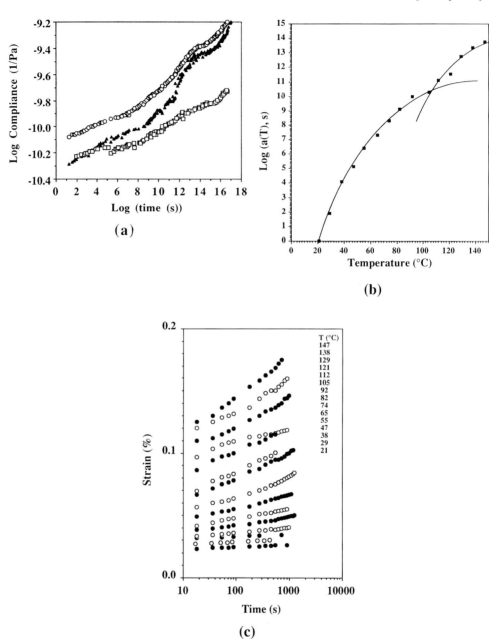

Fig. 8.20 (a) Creep master curves; (b) corresponding empirical horizontal shift factors; (c) strain-time data for the unfilled HBA-HNA.

common failure envelope, or 'Smith' plot [86] as shown in Fig. 23a (although here the stresses have not been normalized with respect to temperature). Finally, Fig. 23b shows an attempt to superpose failure strain data at different strain rates and temperatures using the empirical shift factors obtained from the creep measurements. In spite of the scatter, there appears to be some evidence for time-temperature equivalence, the strain to fail decreasing with decreasing strain rate for T > 100°C (poor superposition at low temperatures might be due to the influence of the β relaxation).

Correlations of this type were originally proposed for failure of rubbers, but have since been extended to other types of polymer, and are generally suggestive of a fracture mechanism in which viscoelasticity has a dominant role [87].

3.6 Impact Testing and Fracture Toughness

In general, impact tests perpendicular to the flow direction at room T show the unfilled mouldings to display very much superior toughness to the filled mouldings. [12] for example gives IZOD toughness values (ASTM D 256) of 520 and 70 Jm^{-1} for unfilled HBA-HNA (Vectra A950) and HBA-HNA/40 wt% wollastonite respectively. These are reasonably typical of glassy semi-crystalline polymers.

In un-notched samples of the unfilled mouldings, the maximum in the force displacement curve coincides with skin failure, beyond which point, impact energy is absorbed by the propagation of L-shaped cracks along the fibrillar interfaces, and ductile deformation of the inner part of the sample. Both mineral and glass fibre filled mouldings on the other hand are found to be macroscopically brittle, that is, the force-displacement curve drops discontinuously to zero beyond its maximum. Some multiple L-shaped cracking occurs in the surface layers, but the remainder of the sample fails by propagation of a single crack roughly perpendicular to the sample length.

It is difficult to draw more quantitative conclusions from impact data owing to sample geometry effects and notch sensitivity, factors which are compounded in the case of inhomogeneous samples such as TLCP mouldings. For example, a notched impact test on a moulded sample in which the notch has subsequently been cut into the surface, is likely to be highly misleading, in view of the role of the skin. Indeed, notched samples of unfilled samples often simply bend on impact. This contrasts with the behaviour of other thermoplastics such as PES and PEEK, for which unnotched samples tend to be most ductile under impact conditions.

Measurements of fracture toughness are also difficult since in general cracks do not propagate perpendicular to the flow direction. Thus in mode I tests, crack propagation perpendicular to the load is only observed for samples loaded perpendicular the to flow direction [65, 67]. *Davies* [89] has obtained room temperature K_q values for HBA-HNA and HBA-HNA/30 wt% glass fibre of 3 and 6.5 MNm$^{-3/2}$ respectively for cracks propagating in the flow direction in single edge notched specimens cut from 5 mm thick injection moulded plaques (the samples in this case did not fully satisfy the requirements for valid K_c testing [89]). This compares with values of 3.8 and 4.1 MNm$^{-3/2}$ obtained by *Davies* and *Moore* at −50°C [67] suggesting a

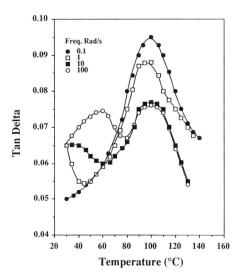

Fig. 8.21 Tan delta as a function of frequency for 6 × 3 mm samples of HBA-HNA in three-point bend tests (measured using the Rheometrics RSA II dynamic mechanical analyzer).

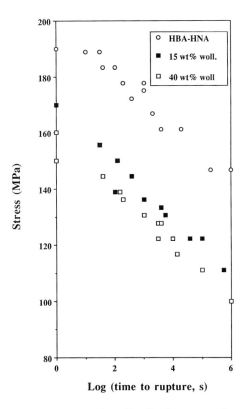

Fig. 8.22 Room temperature creep rupture for 3 × 6 mm samples as a function of filler content.

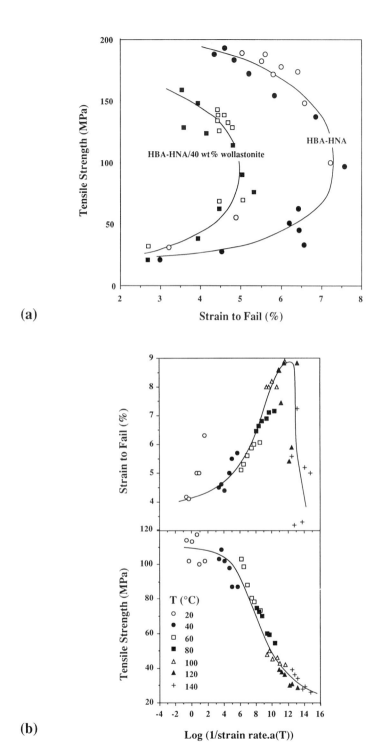

Fig. 8.23 (a) Smith Plot for 3 × 6 mm samples: filled symbols, static data; open symbols, creep rupture data; (b) Strain to fail and tensile strength against strain rate for unfilled 4 × 10 mm samples reduced to 20°C, using shift factors from Fig. 20.

general improvement in crack resistance transverse to the flow direction on glass fibre addition.

However, such simple application of linear fracture mechanics tends not to take into account the influence of sample inhomogeneity, both on the local stresses, and on extent to which data can be considered to be true materials parameters, even where the tests conform to accepted standards. A possible approach based on measurements of modulus profiles, coupled with FEA methods, to the problem of the influence of structure in TLCP mouldings has been suggested by *Brew et al.* [90].

4 Discussion

4.1 Microscopic Interpretations of Stiffness in TLCP Mouldings

4.1.1 Unfilled Mouldings

The rigid rod nature of TLCP molecules makes TLCPs an ideal application for the Ward aggregate model [91] which was originally developed to look at stress strain curves in semi-crystalline polymers (to which it is arguably less suited). The aggregate model averages the elastic constants of sub-units, which may correspond either to individual molecules, or sub-regions of perfect alignment, over a given, or measured orientation distribution to give the macroscopic compliance/stiffness tensor. In general it is found that better results are obtained if the average is performed assuming a constant stress condition (Reuss average). This results in expressions of the form

$$S_{ia} = \sum A_{ij} S_j, \qquad (4)$$

where S_j are the components of the compliance tensor for the sub-units, S_{ia} are the components of compliance tensor for the aggregate and A_{ij} are linear combinations of $\langle \cos^2 \Theta \rangle$ and $\langle \cos^4 \Theta \rangle$ corresponding to the orientation distribution of the sub-units. The unit compliances required for the model calculation are [69]

$$S_{33} = 1/E_3$$
$$S_{11} = 1/E_1$$
$$S_{13} = -v_{13} S_{33}$$
$$S_{13} = -v_{21} S_{11}$$
$$S_{44} = 1/G$$

where E_3, E_1 and G are the axial, transverse and shear moduli of the sub-units.

It has been shown by *Blundell et al.* that by using values for the unit elastic constants estimated from highly oriented samples (see Fig. 24), and X-ray measurements of orientation distributions in injection mouldings, reasonable predictions for the Young's modulus may be obtained by using a combination of the Ward model for individual samples layers, and simple laminate theory [69]. In Fig. 24 the consequences of the model for the room temperature modulus as a function of the orienta-

8. Structure and Mechanical Properties of Polymers   259

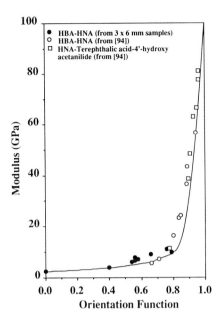

Fig. 8.24 Comparison of Aggregate model predictions for the modulus measured along the orientation direction ($1/S_{33a}$) and data for injection mouldings and extrudates. The following values for the unit elastic constants were assumed following [69]: $E_3 = 100$ GNm^{-2}; $E_1 = 2$ GNm^{-2}; $G = 1$ GNm^{-2}; $v_{13} = v_{21} = 0.4$.

tion for unfilled Vectra are shown in comparison with or own X-ray data and room temperature static modulus data for HBA-HNA and data for extruded TLCPS from [92]. (The model prediction has been interpolated for S between 0.8 and 1, since values of $\langle \cos^4 \Theta \rangle$ corresponding to the data from [92] were unavailable.) Reasonable agreement is obtained, in spite of the possibility of through thickness orientation variations in the layer samples. Moreover it is seen that the range of orientations typically encountered in injection mouldings is one in which the modulus changes relatively slowly with orientation. It is only for higher orientations such as those obtained by extrusion or fibre spinning (see chapter 6) the modulus increases rapidly towards values close to E_{33}.

Use has also been made of the aggregate model to investigate the temperature dependence of highly oriented fibres. Where orientation is high, the expressions of this latter may be simplified to

$$\frac{1}{E_{3a}} = \frac{1}{E_3} + \frac{1}{G} \langle \sin^2 \theta \rangle$$

This expression illustrates clearly how the shear modulus, representing inter-chain slip, plays a crucial role in determining the overall modulus. Indeed, as long as the longitudinal modulus of the fundamental unit, E_3, is very much higher than its shear modulus, G, the final result is insensitive to the behaviour of the former [69], particularly for the relatively low degrees of orientation encountered in injection mouldings. Moreover, one would not expect the longitudinal chain modulus to be strongly T dependent (since conformational changes do not influence the chain length, and the bond stiffnesses are approximately independent of T), whereas G as obtained from torsion measurements will typically drop steeply between ambient T and T_gN. This in turn should account for the strong T dependence of the aggregate modulus. (The work of *Troughton et al.* [93] on oriented fibres of HNA-HBA, nevertheless indicates that there is some contribution from an additional T dependence in E_3.)

In injection mouldings, the relatively low orientation means that the temperature dependence of E_{11} and hence S_{11}, becomes important. By cutting layer samples from injection moulded plaques perpendicular to the flow direction, we have estimated S_{11}' and S_{44}' from DMA measurements in torsion and in tension as shown in Fig. 25, based on a measured orientation of 0.78. Data for S_{33} was taken from X-ray modulus measurements in HNA-HBA [93], and v_{13} and v_{21} were taken to be 0.4 for all temperatures, whence S_{11}' and S_{44}' may be obtained by solving the aggregate model expressions for S_{11a}' and S_{44a}'. In fact, for this degree of orientation S_{11}' and S_{44}', and S_{11a}' and S_{44a}' are respectively quite close, and the calculated values are relatively insensitive to the choice of orientation parameter and data for S_{33}, v_{13} and v_{21}. The values of S_{44}' obtained in this way are consistent with those obtained elsewhere directly from torsion measurements on fibres [93]. (This is not the case where S_{44}' has been calculated from S_{11a}' and S_{33a}' data, with S_{44}' being consistently underestimated at higher T.) The use of the S_{11}' and S_{44}' values from Fig. 25 to predict E_{33a}' is illustrated in Fig. 26. The agreement is reasonable in view of the assumptions on which this approach is based, although the model prediction somewhat underestimates the measured values (cf. also Fig. 24).

So far we have only attempted to model the high frequency dynamic response. The model might nevertheless find useful application for the prediction of long term behaviour, given that direct creep measurements such as those in 3.4 are specific to one orientation distribution only. Given a knowledge of the time dependence of S_{11}' and S_{44}' however, one would be in a position to predict the creep behaviour for an arbitrary orientation distribution, and for different tensile directions. The need for reproducible, relatively uniform samples in order for reliable measurements of the compliance tensor would seem to preclude injection mouldings, but extruded sheets of Vectra are available as a more viable alternative.

4.1.2 The Effect of Fillers

The effect of 30 wt% glass fibre addition to a matrix whose compliance constants are as in Fig. 25 has been modelled using classic 'shear lag' analysis [94], assuming the fibres to be fully aligned, and to have radius 5 μm and aspect ratio of 16 (as measured for the injection moulded samples), and a modulus of 150 GNm^{-2}. As shown in Fig. 27a, the ratio of the predicted fibre contribution to the modulus and that of the matrix increases slightly with T. This tendency appears to be reflected by DMA data, as shown in Fig. 27b for example, and also the creep data in 3.5. In the case of particle reinforcement of the modulus, phenomenological models for reinforcement

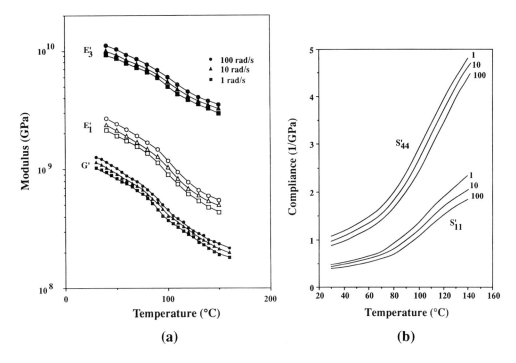

Fig. 8.25 (a) DMA measurements of tensile moduli in in the flow direction and perpendicular to the flow direction, and of the shear modulus (torsion). (b) Calculated values of the real parts of the complex compliances S_{11}' and S_{44}' from DMA measurements (measured using the Rheometrics RSA II dynamic mechanical analyzer).

generally assume combinations of parallel and series reinforcement, in which the shear modulus does not necessarily play a dominant role. Experimentally, we find no improvement in reinforcement at high T in particle filled grades.

4.2 Tensile Strength and High Strain Properties

Since, as suggested in 3.4, heat treatments similar to those used to increase the chain length in fibres do not result in a large increase in the tensile strength of injection mouldings (see 3.4), it is unlikely that existing models for the tensile strength in highly oriented fibres, based on shear-lag analysis of inter chain slip [78], are directly relevant to situations where the orientation is relatively low. In this latter case, it is more reasonable to assume that defects in the domain structure, or else regions where the orientation is locally at high angles to the tensile axis, represent weak links from which microcracks eventually propagate. It seems reasonable to assume that this latter process will be strongly influenced by the extent to which local stress relaxation via viscoelastic mechanisms intervenes for a given set of test conditions (as suggested in 3.5).

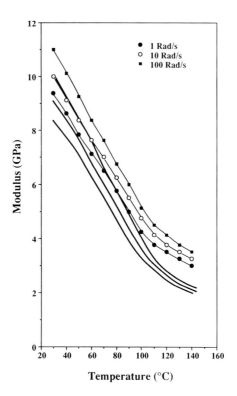

Fig. 8.26 Comparison of experimental and predicted longitudinal modulus using the aggregate model and the compliance constants from Fig. 25b. (the bold lines represent the model predictions).

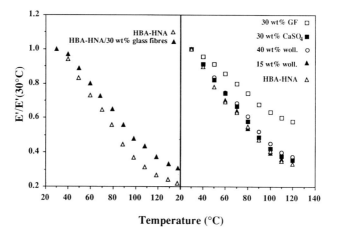

Fig. 8.27 (a) Predicted effect of 30 wt% glass fibre reinforcement on E_a' normalized with respect to the modulus at 30°C, at 1 Hz for HBA-HNA (value of orientation function taken to be 0.78); (b) bend modulus data for various 3 × 6 mm samples at 1 Hz normalized with respect to the modulus at 30°C.

8. Structure and Mechanical Properties of Polymers   263

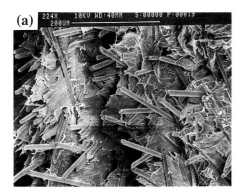 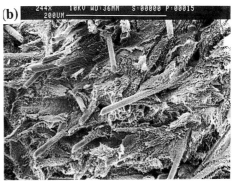

Fig. 8.28 Fracture surfaces in glass fibre filled Vectra: (a) skin region; (b) core region (showing slightly more matrix adhesion).

In the injection mouldings the situation is complicated by the fact that weak links also occur at the macroscopic level as discussed in 3.2.3, making detailed interpretation difficult. Nevertheless, the fact that addition of particle fillers appears to lower the strength for a given orientation seems reasonable if one assumes the particle matrix interface itself to be a source of weakness. Certainly from SEM of fracture surfaces, the bonding appears poor, with little matrix adhesion to the filler particles. The fact that particle fillers also break up long-range structure in the matrix (2.1.2) seems to preclude any beneficial effect from increased delocalization of the crack front, or crack blunting, as is often the case for particle addition to conventional amorphous polymers. Indeed, acoustic emission techniques have been used to demonstrate directly increased damage localization in notched samples on glass fibre addition to TLCPs [95].

Our own acoustic emission measurements, have been limited to attempts to monitor debonding, as shown in Fig. 29 for different filler contents and for the short glass fibre filled material (the threshold in this case was 50 dB). In the particle filled grades a sharp increase in acoustic emission may be identified with a considerable loss of linearity in the stress-strain curves at relatively low strains, as evidenced by the compliance-stress data given in Fig. 29. For the unfilled material, substantial emission is only measurable close to the point of rupture, reflecting the greater linearity of the stress-strain curve. Indeed, at high strains, the modulus of the mineral filled grades tends to fall below that of the unfilled material. That the damage in the particle filled materials is irreversible may be checked by cycling to different stresses; generally the threshold of acoustic emission shifts to the previous highest stress level.

The behaviour of the short glass fibre filled material is somewhat different to that of the particle filled materials. Although acoustic emission commences at low strains, the stress-strain curve remains approximately linear until close to failure, as in the unfilled case. Thus while damage is clearly taking place, it does not greatly affect the reinforcing effect of the fibres even up to large strains, consistent with [67]. Fracture surfaces (Fig. 28(a, b)) nevertheless appear to indicate a high degree of fibre pullout, particularly in the outer regions [65].

Conclusions

The commercialization of TLCPs has proceeded relatively rapidly, and most current research tends to centre on their technological development. Certainly, where mechanical properties of injection mouldings are discussed in the literature it is usually within the context of phenomenological studies of the effect of processing variables. Moreover, the extreme heterogeneity of injection mouldings presents problems in terms both of data scatter, and interpretation, most especially in the case of extensive properties such as strength, whence, not surprisingly, most fundamental work on finished articles has concentrated on melt-spun fibres.

In our own work we have attempted to address a range of questions, each of which merits further attention. Simple scaling of the time and temperature dependence of various mechanical properties, for example, would have obvious practical implications. However, the absence of corroborative evidence in the literature, the limited extent of our experimental data, and the inherent unsuitability of injection moulded samples to detailed study, we can only advance this as a possibility. Extruded sheets, for example, with more controlled orientation would provide a more attractive subject for further work.

As for injection mouldings themselves, we can draw a number of conclusions. First particle addition will tend to reduce orientation to some extent, but particle filled samples nevertheless retain a high degree of anisotropy. Glass fibre is more effective in reducing the anisotropy of plaques, but this appears to be largely due to the presence of transverse fibre orientation in the inner regions of the samples.

Perhaps a more significant effect of filler addition is that it tends to result in the evening-out of fluctuations in orientation, both within a given cross-section, and in the flow direction. In unfilled samples, high orientation differences between skin and core lead to stress concentrations in the skin, which is relatively stiff. Thus, even if the skin strength is high, overall failure (which initiates in the skin) may occur at relatively low loads. This will be compounded by the presence of weak regions of relatively low orientation within in the skin itself.

With fillers, there is better load distribution within the sample cross-section, owing both to lower orientation variations and also to the fact that the high strain behaviour of the stress-strain curves is very similar, in spite of orientation variations, owing to the contribution of debonding to the shape of the former. Thus whilst optimized moulding conditions may lead to higher tensile strengths for unfilled samples, owing to the fact that particle addition will lower the tensile strength for a given orientation, filled grades will tend to perform better where moulding conditions are not optimized (the effect of filler on the tensile strength is in practice small, but for 40 wt% filled grades the room temperature strength appears limited to about 140 MPa). This allows more flexibility in mould design and turnover rates, where tensile strength is an important consideration. In particular, it allows exploitation of possibility offered by TLCPs of fast cycle times - to obtain the best performance from unfilled TLCPs low fill rates are required.

This is quite apart from the other advantages of filler addition, and in particular a systematic increase in modulus in the flow direction with filler content. Filled

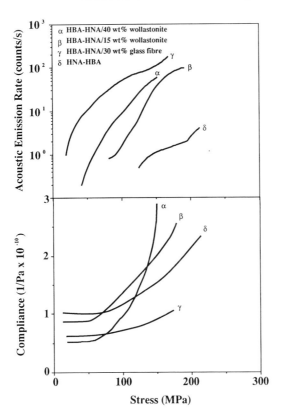

Fig. 8.29 Acoustic emission for filled and unfilled Vectra (crosshead speed 2 mm/minute).

materials are nevertheless more brittle in impact tests, whereas the toughness of the unfilled resin as measured in IZOD tests perpendicular to the orientation direction is considerably higher than most conventional thermoplastics. Whether this manifests itself in other types of test, such as the falling weight test is however highly open to question, and is an area in which we are currently working. Particle reinforcement also tends to be ineffective at large strains, owing to poor matrix particle adhesion. This is not true of glass fibre, where reinforcement is maintained up to strains close to the failure strain, and even improved at high temperature, or long times. Glass fibre fillers are certainly to be recommended where creep properties are important, and indeed may out perform highly creep resistant conventional polymers such as PES (the heat deflection temperature (DIN 53 461 A) for 30 wt% glass fibre filled Vectra is given as 232°C). However, the tensile strength and modulus in all grades will drop off rapidly as T approaches 100°C.

Assuming design allows exploitation of the mechanical properties in the flow direction, TLCPs as with conventional high-performance thermoplastics, offer an alternative to metals, ceramics, with possible weight saving (fuel economy in cars). Although in conventional injection moulding, the need to avoid knit lines limits design flexibility, specialized injection moulding techniques such as mentioned in the introduction [8, 9], are under development with a view to minimizing this problem. As suggested by Fig. 30, for example, the 'Gegentakt' injection moulding apparatus (Klöckner Ferromatik, Malterdingen) which permits the melt to be passed repeatedly through the melt cavity prior to solidification may be used to obtain a substantial improvement in properties compared with conventional injection moulding equipment, even where knit lines are not a problem. The greater control of the melt flow, and the possibility of multiple gating offered by such systems also greatly increases the range of possible approaches to complex design problems [8, 9].

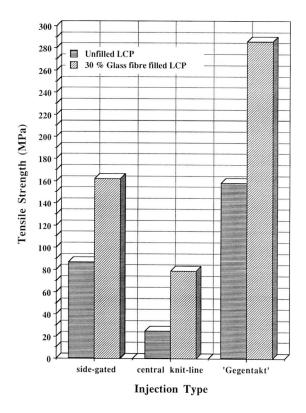

Fig. 8.30 Comparison of tensile strengths for a dumbbell-shaped 10×4 mm cross-section tensile test bar (DIN 53455) moulded from a commercial unfilled and glass fibre filled LCP using different injection techniques. The single side gated bars are similar to those we have used in our own experimental work; the bars containing a central knit line were gated at either end so that two melt streams met in the centre of the sample barrel; the 'Push-pull' (Gegentakt) injection moulded bars were also double end-gated, but the melt was passed back and forth through the mould cavity eight times.

Acknowledgments

Financial support from the Swiss Commission pour l'Encouragement de la Recherche Scientifique (CERS) is gratefully acknowledged. This work was carried out in close collaboration with B. Zülle and A. Demarmels of ASEA Brown Boveri, Baden-Dättwil, and with P. Davies and Y. Wu of the EPF Lausanne. Special thanks are also due to H. Terwyen (Hoechst AG, Frankfurt am Main) and S. Fleury (EPF Lausanne) for their assistance with the injection moulding of some of the specimens, and to Professor W. Kaiser and M. Diez (ETH Brugg) for many useful discussions.

References

1. Cottis, S.G., Economy, J., Nowak, B.E. (Carborundum), US Patent 3,600,350 (1972).
2. Kuhfuss, H.F., Jackson, W.J. (Eastman-Kodak), US Patent 3,778,410 (1973).
3. Calundann, G.W. (Celanese), US Patent 4,161,470 (1979).
4. Calundann, G.W., Jaffe, M., Proc. "The Robert A. Welch Foundation Conferences on Chemical Research, XXVI, Synthetic Polymers", Houston, Nov. (1982).
5. Kirsch, G., Terwyen, H., Kunststoffe **80**, 1160 (1990).
6. Edwards, P., Modern Plastics International, March (1989), 32.
7. Engberg, K., Knutsson, A., Werner, P.-E., Gedde, U.W., Poly. Eng. & Sci., 30 1621 (1990).
8. 'Multilife feed injection moulding' Brunel University, Uxbridge, Modern Plastics International, Jan (1990).
9. 'Gegentakt-Spritzgiessen' Klıckner Ferromatik, Malterdingen, Modern Plastics International, Jan. (1990).
10. Weatherall, J.M., Ono, A.T., Proc. Symp. "Recent Advances in Polyimides and Other High Performance Polymers", ACS Div. Poly. Chem., San Diego, Jan. (1990).
11. Folkes, M.J., in "Short Fibre Reinforced Plastics, Wiley, UK (1982).
12. Hoechst-Celanese, Vectra™ Technical Brochure.
13. Onsager, L., Annals, New York Academy of Science **51**, 627 (1949).
14. Flory, P.J., Proc. Roy. Soc. A234, 73 (1956).
15. DeMarzio, E.A., J. Chem. Phys. **35**, 658 (1961).
16. Flory, P.J., Abe, A., Macromolecules **11**, 1119 (1978).
17. Doi, M., J. Poly. Sci.- Poly. Phys. Ed. **19**, 229 (1981).
18. Demus, D., Demus, H., Zaschke, H., "Flussige Kristalle in Tabellen", VEB Verlag, Leipzig (1976).
19. Kleinschuster, J.J., US Patent 3,991,014 (1974).
20. Jackson, W.J., Kuhfuss, H.F., J. Poly. Sci.- Poly. Chem Ed. **14**, 2093 (1976).
21. Roviello, A., Sirigu, A., Europ. Poly. J. **15**, 61 (1979).
22. Jin, J.I., Antoun, S., Lenz, R.W., Brit. Poly. J. **12**, 132 (1980).
23. Jackson, W.J., Macromolecules **16**, 1027 (1983).
24. Griffin, B.P., Cox, M.K., Brit. Poly. J. **12**, 147 (1980)
25. Demartino, R.N., J. Appl. Poly. Sci. **28**, 1805 (1983).
26. Cao, M.Y., Wunderlich, B., J. Poly. Sci. - Poly Phys Ed. **23**, 521 (1985).
27. Economy, J., Volksen, W., Viney, C., Geiss, R., Siemens, R., Karis, T., Macromolecules **21**, 2777 (1988)
28. Mühlebach, A., Lyerla, J., Economy, J., Macromolecules **22**, 3741 (1989).
29. Duska, J.J., Plastics Engineering **42**, 39 (1986).
30. Frayer, P.D., Polymer Composites **8**, 379 (1987).

31. Bedford, E., Windle A., Polymer **31**, 620 (1990).
32. Friedel, G., Ann. Phys. (Paris) **18**, 273 (1922).
33. Kenig, S., Polymer Composites **7**, 50 (1986).
34. Garg, S.K., Kenig, S:, in "High Modulus Polymers" Zachariades, A.Z., Porter, R.S. eds., Marcel Dekker (1988).
35. Ophir, Z., Ide, Y., Poly. Eng. & Sci. **23**, 261 (1983).
36. Goettler, L.A., Polymer Composites **5**, 60 (1984).
37. Onogi, S., Asada, T., Rheology, Vol. 1, Astarita, G., Marrucci, G., Nicolais, L., Eds., Plenum Press, NY (1980).
38. Alderman, N.J., Mackley, M.R., Faraday Discuss. Chem. Soc., **79**, 149 (1985).
39. Marrucci, G., Maffettone, P.L., Macromolecules **22**, 4076 (1989).
40. Marrucci, G., Maffettone, P.L., J. Rheol. **34**, 1217, 1231 (1990).
41. Cenzig Altan, M., J. Thermoplastic Composites **3**, 275 (1990).
42. Becraft, M., PhD, University of Delaware (1988).
43. Menges G., Schacht, T., Becker, H., Ott, S., Internation Polymer Processing **2**, 77 (1987).
44. Dobb, M.G., Johnson, D.J., Saville, B.P., J. Poly. Sci.- Poly. Symp. **58**, 237 (1977).
45. Sawyer, L.C., J. Poly. Sci.- Poly. Lett. Ed. **22**, 347 (1984).
46. Tharpar, H., Bevis, M., J. Mat. Sci. Lett. 2, 733 (1983).
47. Sawyer, L.C., Jaffe, M., J. Mat. Sci. 21, 1897 (1986).
48. Joseph, E.G., Wilkes, G.L., Baird, D.G., Polymer **26**, 689 (1985).
49. Joseph, E.G., Wilkes, G.L., Baird, D.G., Poly. Eng. Sci. **25**, 377 (1985).
50. Viola, G.G., Baird, D.G., Wilkes, G.L., Poly. Eng. Sci. **25**, 888 (1985).
51. Weng, T., Hiltner, A., Baer, E., J. Mat. Sci. **21**, 744 (1986).
52. Kenig, S., Trattner, B., Andermann, H., Polymer Composites **9**, 20 (1988)
53. Pirnia, A., Sung, C.S.P., Macromolecules **21**, 2699 (1988).
54. Suokas, E., Sarlin, J., Törmälä, P., Mol. Cryst. Liq. Cryst. **153**, 515 (1987).
55. Boldizar, A., Plastics and Rubber Processing and Appl. **10**, 73 (1988).
56. Suokas, E., Polymer **30**, 1105 (1989).
57. Diamant, J., Keller, A., Baer, E., Litt, M., Arridge, R.G.C., Proc. Roy. Soc. London **93**, 293 (1972).
58. Dobb, M.G., Johnson, D.J., Saville, B.P., J. Poly. Sci. **15**, 2201 (1977).
59. Viney, C., Donald, A.M., Windle, A.H., J. Mat. Sci. **18**, 1136 (1983).
60. Graziano, D.J., Mackley, M.R., Mol. Cryst. Liq. Cryst. **108**, 73 (1984).
61. Donald, A.M., Windle, A.H., J. Mat. Sci. **19**, 2085 (1984).
62. Donald, A.M., Windle, A.H., in "Recent Advances in Liquid Crystalline Polymers", Chapoy L.L., Ed., Elsevier, NY (1985).
    Navard, P., J. Poly. Sci.- Poly. Phys. Ed. **24**, 435 (1986).
63. Viney, C., Donald, A.M., Windle, A.H., Polymer **26**, 870 (1985).
    Navard, P., J. Poly. Sci.- Poly. Phys. Ed. **24**, 435 (1986).
64. Stell, D.T., Donald, A.M., MacDonald, W.A., Proc. Sixth Annual Meeting, Polymer Processing Society, Nice, April (1990).
65. Voss, H., Friedrich, K., J. Mat. Sci. **21**, 2889 (1986).
66. Cao, M., Wunderlich, J. Poly. Sci.- Poly. Phys. Ed. **23**, 521 (1985).
67. Chivers, R.A., Moore, D.R., Polymer **32**, 2190 (1991).
68. Tomlinson, W.J., Morton, P.E., J. Mat. Sci. Lett. **10**, 154 (1991).
69. Blundell, D.J., Chivers, R.A., Curson, A.D., Love, J.C., MacDonald, W.A., Polymer **29**, 1459 (1988).
70. Mitchell, G.R. , Windle, A.H., Dev. Cryst. Polym. **2**, 115 (1988).
71. Pirnia, A., Sung, S.P.S., Macromolecules **21**, 2699 (1988).
72. Bright, P.F., Crowson, R.J,, Folkes, M.J., J Mat.Sci. **13**, 2497 (1978).
73. Boldizar, A., Plast. & Rubber Proc. & Appl. **10**, 73 (1988).
74. Yamoaka, I., Kawabe, M., Kimura, M., Higuchi, M., Sunago, H.,Tomioka, T., Proc Sixth Annual Meeting, Polymer Processing Society, Nice France, paper 10-06 (1990).
75. Zülle, B., Demarmels, A., Plummer, C.J.G., Kausch, H.-H., to be submitted to J. Mat. Sci. Lett.

76. Student report from ABB
77. Luise, R.L., US Patent 4,183,895 (1980).
78. Yoon, H.N., Colloid Polym. Sci. **268**, 230 (1990).
79. Blundell, D., Polymer **23**, 359 (1982).
80. Hanna, S.,Windle, A.H., Polymer **29**, 207 (1988).
81. Economy, J., Angew. Chem. **29**, 1256 (1990).
82. Cheng, S.Z.D., Janimak, J.J., Zhang, A., Zhou, Z.L., Macromolecules **22**, 4240 (1989).
83. Butzbach, G.D., Wendorff, J.H., Zimmermann, H.J., Polymer **27**, 1337 (1986).
84. Lin, Y.G., Winter, H.H., Macromolecules **24**, 2877 (1991).
85. Nielsen, L.E., "Mechanical Properties of Polymers and Composites", Vols. 1 & 2, Marcel Dekker, NY (1974).
86. Smith, T.L., J. Poly. Sci. **20**, 447 (1956).
87. Ferry, J.D., "Viscoelastic Properties of Polymers, Wiley, NY (1980).
88. Yujun, W., Unpublished results, Lausanne (1990).
89. Davies, P., Unpublished results, Lausanne (1990).
90. Brew, B., Sweeney, J., Duckett, R.A., Ward, I.M., Proc. 8th International conference on Yield and Fracture of Polymers, Cambridge, April 1991, paper 18.
91. Ward, I., Proc. Phys. Roy. Soc. **80**, 1176 (1962).
92. Chung, T-S, J. Poly. Sci.- Poly. Lett. **24**, 299 (1986).
93. Troughton, M.J., Davies, G.R., Ward, I.M., Polymer **30**, 58 (1989).
94. Cox, H.L., Brit. J. Appl. Phys. **3**, 72 (1952).
95. Weng, T., Hiltner, A., Baer, E., J. Composite Materials **24**, 103 (1990).

PART III
Processing and Joining

Chapter 9
Processing of Thermoplastic-Based Advanced Composites

J.-A.E. Månson

Abstract

Large numbers of novel material preform types and processing techniques have been developed in recent times for the manufacture of thermoplastic-based advanced composite parts. Different types of thermoplastic-based preforms and their production techniques are reviewed, and various manufacturing techniques for shaping and consolidation are presented. Of these techniques, the most widely applied, using preconsolidated sheets and preimpregnated tows, are described. The influence of processing conditions on morphology and internal stress generation is discussed. It is demonstrated that high cooling rates during solidification cause considerable morphological and residual stress distributions through the thickness of a laminate. Finally, the concept of a processing window with respect to solidification rate and forming pressure is introduced and briefly discussed.

1 Introduction

As many industries have been focusing an increased interest on thermoplastic-based composites, large numbers of new and unique processing and manufacturing methods have been developed over the last decade [1–5]. The primary goal of these techniques has been cost effectiveness through simplified and rapid net-shape production, with a minimum tradeoff in performance of the final composite part [6].

The introduction of advanced thermoplastic-based composites has also led to novel requirements being imposed on the processing techniques for the manufacture of high quality pre-pregs. For example, thermoplastic resins have in general caused many more difficulties than thermoset-based resins with regard to fibre wetting and impregnation, due to the higher melt viscosities involved. On the other hand, thermoplastic-based composites have enabled a considerable reduction in manufacturing cycle times by eliminating the time-consuming curing step necessary for thermosets.

Jan-Anders E. Månson, Laboratoire de Technologie des Composites et Polymères (LTC), Ecole Polytechnique Fédérale de Lausanne, EPFL, CH-1015 Lausanne, Switzerland

In parallel to the development of novel manufacturing techniques, new product forms (pre-pregs) have been introduced in order to match the requirements of these production techniques [7–9]. This has led to a wide variety of custom-made manufacturing techniques more or less designed for a particular resin type, fibre configuration and/or part configuration [10].

The implementation of these novel material/process combinations in production has called for a more fundamental understanding of the effects of processing conditions on the resultant composite properties, structure, and morphology [11–15]. In addition to more pronounced non-isothermal solidification and more intricate rheological conditions, the novel material systems introduced often exhibit reduced possibilities for diminishing process-induced effects. The dynamics of pressure and temperature applied during processing, combined with the solidification kinetics and viscoelastic nature of the materials may determine to a considerable extent the morphological features and the internal stress generation in the manufactured parts [12, 16, 17].

The intention of this chapter is to illustrate the dynamic development of this area of thermoplastic composites processing, and to demonstrate the close interplay between the material and the process. Fig. 1 gives an overview of the alternative manufacturing routes, for thermoplastic-based advanced composites, going from resin and fibre to a finished part. This figure serves as an outline for the first part of the chapter.

A further goal is to demonstrate the importance of considering process-induced effects which may considerably influence the durability and long-term performance

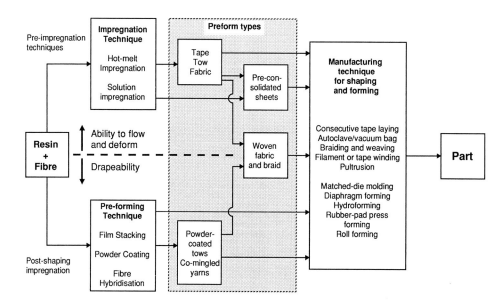

Fig. 9.1 Alternative manufacturing routes for thermoplastic-based advanced composites, going from resin and fibre to a finished part.

of a manufactured part. This is of great importance, due to the large variety of product form and processes invented for unique applications and for which limited practical experience is available. Whereas a general overview is provided in this chapter, various specific examples of design and fabrication of load-bearing machine elements are given in chapter 10.

2 Aspects of Material Processability and Properties

2.1 Thermosets versus Thermoplastics

Nowadays both thermoplastic and thermosetting resins are used as matrices for composites. Each type exhibits particular advantages and disadvantages with respect to processability and service performance, as illustrated in Table 1 [18, 19]. Although a wide range of different chemistries exists within each type, some general features can be distinguished, which have determined their area of application.

In general the crosslinked structure of thermosetting polymers provides potential for higher *stiffness* (E-modulus) and *service temperatures* than thermoplastics. The upper limit of service temperature for advanced composites is most often determined by the glass transition temperature.

On the other hand, *toughness* and *elongation to break* may be considerably higher for thermoplastic resins. This may be a particular advantage in applications where impact strength is a major requirement. Most high-performance thermoplastics offer outstanding interlaminar fracture toughness and acceptable post-impact compression response. This feature of thermoplastic materials has been the major reason for their increased use in composite structures.

From a processing viewpoint, the high *melt viscosities* of thermoplastics generally create considerable difficulties during fibre wet-out and impregnation. Thus, thermoplastic-based composites generally require higher processing temperatures and pressures to ensure sufficient flow during the final forming process.

The higher *processing temperatures* and *pressures* needed for the forming of thermoplastic-based composites generally impose stricter requirements on the processing equipment, and more advanced engineering is needed for tool construction. The

Table 9.1 Property / Process Characteristics for Thermoplastic and Thermosetting Matrix Systems

Property	Thermoset	Thermoplastic
Modulus	high	low
Service temp	high	low
Toughness	low	high
Viscosity	low	high
Processing temp	low	high
Relaxation time (at proc.)	long	short
Conversion costs	high	low
Recyclability	limited	good

higher processing temperatures may also induce considerable difficulties in mismatch of thermal contraction between the matrix and fibres during the processing cycle.

The longer *relaxation times* for thermosetting materials may be a disadvantage, due to a reduced ability to relax process-induced internal stresses. In anisotropic composites in particular, the potential of the polymer to relax internal stress fields is important for the elimination of process-induced defects. Such defects, in the form of voids, microcracking, fibre buckling, warpage, and residual stresses may diminish the durability and long-term performance of the composite [16].

Thermoplastic-based composites offer potential for lower *conversion costs* from intermediate material forms into final end-use parts by process automation. Furthermore, thermoplastics also offer the advantage of having almost indefinite storage life, which facilitates the logistics of the manufacturing procedure.

Finally, thermoplastics may be post-formed and/or reprocessed by the reapplication of heat and pressure, which gives a potential for *recyclability*. The increased awareness, in these last years, about material recyclability has brought about a heightened interest in thermoplastic matrix composites, especially in large volume areas such as the automotive industry.

2.2 Reinforcing Materials

The reinforcing material may be characterised by its intrinsic properties, shape and configuration, as shown in Fig. 2. The choice of a reinforcement is usually a balance between the performance advantage, processability and cost [10]. The reinforcing efficiency is strongly influenced by the aspect ratio of the reinforcement; it improves with increasing aspect ratio, reaching a maximum level for continuous fibres

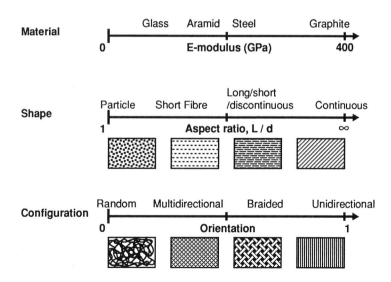

Fig. 9.2 Characterisation of reinforcing materials: properties, shape, configuration.

[20–23]. Increased reinforcement length may also improve the impact strength of the composite, but the potential to flow in a fabrication process is often considerably reduced [24–26]. Control of fibre orientation allows selective reinforcement of the composite. Woven or braided fibre preforms allow a good control of fibre orientation in 2 and 3 dimensions, but most often lead to an increased cost of the final composite structure. The use of a woven fibre configuration can reduce the potential for process-induced fibre misalignment during resin impregnation of the preform.

As the aspect ratio of the reinforcement increases, the impregnation and fabrication of composite parts must be carried out with more care, since vigorous mechanical action during impregnation can greatly reduce the aspect ratio of the longer fibre materials. Therefore wet processes with thermosetting resins are prevalent for long and continuous fibre composites. For advanced thermoplastic composites solvent impregnation, powder coating, or special melt impregnation methods must usually be employed to ensure sufficient pre-preg quality.

3 Basic Product Forms

One of the most critical steps in the preparation of composites is the impregnation of the fibres with matrix resin, as this creates the reinforcing synergy in the final structure. The impregnation can be carried out as a part of the fabrication step or in a preliminary preimpregnation process. In "wet" processes, such as filament winding and resin transfer moulding of thermoset-based materials, the impregnation and final shaping are carried out together. Dry fibre strands or fibre preforms are impregnated with a liquid resin and cured into the final shape in a single step. For thermoplastics, however, the high viscosity of the resins renders wet processes unsuitable, unless specially-designed polymer systems which allow a final polymerization reaction after the impregnation step are used.

In "dry" processes, the composite raw material is prepared prior to the fabrication process, in a separate impregnation step providing favorable conditions for wetting and impregnation. This additional step adds to the raw material cost. Nevertheless, for thermoplastic-based advanced composites, the two-step procedure predominates, due mainly to the difficulties associated with the impregnation process with these high viscosity materials.

As mentioned earlier, the basic problem in producing a high-performance thermoplastic composite is the same as for thermoset composites, that is to ensure good wetting and adhesion under conditions which have a minimum impact on the integrity of the fibres and/or the matrix material. The impregnation of fibres with thermoplastic resins is generally much more difficult than with thermoset resins due to the much higher viscosity. The influence of the resin viscosity on the resin flow through a fibre bed can be demonstrated by D'Arcy's law (see Fig. 3), describing viscous flow through porous media [27–31].

$$\text{Pressure} = \frac{\text{viscosity} \times \text{velocity} \times \text{thickness}}{\text{permeability constant (D'Arcy)}}$$

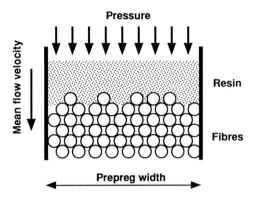

Fig. 9.3 Schematic representation of resin flow through a fibre bed.

D'Arcy's permeability constant for a typical fibre bed for advanced composites is in the order of 10^{-13} m^2.

The dynamic viscosity of commonly used thermoplastic matrix materials at processing temperatures is very high. Typical values may be in the range of 200–4000 Pas even at temperatures between 300 and 400°C. Corresponding values for thermoset resins are approximately 1–10 Pas. [18]. Most commonly, hot-melt prepregging at high shear rates or solvent prepregging are employed in order to achieve a sufficiently low viscosity to guarantee wet-out of the filaments in the fibre bundles.

Several methods have been proposed to solve the problem of proper resin impregnation, resulting in the development of a large range of product forms for thermoplastic-based composites [1, 32, 33]. The different product forms used for these composites can be classified into two basic types according to the sequencing of the impregnation and shaping operation:

- preimpregnated product forms
- product forms for post-shaping impregnation

3.1 Preimpregnated Product Forms

The main advantage in using a separate preimpregnation step is that the critical impregnation and wetting stage can be carried out under optimal conditions to ensure high quality prepreg. The main disadvantages are the additional costs and the boardiness of these product forms, which drastically reduce design freedom in the lay-up phase during manufacturing. Two different techniques are used for producing preimpregnated product forms, namely hot-melt impregnation and solution impregnation.

Conventional *hot-melt impregnation*, as used for thermoset resins, is most often not feasible for high-performance thermoplastics due to the high melt viscosity of the resin [34–36]. However, it is possible in some cases to obtain an acceptable wet-out and impregnation of the fibres if high shear rates and/or pressures are applied during the impregnation stage.

Solution impregnation is primarily used for amorphous thermoplastics [37–40]. By dissolving the polymer in a solution, the viscosity is reduced to an acceptable level for proper wetting and impregnation. A critical step in this process is to ensure the complete removal of the solvent from the prepreg after the impregnation step [41]. Small amounts of residual solvent in the prepreg may have a detrimental influence on the final product performance.

3.2 Product Forms for Post-Shaping Impregnation

By using product forms which allow post-shaping impregnation, a high preform flexibility may be maintained, thus ensuring a good drapeability during lay-up. The fibres and the resin are merged in the preform without any pre-consolidation. This produces a preform with weak interaction between the resin and the fibres. The resin can be in the form of thin films consecutively layered with the fibre tows or as a dispersed polymer powder or polymer fibres in the fibre tow. By introducing the resin into the fibres bundles, the flow distance during the impregnation stage can be diminished, thereby facilitating impregnation and consolidation during the final shaping process. The disadvantage of this product form is the very critical impregnation and wetting stage to be performed simultaneously with the shaping operation in the mould. Different product forms used for post-shaping impregnation are illustrated in Fig. 4.

In the *film stacking* technique, consecutive layers of fibres (individual fibre tows or fabrics) and polymer films are placed on top of one other. The final consolidation of the stack requires high pressures, to ensure a high quality of the laminate [1, 42, 43].

Another method of fibre impregnation, *powder coating*, uses a fine suspension of polymer particles in the fibre tow [7, 8, 44, 45]. The particle size distribution of the polymer powder normally lies in the range of 5–200 µm. The polymer particles are electrostatically deposited onto the fibres using a fluidized bed process. The powder-impregnated tow may be covered by a thin film in order to prevent loss of powder during further handling. A wide range of polymers may be processed by this technique as long as they can be obtained in a fine powder form.

In the *fibre hybridization* technique the reinforcing fibres and polymer fibres are mixed into a hybrid yarn [46–49]. The mixing of the different fibres is usually accomplished by co-mingling or co-weaving. With this technique, as with the powder techniques, impregnation is facilitated by introducing the polymer closer to the reinforcing fibres. A notable advantage is the excellent drapeability of cloth woven from this material, which provides an improved design freedom.

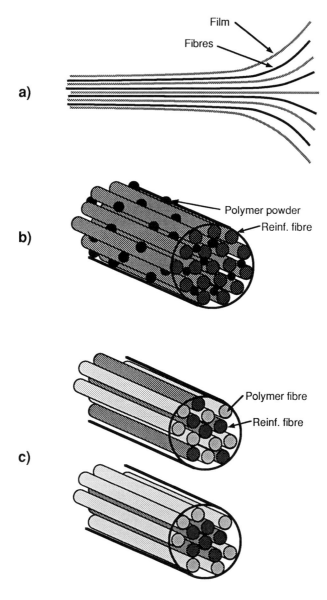

Fig. 9.4 Product forms for post-shaping impregnation.

4 Forming Mechanisms

Two principal mechanisms may be used for the forming and consolidation of a thermoplastic-based component, as was demonstrated in Fig. 1. When using preimpregnated material, the *formability* is determined by the material's potential to flow and deform during the shaping process. When the ability to be formed is determined by the *drapeability* of the preform, the fibre configuration (e.g. weave, stitch, braid) becomes an important parameter. A post-shaping impregnation is normally used for woven thermoplastic preforms.

Fig. 5 describes the different possible modes of flow and the deformation mechanisms contributing to the forming capability of continuous fibre-reinforced preforms [50].

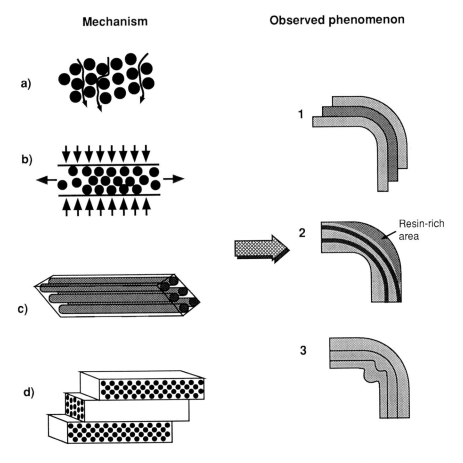

Fig. 9.5 Modes of flow and deformation mechanisms for continuous fibre preforms [50].

The *percolation* of resin through the fibre bed plays an important role in eliminating interlaminar as well as intralaminar flaws, and thus influences the overall laminate quality. The anisotropy of the fibre bed causes a strong orientation dependence of the percolation flow. For a high performance composite containing approximately 60 volume percent of fibres, the permeability transverse to the fibre direction is roughly 3–4 times that in the longitudinal direction. Percolation flow is considered to be of only marginal importance for the overall fibre orientation.

The *transverse flow* (squeeze flow) of the resin/fibre bed is a major source of deformation leading to a uniform compaction of the stacked preforms during the shaping operation. In braided and woven preforms this deformation mechanism is especially important for the homogenisation of the open volume between the yarns.

Intralaminar shear is the deformation of an individual, uniform ply. Intralaminar shear is observed to occur both transversely to and/or along the fibre direction and is necessary for imparting a double curvature to sheets containing unidirectional fibres. This mode of deformation is facilitated by local constraints present at the ply surfaces from, for example, the mould surface or adjacent plies containing different fibre orientations. Of course, any nonuniformity in this deformation mechanism causes unpredictable fibre orientations within the laminate, and as a result gives rise to undesired effects on the dimensional stability of the final product.

Interlaminar shear or slip between adjacent, individual plies is a natural deformation mechanism during the forming of curved shapes. By facilitating the slip between the individual plies, fibre wrinkling and/or ply buckling may be avoided.

It is important to keep these different modes of deformation in mind when choosing combinations of preform configurations and processing techniques.

5 Shaping and Consolidation Techniques

Different manufacturing techniques may be employed for the final forming of the composite part. These depend on the preform configuration and/or the degree of shaping complexity. The manufacturing processes used for advanced thermoplastic composites may be divided into different categories, according to the state of preimpregnation and preconsolidation of the preforms. Among the manufacturing techniques currently used, three main categories may distinguished, namely processes using:

- preimpregnated tape and tow
- preconsolidated sheet
- post-shaping impregnation

Overviews and descriptions of different processing techniques used for thermoplastic-based advanced composites have been provided by various authors [1, 2, 33]. The following gives an overview of some of the most frequently used ones.

5.1 Processes Using Preimpregnated Tape and Tow

The main step in the processing cycle for preimpregnated tape or tow can be broken down into three stages as shown in Fig. 6:

- *Heating and melting*: The temperature of the material is raised, under low contact pressure, up to a level corresponding to the viscosity of the polymer melt required for forming and consolidation. To increase the production speed with preconsolidated sheets, and to avoid large temperature gradients in the mould, the composite may be pre-heated in an external oven, while the mould is kept at a temperature closer to the solidification temperature.
- *Consolidation*: When the desired forming temperature is reached, the consolidation pressure is applied. The temperature and pressure are maintained long enough to ensure void evacuation, healing of flaws in the interlaminar regions, and to provide an overall compaction and consolidation of the laminate.
- *Cooling and solidification*: Following an adequate consolidation time the material is cooled. In order to avoid distortion of the moulded part, the consolidation pressure should be released only after the temperature has dropped below a certain level, usually determined by the glass transition temperature of the resin.

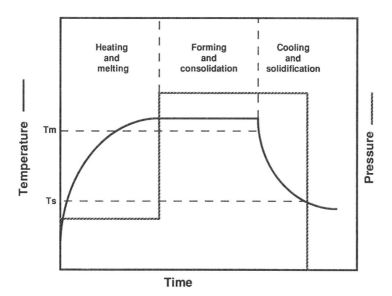

Fig. 9.6 Principal stages of pressure and temperature in the processing of pre-impregnated tape or tow (unconsolidated preforms).

The following gives some examples of processing techniques currently used for preimpregnated tapes and tows.

Consecutive Tape Laying One of the more basic methods for forming preimpregnated tape into a flat part is tape laying (Fig. 7). In contrast with stacking, in which all the layers are set in place at once and consolidated by application of pressure and heat, each layer is applied separately. The tape is rolled off the feed and, as it is positioned onto the substrate, a local heating element melts the matrix and a roller consolidates the tape onto the underlying surface. A thin polymer film can be placed between two consecutive layers of fibre-reinforced material in order to reduce the occurence of voids and thus to ensure a good continuity of the material. Heating can be carried out by a hot air gun or by infrared heating elements. This method can only be used for relatively simple part shapes and, unless it is automated, may lead to inconsistencies in the quality of the final part.

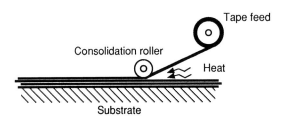

Fig. 9.7 Schematic of the tape-laying process.

Autoclave/Vacuum Bag Autoclaving and vacuum bagging refer to processing methods in which pressure is applied via a membrane to press the material into a mould (Fig. 8). Several layers of preimpregnated tape are stacked in the mould, covered with various layers of bagging material (a bleeder, a release ply, a pressure plate, a breather, and a vacuum bag that is sealed to the mould) and put under pressure in an autoclave. The autoclave pressure is usually supplemented by drawing a vacuum inside the bag. The preimpregnated tape can be heated before layup or pre-shaped to facilitate the stacking of more complicated shapes. The method has several drawbacks, among which the longer cycle times associated with heating the entire processing system, the surface finish on one side of the part, and limitations on the shape of the part. Furthermore, the pressure in vacuum forming may be inadequate for the deformation and consolidation of the material required to form the part. The principal advantage is the lowered tooling cost, since only one mould half is needed.

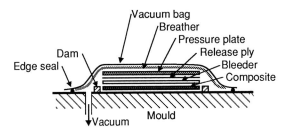

Fig. 9.8 Schematic of the autoclave/vacuum bag process.

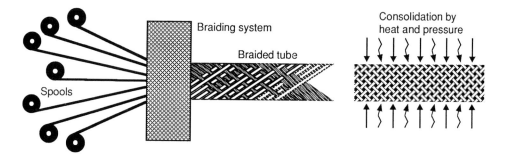

Fig. 9.9 Schematic of the braiding process.

Braiding and Weaving Tows of relatively small diameter may be braided into tubes or woven into flat fabrics (Fig. 9). Various textile manufacturing methods are applied to produce a given weave pattern. The fabric is then draped onto a mould and consolidated by application of heat and pressure, for example in an autoclave. A honeycomb material may be inserted between two layers in order to increase the static height of the section, so as to obtain a higher bending stiffness. Weaving is subject to some limitations. First, care must be taken not to exceed lateral pressures beyond which the fibres are damaged by crimping. Secondly, some loss of in-plane stiffness in comparison to the first two methods mentioned in this section may be observed. An advantage of the method is that a good drapeability of the material can be achieved by an appropriate choice of weave pattern.

Filament or Tape Winding A preimpregnated tow or tape can be wound onto a mandrel with shapes of varying complexity in order to produce tubular elements such as shafts and pressure vessels (Fig. 10). The method is a more complex form of consecutive tape laying, in which the tape is wound helicoidally about the mandrel. The tape coming from a tape feed passes through a guide which heats the tape just before depositing and consolidating it onto the heated mandrel. The tow or tape can be heated by hot air, laser, or infrared radiation. The advance of the tow or tape guide determines the angle of the fibres with respect to the main axis (or axes) of the part being produced. A relatively high adhesion between the successive layers makes it possible to produce parts with concave surfaces. The programming of the layup

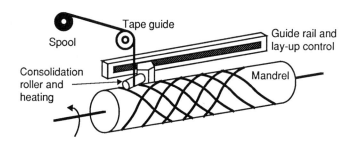

Fig. 9.10 Schematic of the tape winding process.

sequence may require fairly involved calculations to achieve a desired lamination sequence, especially for parts with complex geometries.

Pultrusion Pultrusion is used to produce parts with constant and relatively small cross sectional areas in quasi infinite lengths (Fig. 11). Hollow parts as well as solid cross sections are possible. The preimpregnated tow or tape is taken off the spools (also called creels), fed through guides, and through a preheater, a heated die, and a cooling unit. A drive system that pulls the finished section out of the die ensures that the material is always under tension as it moves through the production machine. The guides control the distribution and the alignment of the tows. The preheater melts the thermoplastic matrix before the material enters the die, which imparts the desired cross-section. The material is cooled by the cooling unit as it exits the die and before it is caught by the pulling mechanism. This method allows a very high fibre content with unidirectional materials and thus high axial stiffness and strength. Another advantage is that the process can be highly automated, which makes pultrusion one of the least expensive processing methods for composites. A difficulty resides, however, in the precision of adjustment of the processing parameters that is required to avoid internal flaws in the finished product.

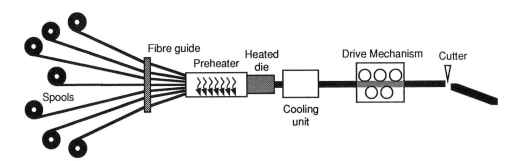

Fig. 9.11 Schematic of the pultrusion process.

5.2 Processes Using Preconsolidated Sheet

The main difference in using preconsolidated sheets for the shaping process, as opposed to preimpregnated tape or tow, is that the time-consuming consolidation step may be reduced or eliminated, as demonstrated in Fig. 12.

The preconsolidated sheet may be preheated in an external oven before being transferred to the mould, which may be kept at a temperature below the solidification temperature of the polymer. In the case of a semicrystalline material this is between the glass transition temperature and the crystallisation temperature; for an amorphous polymer, it is just below the glass transition temperature. By this sequence of operations the cycle time may be considerably reduced, as the time-consuming ther-

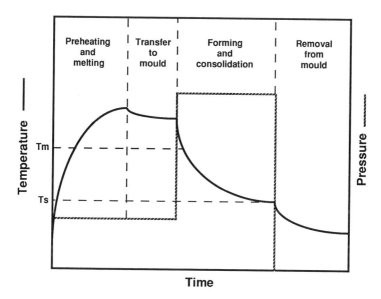

Fig. 9.12 Principal stages of pressure and temperature in the processing of pre-consolidated sheets.

mal cycling of the mould is eliminated. However, the higher thermal gradients imposed on the composite moulding may increase the risk of process-induced defects in the composite (see section 9.6). Furthermore, higher forming pressures may be required to ensure an accurate shaping operation.

The following gives some examples of processing techniques using preconsolidated sheets [1, 2, 5, 33, 52].

Matched-Die Moulding Matched-die press forming is a variant of conventional press forming in which the moulds used are generally designed with a fixed gap having very close tolerances (Fig. 13). The dies can be cooled or heated and high pressures can easily be applied to the part. The pressure must be chosen so as to prevent damage of the fibre bed. In practice, the preform is often preheated to shorten the moulding cycle and to promote early flow in the mould.

This process allows the fabrication of complex shapes at high pressures and achieves close tolerances and a good surface appearance of the moulded part. Matched-die forming has some disadvantages, such as a non-uniform pressure state induced by the thickness mismatch between the formed part and the cavity, and friction at the die interfaces. In the case of non-preheated preforms the process requires long heating/cooling times and thus leads to higher fabrication costs.

Diaphragm Forming In diaphragm forming the charge of pre-preg (either preconsolidated pre-preg or stacks of lamina) is laid up between two plastically deformable sheets. Both aluminium and polyimide have been successfully used as diaphragm

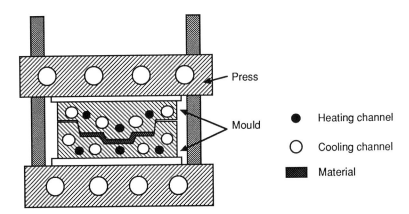

Fig. 9.13 Schematic of the matched-die molding process.

sheets in the forming of high performance composites (Fig. 14). The workpiece is then clamped around its edges by a frame, which maintains biaxial tension on the laminate during deformation. This biaxial tensile stress state prevents the laminate from wrinkling and splitting even when forming parts of relatively complex shapes.

Diaphragm forming does not require high pressures. Usual working pressures are in the order of 0.4–1.0 MPa. The process permits a good fibre placement control. Drawbacks lie in that this process needs a long heating cycle and that the sheet must be clamped by hand. The fixation system of the plate on the tool produces a heat flow at the edge of the part. There are also some temperature limitations.

Hydroforming Hydroforming is similar to diaphragm forming, the principal difference being that the pressure is applied to the composite part by means of a fluid medium acting on a rubber membrane. A hydraulic fluid, behind the rubber diaphragm, is pressurized and deforms the material against a male or female tool half (Fig. 15). The rubber bladder in hydroforming is a permanent part of the system.

Hydroforming enables interply slip deformation in the material. This prevents fibre wrinkling in areas being stretched. Furthermore, the rubber bladder does not contact the part before and after moulding. A problem with the system as described above is that it is suitable only for low temperatures; an extra rubber sheet must be placed on the part if the processing temperature is high. An additional rubber sheet must be placed on the part to prevent the rupture of the diaphragm on sharp edges of the tool for certain mould geometries.

Rubber-Pad Press Forming Rubber pad press forming is very similar to hydroforming, with the exception that the fluid is replaced by a permanent block of rubber (Fig. 16). The rubber part is protected by a wear pad and may be profiled to the tool geometry. Another very similar system is rubber matched-die tooling. This process provides high forming pressures under almost hydrostatic pressure condi-

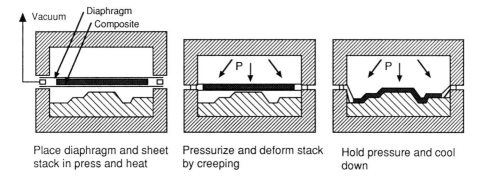

Fig. 9.14 Schematic of the diaphragm forming process.

tions. The bulk of the rubber pad is made of polyurethane foam with a cover pad of silicone rubber, to allow high temperature forming. The advantages and the disadvantages correspond very closely those of hydroforming.

Roll Forming Roll forming is a continuous forming process for composites that is derived from metal working techniques. In this process, a sheet is preheated above its melting temperature and fed into a series of cold roll sections (Fig. 17). The shape is progressively imparted to the material as it moves through the matching rollers. In some cases, it suffices to heat the areas which will be deformed. Line speeds in the order of 10 m/min have been reported.

Roll forming is well adapted for straight channel or hat sections and has also been used for the manufacture of curved beam sections. The process presents restrictions on the complexity of the shapes produced, but its great advantage lies in its continuity which allows the production of profiles in quasi infinite lengths.

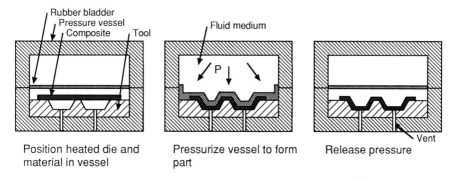

Fig. 9.15 Schematic of the hydroforming process.

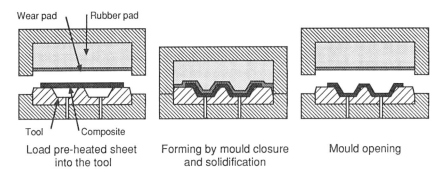

Fig. 9.16 Schematic of the rubber-pad press forming process.

5.3 Processes Using Post-Shaping Impregnation

Most of the above-mentioned processing techniques may be used for product forms requiring post-shaping impregnation. However, it should be emphasised that these product forms contain a considerable amount of air. Vacuum-assisted forming is therefore desirable in order to minimise the porosity of the final product. It should also be noted that the combined shaping and impregnation step is critical in obtaining uniform wetting and impregnation of the fibres [15]. This demands a closer control of the processing parameters.

6 Process-Structure-Property Relations

Many of the recent developments in product forms and processing techniques for thermoplastic composites have enabled higher production efficiency and forming flexibility. In addition to more pronounced non-isothermal solidification and intricate

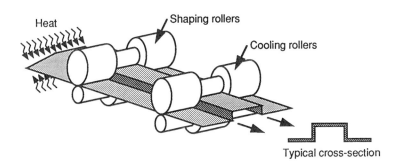

Fig. 9.17 Schematic of the roll forming process.

rheological conditions during the forming stage, the novel high-temperature materials in many cases exhibit an increased sensitivity to processing conditions [12, 13, 54, 55]. The dynamics of temperature and pressure applied during processing, combined with the solidification kinetics and viscoelastic properties of the material, will determine the morphology and internal stress distribution in the manufactured parts [13, 17, 56–60]. Thus the choice of processing conditions may be the main factor influencing process-induced defects such as voids, microcracking, fibre disorder, warpage and/or residual stresses in the moulded part [16]. It is evident that process-induced microstructure and morphology may considerably influence the quality and durability of the final part.

The following section reviews morphological features and properties of advanced thermoplastic composites that are primarily induced by processes imposing considerable thermal gradients on the composite material during the processing step. The general trend has been to apply such techniques for thermoplastic-based composites, given their potential for rapid net-shape manufacturing.

6.1 Process-Induced Morphology

A detailed experimental and theoretical description of the influence of cooling rate on the degree of crystallinity for PEEK-based composites has been presented by many authors. The relationship is illustrated in Fig. 18 [61].

In press forming processes where the composite may be exposed to high cooling rates due to abrupt contact with a cold mould surface, large crystallinity variations may be obtained between the surface and the centre of the laminate. In an autoclave process the composite material is normally exposed to much lower cooling rates,

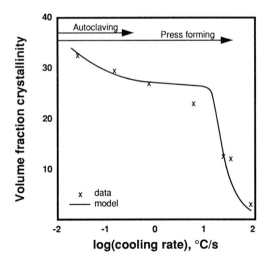

Fig. 9.18 Relation between cooling rate and volume fraction crystallinity for APC2 composite. From *Velisaris* and *Seferis* [62].

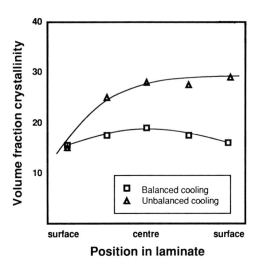

Fig. 9.19 Crystallinity distribution over the laminate thickness (5 mm) for two different cooling rates for APC2 composite. From *Månson, Schneider* and *Seferis* [13].

giving an almost constant degree of crystallinity over the laminate thickness. However, even at the lower cooling rates different morphological structures may be observed across the laminate thickness. For regions exposed to lower cooling rates, larger spherulites and a more highly developed transcrystalline region near the fibre surface have been reported. This demonstrates the importance of relating process-induced morphologies not only to the degree of crystallinity but also to the morphological structure, as their influence on mechanical performance may be significant [62–74].

It has been shown for press forming of APC2 composite with surface cooling rates of the order of $40°K/s$, that the resulting crystallinity is constant over the thickness, at a level of approximately 28% by volume [12, 13]. It should be noted that a slight increase in the cooling rate resulted in a considerable drop in crystallinity closest to the surface (Fig. 19).

If unbalanced thermal conditions are present during the cooling stage considerable morphological differences may be obtained between the two surfaces, resulting in an undesirable distortion of the moulded part [12]. Unbalanced solidification conditions may be present in processing techniques involving a rubber pad on one of the mould surfaces or when pre-heated preforms are loaded into "cold" mould surfaces before mould closing and pressurizing.

6.2 Internal Stress Generation During Composite Processing

It is well known that internal stresses generated during processing, by non-uniformity in thermal contraction and/or thermo-mechanical properties of the material, will have a significant influence on the final quality of the composite material [13].

9. Processing of Thermoplastic-Based Composites   293

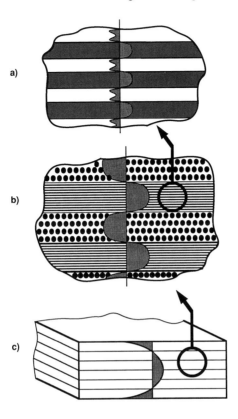

Fig. 9.20 Levels of internal stresses in a fiber-reinforced cross-ply laminate exposed to high cooling rates. From *Månson* and *Seferis* [61].

The internal stresses may act on different levels within the laminate, and thereby have different levels of self-equilibration in the composite structure. As shown in Fig. 20 three different internal stress levels may be identified [17, 56–61, 75, 76].

First, stress generation may be due to the *heterogeneity* of thermal expansivity between the matrix and the fibre. These stresses are primarily generated in the interphase between the fibre and the matrix. It has been demonstrated by using X-ray diffraction techniques that these are of the order of 30 MPa (tensile stresses in the matrix and compressive stresses in the fibre) while cooling a fibre/PEEK composite from the processing temperature down to room temperature [77, 78].

Secondly, stresses are caused by the *anisotropy* in the coefficient of thermal expansion in a unidirectional lamina. In a cross-ply lay-up this will generate stresses with a range of action extending between layers of different orientation. As demonstrated by many authors, these stress fields may be determined by recording the curvature of an unbalanced laminate upon cooling from the processing temperature down to room temperature. Residual stress levels of the order of 40 MPa have be reported for a 0/90 lay-up of APC2 [57–60].

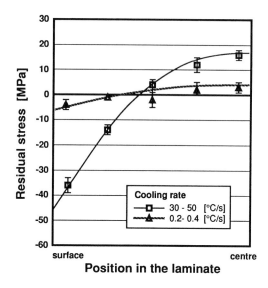

Fig. 9.21 Internal stress through the thickness of a 5 mm thick APC2 laminate, surface cooling rate 30–50°C/s and 0.2–0.4°C/s, respectively. From *Månson* and *Seferis*, [61, 75].

Thirdly, stresses generated by *nonisothermal solidification* may give rise to stress fields acting over the distance from the surface to the centre of the laminate. These thermally induced skin-core stresses are primarily generated by non-isothermal conditions present during solidification, enhanced by contact with the cold mould surface [75, 79, 80]. Fig. 21 shows the residual stress profile as measured by the layer removal technique of a 40-ply laminate, processed at a surface cooling rate of approximately 40°K/s. The compressive stresses in the surface layer reached approximately 40 MPa, while the tensile stresses in the core region of the laminate were in the order of 15–20 MPa [61, 75]. This represents almost 25% of the ultimate tensile strength of the APC2 composite in the 90° direction. The low cooling rates present during autoclaving (0.2°K/s) give an almost negligible contribution from the thermal skin-core stresses.

This shows the importance of considering both the thermodynamic and the kinetic nature of the material and of the process in evaluating internal stress generation during the processing step. It shows, furthermore, the importance of the choice of a characterisation technique for residual stress determination, dependent on whether microscopic or macroscopic stress contributions are to be evaluated.

The principal mechanism of internal stress generation is governed by the kinetics of the process, primarily determined by the cooling rate, combined with the viscoelastic nature of the polymer. Since the solidification of semi-crystalline materials occurs well above the glass transition temperature, the potential for recovery in this temperature range may result in a reduction of the stresses through creep and stress relaxation

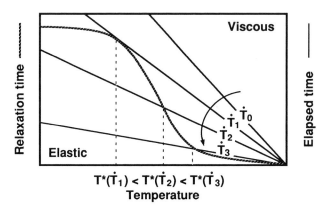

Fig. 9.22 Schematic representation of the viscous to elastic transition temperature for various process-related elapsed times (cooling rates) in relation to the specific relaxation time of the polymer [17].

processes. Given the time-temperature dependence of the viscoelastic recovery, the specific cooling rate during processing or the following heat treatment is of great importance for determining the final stress state. The thermal conditions while passing through the glass transition temperature and down to room temperature are of particularly importance as the relaxation time for the polymer goes through a dramatic change in this temperature range. The large volume contraction for a semi-crystalline polymer during crystallisation has been shown to be of less importance for the internal stress build-up due to the short relaxation time in this stage. The relation between the viscolelastic nature of the polymer, represented by the relaxation time, and the elapsed time given by the process-imposed time-temperature conditions, is shown in Fig. 22 [17, 60, 81]. This figure shows that an elastic build-up of internal stresses occurs in the case when the process-related, available time is shorter than the specific time associated with the relaxation process. If, on the other hand low cooling rates are imposed, the process-related times will exceed the specific relaxation times of the material and relaxation of the thermally generated stresses will be facilitated. It should be stressed that the current trend toward faster process cycles and materials with lower tendencies for stress relaxation may considerably increase the importance of matching these process and material parameters.

The process-induced internal stress profiles may manifest themselves as macroscopic phenomena such as warpage, delamination, microcracking and voids, or may remain in the specimen as residual stresses [13, 16].

For processes such as the different press-forming techniques, which may give rise to a high degree of uncontrolled flow during mould closing, considerable undesirable fibre reorientation normally takes place. It should be emphasised that the above-mentioned process-induced effects may be much less severe in low-pressure and low-flow techniques such as autoclaving.

7 Processing Window

From a manufacturing viewpoint, considering equipment investment and production rate, the goal is to achieve processing at the lowest possible pressures and cycle times. Shortening cycle times requires exposing the moulding to a high cooling rate in some part of the processing cycle. This may occur either while loading a preheated preconsolidated sheet into a "cold" mould or during the cooling phase of the mould when mould temperature cycling is used. As shown in Fig. 23 this may be schematically represented by a processing window defined by the processing pressure and solidification rate (or cooling rate). The x-axis may also be seen as the elapsed time for the process, with decreasing values going from left to right.

The lower limit for the cooling rate is fixed by economic constraints, since long cycle times are costly and therefore undesirable. Furthermore, if the polymer is exposed to excessively high temperatures during an extended time, thermal degradation mechanisms become increasingly important. The upper boundary for the forming pressure will be governed by the practical limit of forces to be applied by the press forming equipment as well as by the increased cost for a mould capable of withstanding these high forces. High forming pressures may also cause a high degree of damage to the fibre bed as well as leading to resin starvation due to a high degree of resin bleed from the mechanically locked fibre bed. However, practical studies have shown that higher pressures may be applied if a faster solidification can be achieved. In this case, the increased viscosity of the resin diminishes the resin bleed-out and the solidified resin carries an increased load.

High cooling rates impose a considerable thermal gradient over the thickness of the composite part and lead to both morphological skin/core effects and thermal skin/core stresses. Obviously, excessive cooling rates or insufficient elapsed times

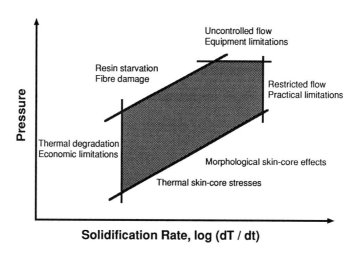

Fig. 9.23 Schematic representation of a pressure / cooling rate processing window.

limit the compaction of the laminate. Relations between different defect mechanisms, most likely governed by the internal stress build-up (voiding, micro-cracking), and the cooling rates in thermoplastic based laminates have been observed. An increased forming pressure, corresponding to the lower right boundary of the diagram, may suppress this defect initiation. This boundary also represents the most desirable processing region.

Thus, an understanding of the relations between the process parameters and the material properties is of great importance for the successful processing of thermoplastic-based advanced composites.

Acknowledgments

The author would like to acknowledge the help of Mr. Patrick Kim in the preparation of the manuscript. The assistance of Messrs. Paul Sunderland and Richard Phillips is also gratefully acknowledged. A particular note of appreciation goes to Professor James C. Seferis, director of the Polymeric Composites Laboratory at the University of Washington, for introducing the author to the field of composite processing science.

References

1. Cogswell, F.N., *Thermoplastic Aromatic Polymer Composites,* ed., Oxford: Butterworth-Heinemann, 1992.
2. Okine, R.K., *Analysis of Forming Parts from Advanced Thermoplastic Composite Sheet Material,* Journal of Thermoplastic Composite Materials, Vol. 2(January): p. 50–76, 1989.
3. Muzzy, J.D., *Processing of Advanced Thermoplastic Composites,* Georgia Institute of Technology, Vol. , 1988.
4. Shukla, J., *Fabrication of Aircraft Structures from Thermoplastic Drapeable Preforms, 21st International SAMPE Technical Conference,* 1989.
5. Mallon, P.J., O'Bradaigh, C.M., and Pipes, R.B., *Polymeric Diaphragm Forming of Continuous Fiber Reinforced Thermoplastics, 33rd International Sampe Symposium,* 1988.
6. Gysin, H., *Machine Industry- A Market for Advanced Thermoplastic Composites?, Verbundwerk 91,* Wiesbaden: 1991.
7. Muzzy, J., Varughese, B., Thammongkol, V., and Tincher, W., *Electrostatic Prepregging of Thermoplastic Matrices,* SAMPE Journal, Vol. 25(5): p. 15–21, 1989.
8. Hartness, T., *Thermoplastic Powder Technology for Advanced Composite Systems,* Journal of Thermoplastic Composite Materials, Vol. 1: p. 210–220, 1988.
9. Olson, S.H., *Manufacturing with Commingled Yarns, Fabrics and Powder Prepreg Thermoplastic Composite Materials, 35th International SAMPE Symposium,* 1990.
10. Lee, W.J. and Manson, J.-A., *Factors Influencing Process Selection and Processing, Polymer Composite Applications for Motor Vehicles,* Detroit, USA: SAE International, SP-850, 1991.
11. Seferis, J.C., *Polyetheretherketone(PEEK): Processing-structure and properties for a matrix in high performance composites,* Polymer Composites, Vol. 7(3): p. 158–169, 1986.

12. Lawrence, W.E., Manson, J.-A.E., and Seferis, J.C., *Thermal and Morphological Skin-Core Effects in Processing of Thermoplastic Composites*, Composites, Vol. **21**: p. 475, 1990.
13. Manson, J.A.E., Schneider, T.L., and Seferis, J.C., *Press-forming of continuous-fiber-reinforced thermosplastic composites*, Polymer Composites, Vol. **11**(2): p. 1–7, 1990.
14. Ko, F., Chu, H., and Ying, E., *Damage Tolerance of 3-D Braided Intermingled Carbon/PEEK Composites*, Proc. 2nd Conf. on Advanced Composites, Dearborn: 1986.
15. Hasselbrack, S.A., Pederson, C.L., and Seferis, J.C., *Evaluation of Carbon-Fiber-Reinforced Thermoplastic Matrices in a Flat Braid Process*, Polymer Composites, Vol. **13**(1): p. 38–46, 1992.
16. Manson, J.-A.E. and Seferis, J.C., *Void Characterization Techniques for Advanced Composites*, Science and Engineering of Composite Materials, Vol. **1**: p. 75, 1988.
17. Chapman, T.J., Gillespie, J.W., Pipes, R.B., Manson, J.-A.E., and Seferis, J.C., *Prediction of process-induced residual stresses in thermoplastic composites*, Journal of Composite Materials, Vol. **24**(June): p. 617–643, 1990.
18. Hancox, N.L., *High Temperature High Performace Composites*, Advanced Materials & Manufacturing Processes, Vol. **3**: p. 359–389, 1988.
19. Johnston, N.J. and Hergenrother, P.M., *High Performance Thermoplastics: A Review of Neat Resin and Composite Properties*, 32nd International Sampe Symposium, 1987.
20. Hill, R., *Theory of Mechanical Properties of Fibre-Strengthened Materials: III. Self Consistent Model*, J. Mech. Phys. Solids, Vol. **13**: p. 189, 1965.
21. Hashin, Z., *On Elastic Behavior of Fiber-Reinforced Materials of Arbitrary Transverse Phase Geometry*, J. Mech. Phys. Solids, Vol. **13**: p. 119, 1965.
22. Halpin, J.C. and Kardos, J.L., Polymer Engineering and Science, Vol. **16**: p. 344, 1976.
23. Folkes, M.J., *Short Fibre Reinforced Thermoplastics*, Research Studies Press, Vol. , 1982.
24. Silverman, E.M., *Effect of Glass Fiber Length on the Creep and Impact Resistance of Reinforced Thermoplastics*, Polymer Composites, Vol. **8**(1): p. 8–15, 1987.
25. Gibson, R.F., Chatuverdi, S.K., and Sun, C.T., *Complex moduli of aligned discontinuous fibre-reinforced polymer composites*, Journal of Materials Science, Vol. **17**: p. 3499–3509, 1982.
26. Gibson, A.G., *Rheology and Packing Effects in Injection Molding of Long-Libre Reinforced Materials*, Plastics and Rubber Processing and Applications, Vol. **5**(2): p. 95–100, 1985.
27. Gutowski, T.G., *A resin flow/fiber deformation model for composites*, SAMPE Quarterly, Vol. **16**(4): p. 58–64, 1985.
28. Gutowski, T.G., Cai, Z., Kingery, J., and Williams, S.J., *Resin Flow/Fiber Deformation Experiments*, SAMPE Quarterly, Vol. **17**(4): p. 54–58, 1986.
29. Dave, R., Kardos, J.L., and Dudukovic, M.P., *A model for resin flow during composite processing: Part 2 - Numerical analysis for unidirectional graphite/epoxy Laminates*, Polymer Composites, Vol. **8**(2): p. 123–132, 1987.
30. Dave, R., Kardos, J.L., and Dudukovic, M.P., *A model for resin flow during composite processing: Part 1- General mathematical development*, Polymer Composites, Vol. **8**(1): p. 29–38, 1987.
31. Lam, R.C. and Kardos, J.L., *The Permeability of Aligned and Cross-Plied Fiber Beds during Processing of Continuous Fiber Composites*, American Society for Composites, 3rd Annual Technical Conference, 1988.
32. Leach, D.C. and Schmitz, P., *Product Forms in APC Thermoplastic Matrix Composites*, 34th International SAMPE Symposium, 1989.
33. Beland, S., *High Performance Thermoplastic Resins and Their Composites*, ed., Park Ridge, NJ, USA: Noyes Data Corp., 1990.
34. Cogswell, F.N., Hezzell, D.J., and Williams, P.J., *Fibre-Reinforced Composites and Methods for Producing such Compositions*, Vol. US Patent 4,549,920, 1981.
35. Cogswell, F.N. and Staniland, P.A., *Method of Producing Fibre Reinforced Composition*, Vol. USP 4,541,884, 1985.

36. Chung, T.S. and McMahon, P.E., *Thermoplastic Polyester Amide-Carbon Fiber Composites*, Am. Polymer Sci., Vol. **31**: p. 965–977, 1986.
37. Leeser, D. and Banister, B., *Amorphous Thermoplastic Matrix Composites for New Applications*, 21st International SAMPE Technical Conference, 1989.
38. Goodman, J.K. and Loos, A.R., *Thermoplastic Prepreg Manufacture*, Proc. Am. Soc. for Composite Materials, 4th Tech. Conf., Technomic, 1989.
39. Peake, S.L., *Processes for the Preparation of Reinforced Thermoplastic Composites*, Vol. US Patent 4,563,232, 1986.
40. Wu, G.M., Schultz, J.M., Hodge, D.J., and Cogswell, F.N., *Solution Impregnation of Carbon Fiber Reinforced Poly(ethersulphone)Composites*, SPE Antec, 1990.
41. Nelson, K.M., Manson, J.A., and Seferis, J.C., *Compression Thermal Analysis of the Consolidation Process for Thermoplastic Matrix Composites*, J. of Thermoplastic Composite Mat., Vol. **3**: p. 216–232, 1990.
42. Hogan, P.A., *The Production and Uses Of Film Stacked Composites For the Aerospace Industry*, SAMPE Conference, 1980.
43. Lee, W.I. and Springer, G.S., *A Model of the Manufacturing Process of Thermoplastic Matrix Composites*, J. Comp. Mats, Vol. **21**(11): p. 1017–1055, 1987.
44. Ganga, R., *Fibre Imprégnée Thermoplastique (FIT)*, Composites et Nouveaux Matériaux, Vol. **5**(May), 1984.
45. Chang, I.Y., *Thermoplastic Matrix Continuous Filament Composites Kevlar Aramid or Graphite Fiber*, Composites Science and Technology, Vol. **24**: p. 61–79, 1985.
46. Taske, L.E. and Majidi, A.P., *Performance Characteristics of Woven Carbon/PEEK Composites*, Proc. Am. Soc. for Composites, 2nd Tech. Conf., Technomic Publishing Co, 1987.
47. Lynch, T., *Thermoplastic/Graphite Fiber Hybrid Fabrics*, SAMPE Journal, Vol. **25**(1): p. 17–22, 1989.
48. Majidi, D.A., Rottermund, M.J., and Taske, L.E., *Thermoplastic preform fabrication and processing*, Sampe Journal, Vol. (January/February): p. 12–17, 1988.
49. Handermann, A.C., *Advances in Commingled Yarn Technology*, 26th Int. SAMPE Technical Conference, 1988.
50. Cogswell, F.N., *The Processing Science of Thermoplastic Structural Composites*, Intern. Polymer Processing, Vol. **1**(4): p. 157–165, 1987.
51. Muzzy, J., Norpoth, L., and Varughese, B., *Characterization of thermoplastic composites for processing*, Sampe Journal, Vol. **25**(1): p. 23–29, 1989.
52. Smiley, A.J. and Pipes, R.B., *Analysis of the diaphragm forming of continuous fiber reinforced thermoplastics*, Journal of Thermoplastic Composite Materials, Vol. **1**(October): p. 298–321, 1988.
53. Cattanach, J.B. and Cogswell, F.N., *Processing with aromatic polymer composites*, Developments in Reinforced Plastics, Applied Science Publishers: London, p. 1–38, 1986.
54. Corrigan, E., Leach, D.C., and McDaniels, T., *The Influence of Processing Conditions on the Properties of PEEK Matrix Composites*, Materials and Processing-Move in the 90's, Proc.10th Int. Eur. Chapter Conf. of SAMPE, Elsevier: p. 121–132, 1989.
55. Kenny, J., D'Amore, A., Nicolais, L., Iannone, M., and Scatteia, B., *Processing of Amorphous PEEK and Amorphous PEEK Based Composites*, SAMPE Journal, Vol. **25**(4): p. 27–34, 1989.
56. Galiotis, C., Melanitis, N., Batchelder, D.N., Robinson, I.M., and Peacock, J.A., *Residual strain mapping in carbon fibre/PEEK composites*, Composites, Vol. **19**(4): p. 321–324, 1988.
57. Nairn, J.A., *Thermoelastic Analysis of Residual Stresses in Unidirectional, High-Performance Composites*, Polymer Composites, Vol. **6**(2): p. 123–130, 1985.
58. Nairn, J.A. and Zoller, P., *Residual Thermal Stresses in Semicrystalline Thermoplastic Matrix Composites*, Fifth International Conference on Composite Materials ICCM-V, San Diego, California, USA: 1985.

59. Nairn, J.A. and Zoller, P., *Matrix solidification and the resulting residual thermal stresses in composites*, Journal of Materials Science, Vol. **20**: p. 355–367, 1985.
60. Nairn, J.A. and Zoller, P., *Residual Thermal Stresses in Amorphous and Semicrystalline Thermoplastic Matrix Composites*, Central Research & Developement Dpt, E. I. du Pont de Nemours & Company, Du Pont Experimental Station, Wilmington, Delaware, 19898 USA: 1986.
61. Manson, J.-A.E. and Seferis., J.C., *Process Simulated Laminate (PSL): A Methodology for Internal Stress Characterization in Advanced Composite Materials*, Journal of Composite Materials, Vol. **26**(3): p. 405–431, 1992.
62. Velisaris, C.N. and Seferis, J.C., *Crystallization kinetics of polyetheretherketone (PEEK) matrices*, Polymer Engineering and Science, Vol. **26**(22): p. 1574–1581, 1986.
63. Velisaris, C.N. and Seferis, J.C., *Heat Transfer Effects on the Processing-Structure Relationships of Polyetheretherketone (PEEK) Based Composites*, Polymer Engineering and Science, Vol. **28**(9): p. 583–591, 1988.
64. Seferis, J.C., Ahlstrom, C., and Dillman, S.H., *Cooling Rate and Annealing as Process Parameters for Semi-Crystalline Thermoplastic-Based Composites*, SPE ANTEC, 1987.
65. Blundell, D.J., Chalmers, J.M., Mackenzie, M.W., and Gaskin, W.F., *Crystalline morphology of the matrix of PEEK - carbon fiber aromatic polymer composites. Assessment of crystallinity*, Sampe Quarterly, Vol. **16**(4): p. 22–30, 1985.
66. Blundell, D.J. and Osborn, B.N., *Crystalline morphology of the matrix of PEEK - carbon fiber aromatic polymer composites II.œCrystallization behavior*, Sampe Quarterly, Vol. **17**(1): p. 1–6, 1985.
67. Blundell, D.J. and Willmouth, F.M., *Crystalline morphology of the matrix of PEEK - carbon fiber aromatic polymer composites*, SAMPE Quarterly, Vol. **17**(2): p. 50–57, 1986.
68. Cebe, P., Chung, S.Y., and Hong, S.-D., *Effect of thermal history on mechanical properties of polyetheretherketone below the glass transition temperature*, Journal of Applied Polymer Science, Vol. **33**: p. 487–503, 1987.
69. Cebe, P., *Application of the parallel Avrami model to crystallization of Poly(Etheretherketone)*, Polymer Engineering and Science, Vol. **28**(18): p. 1192–1197, 1988.
70. Cebe, P., *Non-isothermal crystallization of poly(etheretherketone) aromatic polymer composite*, Polymer Composites, Vol. **9**(4): p. 271–279, 1988.
71. Maffezzoli, A.M., Kenny, J.M., and Nicolais, L., *'Modelling of Thermal and Crystallization Behaviour of the Processing of Thermoplastic Matrix Composites'*, in Materials and Processing- Move into the 90's, Proc. 10th Int. Eur. Chapter Conf. of SAMPE, Elsevier: Materials Science Monographs, 1989.
72. Lee, Y.C. and Porter, R.S., *Crystallization of polyetheretherketone (PEEK) in carbon fiber composites*, Polymer Engineering and Science, Vol. **26**(9): p. 633–639, 1986.
73. Blundell, D.J., *On the interpretation of multiple melting peaks in poly(etheretherketone)*, Polymer, Vol. **28**(December): p. 2248–2251, 1987.
74. Kemmish, D.J. and Hay, J.N., *The effect of physical ageing on the properties of amorphous PEEK*, Polymer, Vol. **26**(June): p. 905–912, 1985.
75. Manson, J.-A. and Seferis, J.C., *Internal stress determination by process simulated laminates*, ANTEC '87, 1987.
76. Manson, J.-A.E. and Seferis, J.C., *Process Analysis and Properties of High-Performance Thermoplastic Composites*, Engineering Application of New Composites ed., Paipetis, S.A. and Papanicolau, G.C., ed., Omega Scientific, 1988.
77. Hank, V., Troost, A., and Ley, D., *Use of X-Ray Diffraction of Measure Lattice Strain and Determine Stress in Carbon Fiber Reinforced PEEK*, Kunststoffe-German Plastics, Vol. **78**(11): p. 41–43, 1988.
78. Young, R.J., Day, R.J., Zakikharic, M., and Robinson, I.M., *Fibre Deformation and Residual Thermal Stresses in Carbon Fibre Reinforced PEEK*, Composites Science and Technology, Vol. **34**: p. 243–258, 1989.
79. Maneschy, C.E., Miyano, Y., Shimbo, M., and Woo, T.C., *Residual-stress analysis of an epoxy plate subjected to rapid cooling on both surfaces*, Experimental mechanics, Vol. (December): p. 306–312, 1986.

80. Chapman, T.J., Gillespie, J.W., Pipes, R.B., Manson, J.-A.E., and Seferis, J.C., *Thermal Skin/Core Residual Stresses Induced during Cooling of Thermoplastic Matrix Composites, Amer. Soc. for Comp., 3rd Tech. Conf.,* 1988.
81. Lawrence, W.E., Manson, J.-A.E., Seferis, J.C., JR., J.W.G., and Pipes, R.B., *Prediction of Residual Stresses in Continuous Fiber Semicrystalline Thermoplastic Composites: a Kinetic-Viscoelastic Approach, Proc. American Society for Composites, 5th Technical Conference,* 1990.

Chapter 10

Conception and Processing of Advanced Thermoplastic Composite Structures for the Machine Industry

J. Müller

1 General Situation for Advanced Composites in the Machine Industry

Because of their excellent mechanical properties combined with low specific gravity high performance thermoset composites are being successfully used in the aircraft industry. There, highly loaded lightweight structures are produced at costs comparable to metal parts.

Until recently this success could not be duplicated in the machine industry, even though a keen interest exists in lightweight, highly loadable structures for high performance production machines. This interest resulted in many costly but unsuccessful projects being carried out in order to develop such structures based on the aircraft technology of the 1970's.

The reasons for this lack of success become apparent when one considers the specific requirements of the machine industry for high performance components and when one evaluates the properties of the thermoset composites mainly used in the aircraft industry against these requirements. An examination of these requirements will also lead to new materials – also those yet to be invented – finding use in the machine industry as well as showing the points to be observed when developing and designing structures made with these materials i.e. their conception.

2 Requirements of the Machine Industry for Advanced Composites

According to *Müller et al.* [1] the machine industry has the following requirements for highly loaded structures:

- good mechanical properties, even under very rapid impact loading conditions
- low weight respectively low energy consumption in service

Dr. J. Müller, Ingenieurbüro für Verbundwerkstofftechnologie, Unterloosstraße 12, CH-8461 Oerlingen

- long fatigue life in cyclic loading at high frequencies *beyond* 100 million cycles in various environments
- resistance against all lubricants used in the machine industry
- good corrosion resistance
- close dimensional tolerances maintained over the entire life period of a structure in different environments
- longlasting and safe joints, also to metal parts
- compatibility with adjacent metal parts especially with regard to thermal properties
- good tribological behavior and wear resistance against metal partners in bearings
- ease of installation, adjustment and disconnection of structures
- production processes and cycles applicable for series of medium to high quantities i.e.
  - cost effective equipment
  - short production cycles
  - steady production rates
  - economic, durable tooling
  - low cost ancillary materials which can quickly be applied
  - conclusive, rapid and simple quality assurance tests.

Besides these technical requirements, the machine industry emphasises assured, consistent material supply at constant quality over long periods as well as easy environmentally safe disposal or reclamation. Competent and prompt customer support by the supplier is also expected during these periods.

In response to these demands, constant and significant quantities will be consumed by the machine industry.

From the foregoing it can be seen that the uneconomic long processing cycles and expensive production equipment necessary for advanced thermoset composites are chiefly responsible for the lack of success of these materials in the machine industry. Other reasons are the lack of high impact and chemical resistance of most thermoset composites. In addition, the knowledge acquired in the aircraft industry is not sufficient to allow their wide application in the machine industry.

3 Significance of Advanced Thermoplastic Composites for the Machine Industry

With the advent of high performance carbon fiber reinforced thermoplastics the situation has changed fundamentally. This material group, which includes carbon fiber reinforced PAS, PEEK, PEI and PEK, among others allows rapid production cycles and the application of simple or already available production equipment. The differences between the production cycles of high performance thermosets and thermoplastic composites are shown in Fig. 1.

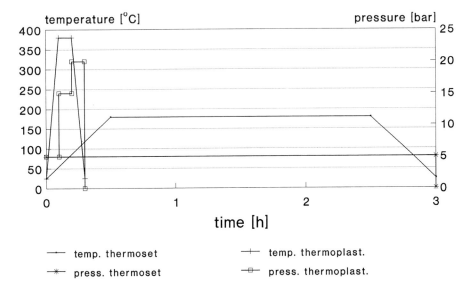

Fig. 10.1 Typical processing cycles of advanced composites with thermoset and modern thermoplastic matrices.

The rapid processing together with high impact resistance and excellent fiber to matrix adhesion has led to a breakthrough as the machine industry became interested in the application of these materials. Development projects were launched which resulted in successful designs. Also the machine industry together with research institutes, is starting to fill the gaps in the knowledge on these materials. However because of high material and production costs application of these materials is economical only for the high performance sector of the machine industry. Further limitations are posed by the relatively low service and high processing temperatures of modern thermoplastic composites. The former in general for different environments can be placed at roughly 100°C. The latter lies between 300°C and 400°C. For economic processing cycles and equipment it effectively sets a size limit as constant temperature distribution in the structure can only be achieved for small to medium sized parts.

One of the branches of the machine industry which has overcome its mistrust of high performance composites resulting from previous failures with thermosets is the textile industry. For its high performance machines it is very much interested in high weaving rates at low energy consumption i.e. in for lightweight structures capable of withstanding high loads at high loading rates. Metal designs have been exploited often to their limits and further improvements can only be realised by using lighter highly loadable materials with which parts in significant quantities of roughly 10000 parts per year can be produced at acceptable costs.

The textile industry recognized the advantages of advanced thermoplastic composites and therefore has started to develop structures with these materials. Their limitations are not significant since the highly loaded parts of a textile machine are of small to medium size and the service temperature of many parts falls well within the temperature limit of these composites. Up to a point, it is acceptable if these structures cost more, provided that they allow significant increases in machine performance. *It should be noted, however, that it is entirely feasible to produce structures at the same costs as competitive metal parts.*

These conditions allow the application of advanced composites for highly loaded structures in high performance machines i.e. in a first step a limited consumption of advanced thermoplastic composites. In order to reach the second step, namely to emulate the substantial significance of composites in the aircraft industry the production and material costs must be reduced while maintaining the level of properties and processing times.

An example of such a successful structure made of advanced thermoplastic composites is the picking lever of a Sulzer Ruti textile machine, shown in Fig. 2. In service, this structure is subjected to high and very rapid loads and has to maintain its resistance under continuous temperatures of up to 80°C in a high humidity and a chemical environment. It has been tested successfully at the design load and in operation for over 100 million load cycles. Compared to the competitive titanium lever it shows a significantly lower moment of inertia around its rotation axis. In addition, its production costs are comparable to those of the titanium lever.

4 Applicability and Limitations, Guidelines for Application

In summary one can reach the following conclusions concerning the applicability and limitations of advanced thermoplastic composites in the machine industry:

- parts subject to high loads and acceleration where low weight is a priority target
- cost expectations are of the order of those for high performance metal structures
- medium to large series, i.e. short production cycles, are required
- small to medium part size
- service temperature does not significantly exceed 100°C.

5 Design Procedure

When designing highly loaded parts made of advanced thermoplastic composites the sequence shown in Fig. 3 [based on *Brünings et al.* [2], adapted for advanced composites by the author] can be followed. As a basis for success, designers must be well schooled in the field of composites in order to be able to decide upon the most efficient type of structure, optimum design and most economical production method. Composites should not be treated like metals. In order to be successful, the conception of composite parts should include knowledge not only of their properties but also of their production methods.

Fig. 10.2 Picking lever of a Sulzer textile machine made of modern thermoplastic composites.

Often one receives the task of designing a component of a machine to compete with or replace an already existing metal structure. The pressure is great to "translate" the metal part into composites, i.e. to duplicate the geometry of the metal in the composite structure and only vary the fiber orientation. In this case one should endeavor to assure as great a degree of geometrical freedom as possible. The smaller this degree of freedom, the lower are the chances for the development of a successful composite part.

Ideally one should strive not to replace metal structures with structures made of composite materials but to replace systems.

5.1 Requirements for Structure

As a first step the requirements for the structure or the system must be established. This step has to be taken very carefully and should include not only the exact determination of mechanical service loads (size, type, direction) but also the service environment such as its climate (temperature, humidity, chemical attack etc.), physical demands (e.g. electrical, tribological), required tolerances or installation requirements.

Furthermore, the available degrees of freedom mentioned above must be established at this point.

As an example the list of requirements for the picking lever shown in Fig. 2 contained:

- reduced moment of inertia around its rotation axis compared to the titanium version

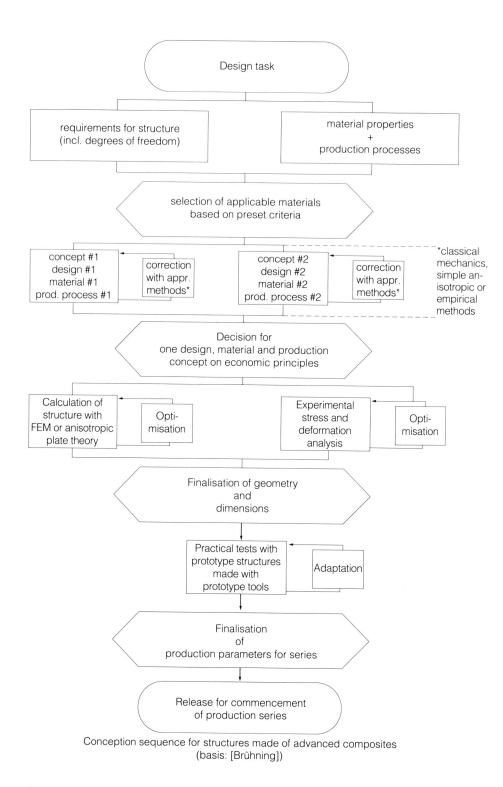

Fig. 10.3 Design and manufacturing sequence for structures made of advanced composites.

- high impact resistance
- fatigue resistance when subjected to impact loading conditions beyond 100 million cycles
- excellent wear resistance beyond 100 million cycles
- service temperature up to 100°C
- resistance against humidity, chemical attack by textile sizings, lubricants etc.
- precise tolerances
- easy, precise machining of the finished part
- ease of installation and adjustment as well as removal in an existing textile machine
- steady production rates with cycles not exceeding 1/2 h
- production costs of the order of the titanium lever.

The degrees of freedom of the composite picking lever were limited not only by the geometrical conditions of the textile machine but also by the very strict conditions of the load applications on its rotation axis and the projectile acceleration. Clearly the establishment of the degrees of freedom and their limitations delineated the space within which the structure could successfully be designed.

5.2 Material Properties and Production Processes

As stated before, it is very important to know the properties of composites not only in relation to the service environment but also as a function of their production processes. With metals, a very high degree of standardisation has been achieved. Properties are fixed by material designations and in most cases will not be changed by the fabrication process of the structure.

This is not the case with composite structures as their properties depend upon their fabrication processes and parameters. Even very detailed material designations still allow variations of properties unless the production process is also fixed. Hence, production processes and their parameters must be considered already at the early stage of material selection.

Especially for advanced thermoplastic composites the processing method and its parameters, e.g. cycle time, consolidation pressure, temperature and cool-down rate, can be of importance.

Another important point to consider is the form in which the material is available. Up to now, advanced thermoplastic composites have mostly been available in unidirectional prepreg tape form. Hence, as with advanced thermoset composites, structures will be made by lamination. This means that for material selection, ply data must be available for different loading conditions and service environments as well as the processing conditions with which they will be achieved. The last point is of great importance for the economy of the structure.

In Fig. 4 the mechanical properties for carbon fiber reinforced PEEK (tradename APC2) and PAS (tradename Radel) are shown as quoted by the material suppliers

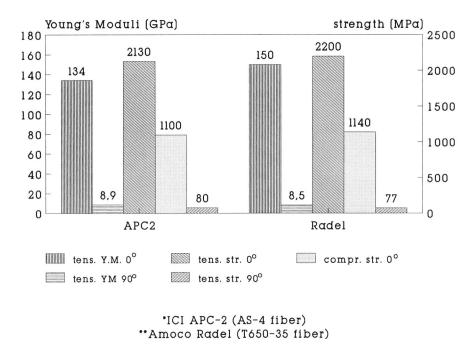

*ICI APC-2 (AS-4 fiber)
**Amoco Radel (T650-35 fiber)

Fig. 10.4 Mechanical data of PEEK-CF and PAS-CF (supplier data).

ICI and Amoco respectively for the cycles shown in Fig. 5. These cycles basically meet the requirements outlined in chapter 2 even though they do not fully exploit the economical potential of these materials. The data allow to carry out the material screening step though.

However, for the material selection process it must be noted that even at this early stage the influence of the service environment on the properties must be known. In addition, often impact resistance and fatigue data or tests and statistical data proving invariance of properties must be available so that a selection can be made against the preset criteria.

As an example, the mechanical data for APC2 are constant in humidity and temperatures up to 100°C for the processing limits shown in Fig. 6.

In the material selection process one should select at least two basically equivalent competitive materials in order to assure supply.

Supplier support and long term availability of the product should also be evaluated and considered as an important part of the material selection process.

5.3 Structural Concepts

Because of the high processing temperature of advanced thermoplastic composites, monolithic types are mainly used as structural concepts. The difficulty lies in the availability of sandwich core materials with sufficient heat resistance. These have

10. Structures for the Machine Industry   311

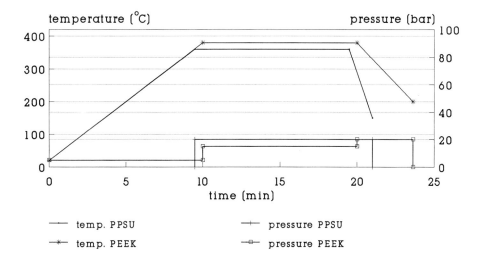

*ICI APC2, cool down rate per Cantwell
**Amoco Radel

Fig. 10.5 Processing cycles of PEEK-CF and PAS-CF: press cycles indicated by supplier.

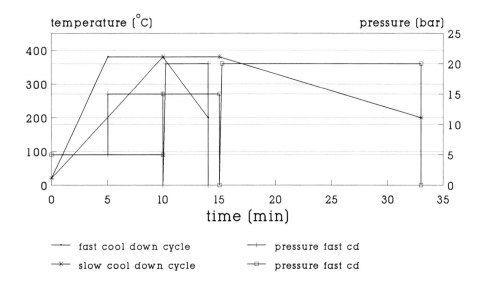

average cool-down rates
per ICI and Cantwell

Fig. 10.6 Cool down processing window for PEEK-CF laminates with thickness of 2 mm.

only become available lately so little experience with the sandwich concept yet exists for use with advanced thermoplastic composites.

As a result, the main type of structures applied are:

- pressed shapes
- stiffened panels
- filament wound tubes and vessels
- constant cross sections.

Hybrid material designs i.e. combination of different fiber types so far are not possible because successful advanced thermoplastic composites are only available with carbon fibers.

For most production processes, size limitations are imposed by the high processing temperature and the precision requirements for the structure which in turn demand a uniformity of processing parameters over the entire structure. (Exceptions will be shown in chapter 6.)

5.4 Preliminary Design: Stress Calculation Methods, Design Data, Failure Criteria

Normal design procedures for advanced thermoset composites, which are based on the thin anisotropic plate theory, can also be applied for advanced thermoplastic composites. The theories shown in *Tsai* [3] and related software were used for the preliminary design of the picking lever.

The ply design data which were used for APC2 are shown in Table 1. The stress free state of the laminate was identified with the glass transition temperature e.g. for PEEK with $T_g = 143°C$.

Safety factors of $S = 1.5$ were taken for both static and fatigue loads, which worked well in practice.

Fatigue strength data for different laminates are discussed in chapter 7.

The failure criterion of *Tsai* and *Wu* is described by the following formula [3]:

$$F_{ij}\sigma_i\sigma_j + F_i\sigma_i = 1 \quad i, j = 1, 2, 6$$

$$\sigma_{i,j} = \text{stresses}, \quad F_{i,ij} = \text{strength factors}$$

For thin orthotropic plates these strength factors are:

$$F_{xx} = 1/XX'; \quad F_{yy} = 1/YY'; \quad F_{ss} = 1/S^2$$

$$F_x = 1/X - 1/X'; \quad F_y = 1/Y - 1/Y'; \quad F_s = 0$$

$$F_{xy} = F^*_{xy}[F_{xx}F_{yy}]^{1/2},$$

where F^*_{xy} is the absolute normalized interaction term. The factor F^*_{xy} can only be determined by biaxial tests. See Table 1 for the definition and the values of the strength factors $F_{i,ij}$.

Table 10.1 Static Ply Design Data for PEEK-CF (AS-4 Fiber)

Property	Value	Units
Modulus E_x	134	GPa
Modulus E_y	9	GPa
Shear modulus S	5.2	GPa
Poisson ratio v_{xy}	0.34	—
Tensile strength X	2130	MPa
Tensile strength Y	110	MPa
Compr. strength X'	1100	MPa
Compr. strength Y'	246	MPa
ILS strength S	113	MPa
normal. int. action F_{xy}*	−0.5	—
Degradation factor	0.1	—
Lin. thermal expansion coefficient α_x	−0.3e−6	/°K
Lin. thermal expansion coefficient α_y	28.1e−6	/°K
Moisture expansion coefficient β_x	0	—
Moisture expansion coefficient β_y	0	—
Density ρ	1600	kg/m^3

The criterion was applied for last ply failure and its results were in good agreement with those of practical tests when using a normalized interaction term of $F^*_{xy} = -0.5$ recommended by *Tsai* [3].

It should be noted that, as a result of their excellent fiber to resin adhesion, advanced thermoplastic composites show better load transfer in areas of stress concentrations than advanced thermoset composites. Also, due to the plastic behavior of the matrix, failures are not catastrophic. This was shown in load bearing tests and fatigue loading of the picking lever discussed above.

5.5 Economic Principles for Decision: Estimated Material and Production Costs

In order to be able to reach a decision on economic principles in the screening phase, cost estimation of the different designs and processes involved must be carried out.

Rough cost estimates for modern thermoplastic composite structures made with diaphragm forming and tape laying processes established by *Harvey* [4] are shown in Fig. 7. The cost factors for the transformation of material into parts are 2.5 for tape laying and 3.0 for diaphragm forming. The factor of 3.0 was also found to be realistic for matched die molding, which was used in the production of the picking lever shown in Fig. 2. These factors can be used if the material price stays more or less at its present level of roughly 250 US$/kg.

Of course, in all cases one must consider also the costs for the additional steps needed in order to achieve the final part such as machining. The figures of *Harvey* were adapted accordingly in order to draw attention to this point.

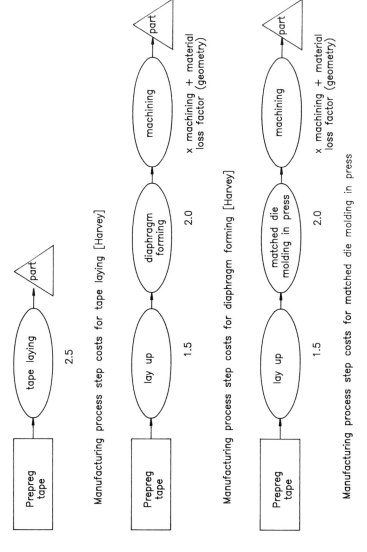

Fig. 10.7 Rough estimates of cost for processing of modern thermoplastic composites.

More precise estimates can be reached by establishing the following costs and production parameters based on the part to be made:

- material costs
- material waste factor, which allows for cutouts and offaxis layers
- preparation costs which are a function of the thickness and the complexity of the part
- ancillary materials
- equipment costs as a function of time.
- time in equipment based on heat-up rate of the equipment to be used, heat transfer to the tool and the material, part complexity and pressure which can be applied. A complex or thick part will obviously require more time in the processing equipment than a simple or thin part. Also low pressures will prolong the processing cycle since otherwise inadequate consolidation will result.
- tooling costs for tools with heat resistance to 400°C.

Definite and detailed cost evaluations can be made when the final design is initiated i.e. after the decisions on type of design, manufacture and material have been reached. For those cost calculations a designer's cost sheet shown by *Noton* [5] (Table 2) can be used.

5.6 Detailed Mechanical Calculation Methods: Finite Element Methods, Airy Functions

Finite Element Analysis Detailed mechanical analysis on relatively complex composite structures with joints and cutouts can be done with the Finite Element Method. The stress distribution in the total structure can also be determined which allows the definitive identification of critical areas and the determination of the critical load. These methods can, of course, also be used for advanced thermoplastic composites.

Disadvantages include the cost of the software and the high requirements for computer performance. In spite of the many available software programs, it seems that no single program satisfies all requirements for the mechanical analysis of anisotropic composites. In view of the cost effect, it is recommended that Finite Element Methods analysis be done only when the geometry is fixed within limits and the choice of lay-ups and fiber orientations has been sufficiently narrowed down. Also it can be stated that in some cases a practical test is more cost effective and conclusive than a Finite Element analysis.

A well known software package which allows calculation with anisotropic layered plate elements is ANSYS.[1] With this software, the stress distribution and the failure load of the picking lever in Fig. 2 were adequately and economically determined after the lay up and fiber orientation options had been narrowed down to two. The

---

[1] Other applicable programs are described and compared by Dropek: Numerical Design and Analysis; Composites, ASTM 1987.

Table 10.2 Designer's Cost Sheet [Based on 5]

Design concept	Recurring cost										Non recurring cost				Program cost		
	(L×LCF + TI&E)×LR = L[$] ; L[$]+M[$]=RC/m; RC/m×DQ=PRC										(NR(T)C+TI&E)×LR=PNRC				(PRC+PNRC)/DQ= costs/machine		
part description no.	Labor MH/PT (1)	Learning curve factor (2)	Labor TI&E MH/PT (3)	Labor rate $/MH (4)	Labor RC $/PT (5)	Material $/PT (6)	REC. cost/ part $ (7)	Parts per machine (8)	Des. qty. (9)	Program RC $ (10)	NR(T)C MH (11)	NR(T)C TI&E MH (12)	Labor rate $/MH (13)	Program NRC $ (14)	Program cost $ (15)	Des. qty. (16)	Cost/ machine (17)
Totals																	

Abbreviations: MH = man hours; PT = part; LCF = Learning curve factor; LR = Labor rate; RC = recurring costs; L = labor; M = material; m = machine; TI&E = test, inspection and evaluation; PRC = program recurring costs; NR(T)C = non-recurring (tooling) costs; PNRC = program non-recurring costs; DQ = design quantity

failure criterion of *Tsai* and *Wu* [3] was also applied here. The result is shown in Fig. 8 [6].

Composite Plate Analysis with Airy Functions Interestingly, the analysis with the anisotropic plate theory of Airy developed for wood can also be applied to advanced thermoplastic composites. As the amount of mathematical work needed for development of the stress distribution formulas is considerable, its application is only cost effective for simple problems such as bolted joints or cutouts. On the other hand, it should be noted that a small computer can be used for the solution of these relatively simple problems with Airy functions.

The details of this method are described by *Sawin* [7]. Its applicability to bolted joints in advanced composites was shown by *Müller* [8].

5.7 Lay-Up Sequences

Of course lay-up sequences will be dictated by the mechanical requirements for the part. Because of the high processing temperature, however, care must also be taken to ensure that the lay-up sequence of differently oriented layers does not cause warpage. This point can be controlled by using the appropriate theory shown by *Tsai* [3].

5.8 Joints

For the efficiency of advanced composite structures, the design of joints is of extreme importance. In order to exploit the potential of advanced composites fully, one must first locate the joints and design them within their geometrical limits. Only then can the rest of the structure be designed. The design of joints in composites includes geometry, fiber orientation, lay up and, of course, economic manufacture. Advanced thermoplastic composites have more joining possibilities than thermosets, as they can be welded.

As bonded and welded joints will be described in chapter 11, this chapter will deal mainly with bolted joints.

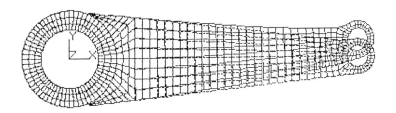

Fig. 10.8 Finite Element analysis for CFRP picking lever: deformation due to design load.

5.8.1 Bolted Joints

In contrast to bonded or welded joints, the strength of laminates cannot be fully exploited with bolted joints. With single bolted joints an efficiency of only 1/3 of the laminate strength can be reached. The efficiency of single bolted joints can be increased by multiple bolted joints but still only to 1/2 of the laminate strength.

Despite this fact, bolted joints are used because they can be disconnected, and allow easy assembly and repair of subcomponents.

Fiber Orientation for Bolted Joints When selecting fiber patterns for bolted joints, as a general rule, it is advisable to remain within the shaded area of Fig. 9 which was established for advanced thermoset composites [9].

Within the shaded area of fiber orientations i.e. for nearly isotropic fiber orientations the strength of bolted joints does not vary greatly also for modern thermoplatsic composites.

However, with sophisticated calculation methods and practical tests one can select within limits fiber orientations outside the shaded area i.e. fiber orientations resulting in composites with an elevated degree of anisotropy. One must bear in mind, however, that the strength of bolted joints in highly anisotropic composites is relatively low.

The most efficient fiber orientation for bolted joints is quasi-isotropic with equal contributions from $0/90/45/-45°$ layers. On the other hand, this fiber orientation is most often the least efficient one in areas without stress concentrations. In view of

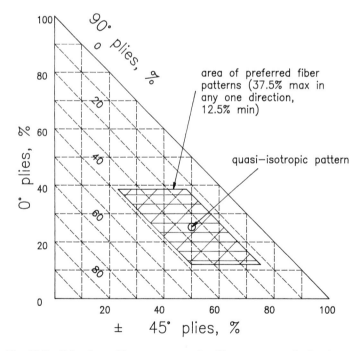

Fig. 10.9 Selection of lay-up pattern for fibrous composite laminates.

Fig. 10.10 Local reinforcements for bolted joints in advanced composites [8].

this evidence, it is fair to say that the selection of the optimum fiber orientation for joint and structure with constant thicknesses is often a compromise.

It is of course possible to strengthen the joint areas locally with appropriate means e.g. addition of quasi-isotropic layers or other configurations shown in Fig. 10 [8]. The selection of these additional reinforcements depends upon the load, its size and direction, as well as the basic fiber orientation. Care must be taken, however, to ensure that the weight of these reinforcements does not exceed the weight saved by the use of composites.

Dimensions of Bolted Joints Bolted joints in advanced composites have optimum dimensions at which theoretical bearing failure occurs simultaneously with tensile failure in the narrow cross sections immediately adjacent to the hole. Increasing these dimensions will not increase the strength of the joint but only add weight.

Practical test results on bolted joints without lateral constraints (pin loaded holes) and with a bolt diameter of 13 mm (Fig. 11) show the optimum dimensions for some laminates with fiber orientations $0/90/45/-45°$ (Fig. 12a, 12b).

Even though the 2222[1] shows the highest bearing strength ($\sigma_{bs} = 330$ MPa) of the illustrated fiber orientations, it can be seen that the bearing strength of the other laminates is not significantly lower ($\sigma_{bs} = 300$ MPa). It is expected that the bearing strength of bolts with lower diameters is higher than the one shown above for D = 13 mm as there is significantly more stress relief for the smaller bolt diameters.

In view of the comparison of theoretical and practical test results, it seems that for this relatively large bolt diameter little stress relief occurs at maximum load and

---

[1] Quantities of layers with fiber orientations $[0/90/45/-45°]$ in 8 layer packages.

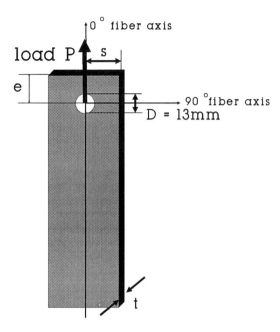

Fig. 10.11 Dimensions of bolted joints: definitions.

linearly elastic predictions apply as in the case of advanced thermoset composites. Increases in the bearing strength of these joints can be achieved by lateral restrictions on the top surfaces which impede deformation perpendicular to the laminate plane and limit material deflection. These lateral restrictions can be provided for by the nut of a screw-bolted joint. According to *Walsh et al.* [10], for diameters of 6.25 mm a bearing strength of 1000 MPa was achieved with a bolt torque of 3.34 Nm in advanced thermoset composites. It must be noted that the control and invariance of this bolt torque will be difficult to achieve in service. It is suggested by *Hart-Smith* [9] that design values for finger-tight bolted joints only be used. In the absence of true comparative values, one can assume these values to be about double those of the pin loaded holes.

There is a marked difference between the behavior of bolted joints in advanced thermoplastic [PEEK-CF] and in thermoset composites *after* maximum load. As shown in Fig. 13 [1] maximum load failures first occurred which could be visibly seen. After more loading this visible failure increased and resulted in slow degradation (Fig. 14). But only when a very high damage level was reached did final failure occur at a low load level. The progress of the bolt through the laminate up to that point was gradual, which is in contrast with the frequently catastrophic failure behavior of advanced thermoset composites. The same failure behavior could be noted in dynamic loading conditions of the picking lever shown in Fig. 2.

This evidence is further supported by *Walsh et al.* [10], who have found a superior behavior of bolted joints in PEEK compared to epoxy composites, which expressed

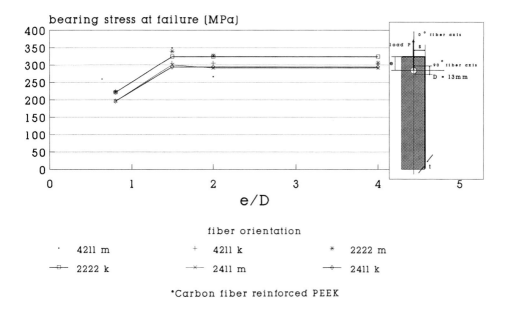

Fig. 10.12a Load bearing strength as a function of dimension e/D for advanced thermoplastic composites.

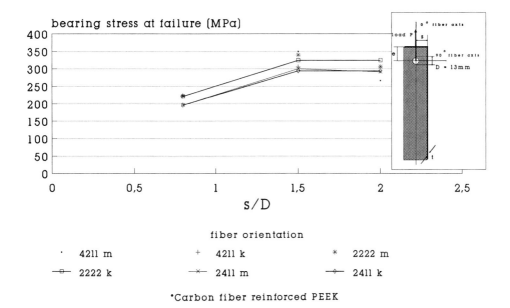

Fig. 10.12b Load bearing strength as a function of dimension s/D for advanced thermoplastic composites.

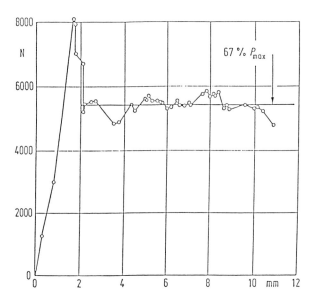

Fig. 10.13 Typical load deformation curve for bolted joints in PEEK-CF with multidirectional fiber orientation.

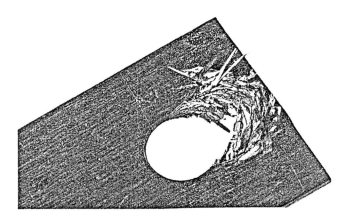

Fig. 10.14 Bearing failure in a bolted joint of a PEEK-CF specimen (see Ref. [1]).

itself not only in better failure characteristics, but also higher bearing strength by about 10% for quasi-isotropic laminates.

Theoretical calculations for bolted joints can be carried out also for cases where one is limited to lower than optimum dimensions with either airy functions or finite element methods. This is possible for bolted joints with different dimensions and fiber orientations. Strengths determined with both methods and evaluated against the failure criterion of *Tsai, Wu* showed good agreement with practical test results.

Simple formulae [9] exist for the preliminary design of bolted joints in *advanced thermoset composites* which fail in tension i.e. with dimensions below the optimum dimensions shown above and nearly isotropic fiber orientations. There, the stress concentration factor K_{tc} valid up to the linearly elastic limits of these materials is determined in the net area adjacent to the hole:

$$K_{tc} = w/D + 1.5 - 1.5*\{w/D - 1\}/\{w/D + 1\}$$

For bolted joints with bolt diameters $D \leq 10$ mm loaded above the linearly elastic limit, stress relief can be expected which is expressed by

$$C \approx (\% \ 0° \text{ plies})/100$$

The effective stress concentration factor for bolted joints in thermoset advanced composites when stress relief occurs then is

$$K_{tC} = 1 + C(K_{te} - 1)$$

However these formulae do not show good agreement with experimental results for advanced thermoplastic composites. This evidence can be explained by the different strength values of the two material groups in compression and tension as well as the different failure behavior.

Until a similarly simple formula is developed for bolted joints in advanced thermoplastic composites, one must either use the theoretical methods mentioned above or run practical tests.

5.8.2 Bonded Joints

Bonded or welded joints in advanced thermoplastic composites are more efficient than bolted joints, as the strength of the bond can, and must, exceed that of the laminate. They do however have the disadvantage of difficult repair and disconnection. Hence for practical purposes one must as in the case of bonded joints in metals or thermoset composites, consider them as non-disconnectable.

Bonding will be discussed in chapter 11 so the details of the potential processes will not be discussed here except for the main points and those concerning applications in the machine industry.

The main possible processes for bonded and welded joints are among others:

- bonding:
  - with thermosetting adhesives
  - aromatic or diffusion bonding

- welding or fusion bonding
  - hot plate welding
  - resistance heating
  - induction heating
  - dielectric/micorowave heating
  - infrared/laser heating
  - ultrasonic welding.

In the bonding processes the temperature in the contact areas is below the melting point of the laminate matrices. This has the advantage that the adherends are not exposed to the danger of thermal deformation due to the bonding process. However in the case of bonding with thermosetting adhesives the surface preparation is quite demanding and must be done precisely in order to reach a satisfactory level of joint strength. Also in the case of most adhesives applicable to modern thermoplastics lengthy bonding processes must be carried out.

Diffusion bonding[1] in the case of PEEK involves interleaves of PEI film which diffuse into the PEEK surfaces under pressure at temperatures of about 295°C i.e. well below the melting temperature of PEEK (343°C). Both processes introduce the hazard of low chemical resistance into the bond. It is possible that the strength of the bond under chemical attack becomes weaker than the adherends which in the case of PEEK are quite stable.

This signifies that the resistance of the bond against chemical attack likely to be encountered in service must especially be proven before commitment to that joining technique.

Joining by welding does not pose this problem as the bond has the same chemical resistance as the laminate matrix. But one must at least locally, i.e. in the bonding and contact area of the adherends, exceed the melting temperature of the matrix. This poses the problem of possible thermal deformation due to the melting process. A careful balance between pressure and applied heat energy must be reached in order to minimize this effect.

Of the fusion bonding processes evaluated by *Benatar* and *Gutowski* [11], resistance and induction heating and ultrasonic welding offer the best potential for fusion bonding of advanced thermoplastic composites. They bond the adherends quickly, achieve good bond strength above 30 N/mm^2 and do not distort the adherends.

---

[1] Thermabond process patented by ICI.

6 Manufacturing

The manufacturing techniques for thermoplastic composite parts in the machine industry must exploit the rapid processing potential of these materials in order to achieve the desired cost-effectiveness and flexibility. Manufacturing techniques are described in chapter 9. Therefore in this chapter only the main points of the different manufacturing techniques, with practical experiences and the necessary tooling and machining techniques, will be described.

6.1 Manufacturing Techniques and Tooling Requirements

6.1.1 Matched Die Molding

The matched die molding process is the most attractive for the machine industry as most of their trusted suppliers in the plastic industry use presses for other processes, such as SMC molding. Also, relatively precise parts can be made with this process. It does have the disadvantage of being a slow process if proper precautions are not taken. Until more development work is done with more elaborate tooling, this process is limited to relatively simple, flat or slightly curved structures.

It is possible to achieve rapid process times with *two presses*, shown in Fig. 15. This has the disadvantage of blocking two presses for essentially one process which can result in cost and time problems. The tool and the material to be formed are heated up in one press above the melting point of the matrix to the optimum processing temperature. The part is consolidated in the second press. The typical cycle is shown in Fig. 16.

A dwell time should be observed in order to ensure that the material has reached the required temperature and it is constant throughout the part. Because of the high processing temperature, this point is important because uneven temperature distribution along with uneven cool-down rates will cause part warpage and mechanical problems. The dwell time depends upon the thickness of the part. For thin parts in the range of 2 mm thickness it generally is five minutes. Thicker parts must have a correspondingly longer dwell time in order to assure complete through heating. In view of the economy of the process, optimum consolidation, and possible forming and bonding cycles, the dwell time should not exceed 15 minutes.

In order to achieve rapid cooling after the dwell time the tool is transferred to the second press with a platen temperature of 200°C where the part is consolidated and cooled down to 200°C with a pressure of 20 bar or more. This can take 5 to 10 minutes if the tool is not too massive. According to *Cantwell et al.* [12] the average cool down rate should not exceed 50°C/min in order to avoid microcracking. On the other hand, cool down should not be slower than 5°/min as then the fracture toughness of PEEK is reduced due to excessive crystallinity. After the part has reached about 200°C it can be demolded.

When using only *one press* [1] the part can be heated outside the press by other means than heated press platens such as ovens, microwaves, infrared energy etc. in an ancillary tool, which has to have either good thermal conductivity or permeability

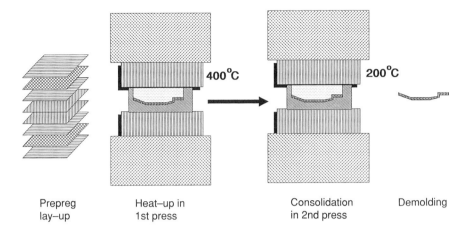

Fig. 10.15 Matched die molding of parts made with modern thermoplastic composites: procedure with two presses according to ICI.

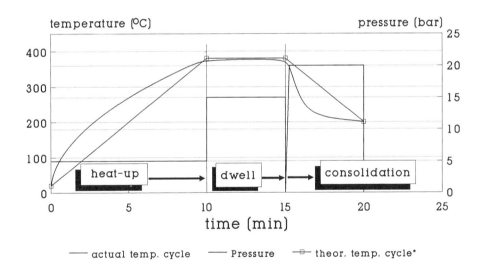

laminate 2mm thick, ICI recommendation,
average max. cool down rate per Cantwell

Fig. 10.16 Typical processing cycle of PEEK-CF in matched die molding technique with two presses.

(Fig. 17). Light pressure can be applied by simple means. When the entire material has reached the processing temperature and after the dwell step the ancillary tool will be transferred into a consolidation tool in the press. This processing cycle (Fig. 18) is similar to the one shown in Fig. 14. It has a relatively long heat-up period due to the greater thickness of 10 mm of the part made with it. The temperature of the consolidation tool is again 200°C in order to reach optimum properties especially with respect to impact and chemical resistance.

The tool requirements for both methods are:

- heat resistance up to 400°C
- good thermal conductivity or permeability
- thermal expansion relatively compatible to that of the part.

The heat resistance of the tools can be such that they have good durability for a large number of cycles. Steel tools will meet this requirement. However in order to have good heat transfer they must be made as light as possible, otherwise a long processing cycle results. For complex parts, steel tools can be quite expensive due to extensive machining. It is important that the difference in thermal expansion between the part and tool be considered in order to avoid inaccuracies or warpage.

It is of course possible to use throw-away tooling made of e.g. organic materials (silicone, polyimides) which have the potential of being shaped to the part contour. One must carefully evaluate the cost of this approach when comparing it with more durable tooling since these materials usually are also costly and, depending on the part, there may be an expensive shaping process involved before they can applied in form of a tool or tooling aid.

The ideal manufacturing process, of course, would be one where only the laminate is heated-up and the tool only consumes a minimum amount of energy in the process.

The energy consumption of a tool can also be held low by good thermal conductivity when using conductive heat-up processes. This can be achieved with light tools made of materials with good thermal conductivity or appropriate processes described above where only the tool is used for the consolidation step.

In other processes where energy is transferred into the laminate and converted into heat the tools must have good energy permeability.

6.1.2 Diaphragma Forming

The lay-up of modern thermoplastic composite prepreg layers is somewhat difficult as a result of the absence of tackiness and very limited drapability. Conventional autoclave lay-up work for parts which are not flat or slightly curved is therefore uneconomic.

A process which allows the structural part to be formed from flat lay-ups is diaphragm forming. The process is described in [13] and [4] and is shown in Fig. 19.

A flat unconsolidated lay-up is placed between two plastically deformable diaphragms. The lay-up is then heated up in an autoclave to the optimum processing temperature, and simultaneously consolidated and formed against a one sided tool

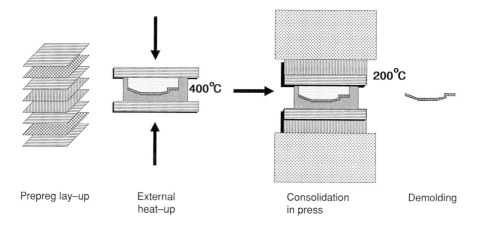

Fig. 10.17 Matched die molding of parts made with modern thermoplastic composites: procedure with external heatup and one press.

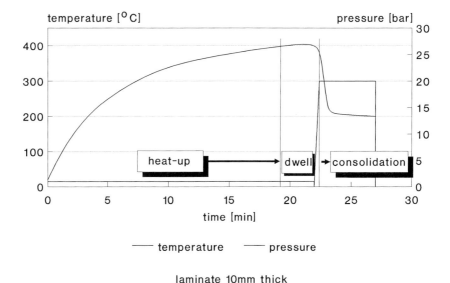

Fig. 10.18 Typical processing cycle of PEEK-CF in matched die process with one press: external heat-up and consolidation in press.

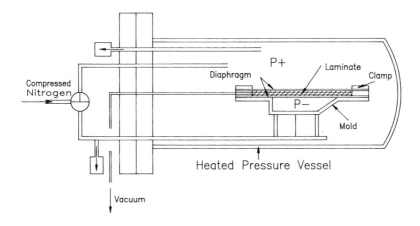

Fig. 10.19 Heated diaphragm forming inside an autoclave.

by the action of gas and or mechanical pressure in an autoclave. Because of the high forming temperature associated with modern thermoplastic composites, it is strongly advisable to use a nitrogen atmosphere in the pressure vessel as the fire hazard in pressurized air at 400°C is very high [13]. Cycle times of 30 to 60 minutes have been achieved with this process using the equipment shown above. The ancillary material consumption, as well as tool weight and costs are low. Due to the relatively low pressure which can be applied, the limited strength of the diaphragm and the stiffness of the material to be formed, only relatively thin parts – typical thickness is 2 mm – can be processed. Also, it is difficult to produce parts with variable thicknesses and complex structures. Subcomponent production and subsequent bonding can overcome this problem however.

The typical cycle of 30 to 60 minutes for an autoclave does not fully exploit the rapid processing potential as pressure vessel, tool and part are heated up and cooled. In order to achieve rapid cycles with this process, local heating and cooling of tool and part should be envisaged which could result in cost effective processing times of 10 to 20 minutes. An apparatus of this type would be a pressclave in which only the tool and its contents would be heated up and cooled down.

6.1.3 Pultrusion

Pultrusion of different shapes with modern advanced thermoplastics is a process which allows the cost effective production of parts with constant cross sections over their length. The difficulty lies in the high processing temperature and the high melt viscosity of the matrix for which appropriate tooling must be used.

Since the material suppliers refuse to supply the matrix resin, prepreg must be used. In consequence one processing step is added and one of the major advantages of the pultrusion process i.e. the combination of resin and fiber in its original form to make a structure is eliminated, thereby reducing the economic advantage of the process.

However, because of the availability of off-angle prepreg tape or braid the process lends itself to the production of subcomponents with constant cross sections such as stringers which can be bonded to panels with the described processes.

6.1.4 Filament Winding and Tape Laying

Filament winding also faces the same difficulties as pultrusion because of the nonexistent tack and low drapability of the prepreg as well as the unavailability of fiber and resin in its basic forms.

Possibilities of filament winding or tape laying with modern thermoplastic composites are shown by *Harvey* [4]. The process involves local application of heat, cool down and consolidation with a roller Fig. 20. The local application of heat must be well controlled to avoid exceeding the processing window of PEEK. Also a rapid cool down must be done under relatively high pressure in order to have satisfactory processing rates. This condition requires elaborate tape rolling and consolidation equipment so that the cost effectiveness of this process must be carefully examined. On the other hand this process allows the production of large structures.

6.1.5 Ancillary Materials

As a result of the high processing temperature and pressure involved in the manufacture of parts made with modern thermoplastic composites, ancillary materials must meet high standards in order not to compromise the production and its equipment. It is strongly suggested by the author that this aspect be carefully considered otherwise one might experience unpleasant surprises. It must be proven with certainty that the ancillary materials, which are relatively inexpensive in comparison to the part

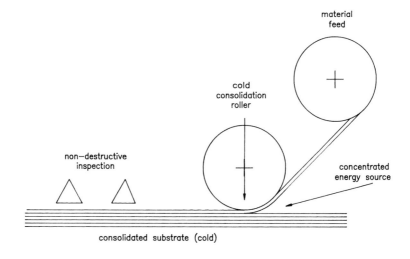

Fig. 10.20 Automated tape laying process with modern thermoplastic composites [4].

and its tooling, will sustain the processing conditions before their application and will not create fire, health or manufacturing hazards.

For processes which involve forming under vacuum, special polyimide films until now must be used as bagging materials. These materials are relatively stiff and brittle so that extreme shaping is difficult to achieve, and then only with major material and labor outlays. To this is added the problem of sealant tapes with relatively low tack. As release agents, usually those used for 180°C epoxy curing can be used.

6.1.6 Manufacturing Tolerances

When manufacturing composite structures especially for the mechanical industry the problem of dimensional tolerances will have to be faced. Prepreg thickness tolerances of modern thermoplastic composites can vary as much as ±10% between production lots so the problem can be difficult to solve for thick structures when close tolerances are required. Also, as a result of the high processing temperature, the thermal expansion behavior of complex structures will be difficult to predict and affect tolerances.

The question of tolerances must therefore be discussed from the start of the design process so that it can be followed during the detail design sequence.

There are several solutions to the problem:

- widen tolerance requirements of structure
- adjust layer quantity during manufacture according to thickness of supplied prepreg
- use process variations
- machine critical areas.

Widening the tolerances is the easiest solution, of course, but often impractical because of not only dimensional but also strength or stiffness requirements. It is important, though, that tolerances within acceptable limits are wide enough, in order to avoid a significant increase in manufacturing costs.

The adjustment of the layer quantity in order to reach tolerances will only affect thickness tolerances. Also it can be very demanding on logistics and require qualified personnel making decisions during manufacture.

Process variations such as pressure due to the high viscosity of the thermoplastic matrix will only slightly affect dimensional tolerances.

Machining, is of course, an added step in the manufacturing process but sometimes the only solution for close tolerance requirements. In order to keep costs low and maintain functionality of the structure, one should minimize machining as a finishing step by applying other means mentioned above.

6.2 Cutting and Machining

Cutting and machining of advanced thermoplastic composites can be done relatively easily. Most metal machining processes can be used for carbon fiber reinforced PEEK. Difficulties are caused by the toughness of the matrix, which slows down machining and cutting progress. For drilling and machining, hard metal or diamond tipped tools can be used which must be sharp. Cutting speeds must be high but feed rates must be low in order not to damage the part. Adequate cooling is recommended for the same reason.

Good results have been obtained with waterjet cutting for parts made of advanced thermoplastic composites. Parts with thicknesses above 10 mm have been cut without problems. In order to achieve good cuts, abrasives must be added to the waterjet. The start of the cut must be done with care so that delamination does not occur. For precision, the trailback of the water jet must be considered especially for curvature changes in the cuts.

7 Quality Assurance

The aspect of quality assurance of course is also very important in the manufacturing process of parts made with advanced thermoplastic composites. The following aspects need to be verified:

- quality of the supplied prepreg with regards to physical properties
  - fiber alignment
  - fiber/resin content, areal weight
  - prepreg thickness
  - resin impregnation

  chemical properties
  - processing characteristics
  - consistency of chemical formula
  - crystallinity

  mechanical properties
  - strength data
  - stiffness data
- quality of the part
  - strength, stiffness
  - fiber orientation
  - lay-up
  - consolidation
  - processing conditions.
  - dimensions.

The test methods used for advanced thermoset composites can also be used for their thermoplastic counterparts. Most of the points outlined above are already clear from advanced thermoset composites. Some of them, are, however, of special importance for advanced thermoplastic composites.

Because of the tolerance requirements mentioned above it is especially important to know the thickness of the prepreg when pressed.

In addition, in order to check the processing conditions of the part or prepreg, the crystallinity should be examined as it affects the impact strength of the part. This property can be determined with differential scanning calorimetry (DSC).

For thick parts it is especially important that the correct lay-up sequence be followed due to the high processing temperature. Small errors may be forgiven for strength reasons but may manifest themselves in unnecessary warpage.

8 Cost Effectiveness

As stated in section 1, the application of modern thermoplastic composites can be cost effective for highly loaded structures in high performance machines. If made of metal, these structures are either machined with high costs out of special steels or made of high grade metals such as titanium. The costs of machining can be kept quite low in the manufacture of composite structures so in some cases in the machine industry, composites can compete with steel structures even in costs. The material and processing costs of titanium are also high, so that here again composites can be competitive in costs.

It was proven that the picking lever shown in Fig. 2 could be produced for about the same costs as the competitive lever made of titanium even though here structures were in competition and not structural systems. One of course must also weigh the advantages of the application of composites i.e. its performance or costs in service. In the case of the picking lever, the inertia mass moment around its rotation axis was 25% lower than that of the titanium lever, allowing either less energy consumption or higher weave rates of the textile machine.

9 Service Experiences

It is also very important for the machine industry to have a good service history for the parts it employs, because usually relatively large numbers of structures will be made and maintenance or replacement of unsatisfactory parts can be very costly. The machine industry therefore will first evaluate these structures in tests, then in limited series in service, possibly in machines of selected customers in order to gain that experience. This development is relatively slow but needs to be undertaken in order to keep the cost effects low for structural changes and their introduction onto the market.

As an example a prototype of the picking lever shown in Fig. 2, was first tested in a pulser at triple design load. After one lever reached 40 million cycles without damage, another one was evaluated in a textile machine for about 100 million cycles and again tested in the pulser for another 40 million cycles at triple design load. In service it showed not only energy advantages and sufficient strength and stiffness, but also good resistance against wear, lubricants and chemicals.

One other positive aspect demonstrated by one of the prototype levers evaluated at limit design load in the pulser was its controlled failure behavior (Fig. 21) [1]. This lever showed visual enlargement of the bolt hole after 28 million cycles. When the loading of the same level and high frequency was continued it did not fail catastrophically. Instead the bolt hole was enlarged progressively during a further 8 million cycles, after which the test was stopped.

After these tests with positive results it was decided to evaluate the picking lever in a limited series in production and service.

10 Future Aspects, Requirements for Materials

In summary one must state clearly that even though modern thermoplastic composites have a better applicability in the machine industry than their thermoset counterparts they are limited to the high performance sector of that industry, where they can compete with metal parts in both performance and cost. The extent of their application will increase only when the following improvements have been achieved:

- availability of more favorable intermediate materials
- improvement of processing conditions
- cost reduction of equipment and ancillary materials
- processing methods with even shorter production cycles
- reduction of material costs
- compatibility with other material types.

Up to now only thin prepreg material has been available with nominal thicknesses of about 1/8 mm. This signifies that for a thick structure many plies must be layed-up, which is very time consuming even when automatic lay-up methods are used. Thicker, woven or braided materials will reduce the preparation costs significantly. Also another aspect which could be improved upon is tack and drapability which affects lay-up time and costs.

Processing conditions of modern thermoplastics are quite favorable, the only negative being the high processing temperature. Improvement of this parameter whithout significant decrease in the good mechanical properties of these materials will not only reduce processing times and costs but also thermally induced stresses and other effects.

The equipment and ancillary material costs can also at least partly be reduced through lower processing temperatures.

In order to shorten the production cycles further, other methods should be possible where only the material is heated up for processing but not the tool in order to minimize energy consumption and speed up the processing cycle.

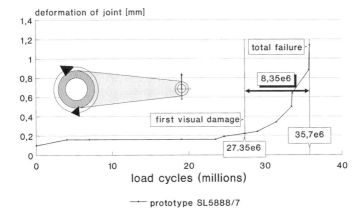

Fig. 10.21 Load deformation curve of prototype picking lever made with modern thermoplastic composites, under fatigue loading.

Reductions in material costs will of course increase the application of modern thermoplastic composites.

Improved compatibility with other materials can also lead the way to wider application. This would allow the cost effective selection of the most economic material for specific load cases in hybrid design which could be joined in situ or in steps subsequent to the manufacture of the components.

One can summarize these requirements by stating that the material to be developed has to have the positive properties of both advanced thermoset and thermoplastic composites.

At the moment, this may look like an extremely demanding request, but in view of the many material developments made so far, it certainly should not be impossible to meet.

References

1. Müller, J., Flüeler, P., Just, Ch., de Kalbermatten, Th.: Faserverbundwerkstoffe - Hochleistungswerkstoffe für den Maschinenbau, Kunststoffe 80 (1990)/9.
2. Brünings et al, Bauteile aus Kunststoff, Maschinenmarkt 64 (1981).
3. Tsai, Composites Design; 4th edition, Think Composites.
4. Harvey: Thermoplastic Matrix Processing; Composites ASTM 1987.
5. Noton: Cost Drivers in Design and Manufacture of Composite Structures; Composites, ASTM.
6. Kütükçüoglu: Finite Elemente Berechnungen am Schlaghebel aus kohlefaserverstärktem PEEK; unpublished internal report Swiss Project 1500 "Composites".
7. Sawin: Stress distributions around holes, Naukova Dumbka Press, Kiev 1968.
8. Müller, J.: Spannungsverteilung an gelochten über einen Bolzen belasteten Stäben aus kohlenstoff-faserverstärktem Kunststoff, Doctorate Thesis, ETH-Zurich, 1975.

9. Hart-Smith: Joints; Composites ASTM 1987.
10. Walsh et. al: "A Comparative Assessment of Bolted Joints in a Graphite Reinforced Thermoset vs. Thermoplastic SAMPE Quarterly July 1989.
11. Benatar, Gutowski: A review of methods for fusion bonding thermoplastic composites; SAMPE Journal 1987.
12. Cantwell, Davies, Kausch: The effect of cooling rate on deformation and fracture in IM6/PEEK composites; Composite Structures 14 (1990).
13. Mallon, O'Brádaigh, Pipes: Polymeric diaphragm forming of continuous fiber reinforced thermoplastics; 33rd International SAMPE Symposium 198.

Chapter 11

Bonding and Repair of Thermoplastic Composites

P. Davies and W.J. Cantwell

1 Introduction

The use of fibre reinforced organic matrix composite materials is now widespread, particularly in all forms of transportation. The high specific properties of these materials are the principal reason for their popularity but other advantages have also promoted their application. For example, the design of carbon fibre reinforced materials with properties tailored to support anisotropic loading is exploited in the aerospace industry while the ease of fabrication and corrosion resistance of glass reinforced polyesters are particularly attractive to boatbuilders. In virtually all examples of composite components however, one question must be answered ; How is the composite to be joined to the rest of the structure ? Thus the transfer of load from a glass/epoxy leaf spring to a steel car chassis, the joint between the composite hull and the composite superstructure of a yacht or between the composite panels and reinforcing stringers in aircraft will be the the critical areas which condition the successful adoption of fibre reinforced materials. Indeed, the importance of the subject has been recognized in several books, [e.g. 1, 2]. There are two possibilities available to the designer using "conventional" composites, i.e. composites based on thermosetting resins such as epoxies or polyesters:

(i) Mechanical fastening
(ii) Adhesive bonding (or co-curing)

The use of thermoplastic polymers as composite matrices, whether high performance (e.g. PEEK (polyetheretherketone), PPS (polyphenylene sulfide), PES (polyethersulphone)) or lower performance (e.g. polyamides, polypropylene) allows a third possibility which takes advantage of the re-meltability of thermoplastics:

(iii) Fusion bonding

P. Davies, Laboratoire des Matériaux, Institut français pour la recherche sur l'exploitation de la Mer IFREMER Centre de Brest, 29280 Plouzané, France
W.J. Cantwell, Laboratoire de Polymères de l'Ecole Polytechnique Fédérale de Lausanne (EPFL), MX-D-Ecublens - CH-1015 Lausanne

On the other hand, for most thermoplastics adhesion bonding is more difficult than for thermosetting resins so that more care is necessary in the choice of adhesive and surface treatment.

In the first and main part of this chapter, the application of these three approaches to thermoplastic composites will be discussed. The properties of some joints will then be presented and the different methods will be compared.

In the second part of the chapter the application of the different bonding methods to repair of composite structures will be discussed. Repair of aircraft structures is a particularly important area and has been the subject of a number of publications (e.g. ([3, 4]). Repair technology also features in a recent review of current US Air Force contractual programmes on thermoplastic matrix composites [5]. The application of fusion bonding techniques is still in its infancy but should be rapidly developed as more thermoplastic composites find applications in primary aircraft structures. It should also be remembered that both the automobile and boat repair industries are quite extensive. In the future, the recyclability of thermoplastics may make their increased use and hence more frequent repair an attractive proposition. Throughout the present chapter the emphasis will be on high performance thermoplastic composites as these have received most attention to date, but most of the methods treated may also be applied to composites based on lower performance thermoplastic materials.

2 Mechanical Fastening

Under this heading are included joining techniques involving principally

- Bolts
- Rivets
- Other fasteners

The use of these fasteners to assemble thermoset matrix composites is described elsewhere and the different types of joint are clearly applicable to thermoplastic composites. *Vinson* gives an overview of the literature on mechanical fasteners and presents the different failure modes [6]. He also emphasizes previous work on the importance of tightening in suppressing failure modes and increasing bearing strength. The design and analysis of bolted and riveted joints is discussed by *Hart-Smith* [7] and empirical rules to aid design are presented.

The common feature of such assemblies is the necessity to prepare a hole, so that in changing from a thermoset to a thermoplastic matrix two aspects of the behaviour of the composite are critical:

(i) Ease of machining
(ii) Holed strength of the material

The machining of composites is a delicate operation and has been described in some detail [8]. Practical details on preparation of holes in graphite/epoxy composites are also given by *Phillips* and *Parker* [9].

Two aspects which can cause problems are overheating and delamination. Firstly, even if local melting does not occur in thermoplastic matrices, changes to the matrix structure may take place around the hole. The high level of residual stresses in thermoplastic composites may also necessitate particular care if the hole tolerances specified for thermosets, typically 0.075 mm on the diameter, are to be met. Secondly, it is very easy to delaminate composites during drilling operations, so that the high delamination resistance of some of the thermoplastic composites is of direct benefit. Nevertheless, a study of the drilling of holes in carbon fibre reinforced PEEK has shown that the success or otherwise of the operation is extremely sensitive to the method used [10].

Once the composite structure has been assembled, the thermoplastic matrix may not offer a significant improvement in strength over a thermoset if stress concentrations at the holes are not relieved by delamination. Thus, paradoxically, the higher delamination resistance of a carbon/PEEK composite compared to that of a carbon/epoxy may then be detrimental unless plastic deformation can take place to allow a redistribution of stresses [11, 12].

Although widely used, a further disadvantage of mechanical fasteners is that they increase weight. For this and the reasons discussed above adhesive bonding is an attractive alternative.

3 Adhesive Bonding

3.1 Historical

Adhesive bonding is a vast subject and has been reviewed recently by *Kinloch* [13]. In general it has been found that unreinforced thermoplastics are more difficult to bond adhesively than thermosets, and much research has been devoted to surface treatments to improve bond strengths. Many publications have described treatments of polyolefins over the last 30 years, but there is still controversy over the mechanisms by which bond strength can be improved in these materials. In addition, a number of quite effective surface treatments for polyolefins are based on mixtures such as potassium dichromate/sulphuric acid/water solutions, which are both carcinogenic and difficult to dispose of without pollution [14], so that this subject is far from exhausted as new treatments are sought. Far fewer studies have treated the high performance thermoplastics such as PEEK and PPS and so it is not surprising that neither adhesives nor surface treatments have been optimised for these materials.

3.2 Adhesive Selection

The choice of structural adhesives available is quite extensive and as, understandably, suppliers are reluctant to give details of formulations objective generalizations about the most appropriate adhesives for different materials and applications are difficult. Most of the published work on thermoplastic bonding has been performed using a small number of modified epoxy film adhesives and this should be borne in mind for the discussion on surface treatments which follows.

3.3 Surface Treatments for Thermoplastic Composites

The influence of surface treatment on the lap shear strength of injection moulded unreinforced PEEK/PEEK joints has been discussed in detail by *Hamdan* and *Evans* [15]. These authors obtained values of 2.3 MPa for abraded and degreased surfaces. This was increased to 4.6 MPa for joints with grit blasted surfaces, (although one of these failed outside the joint), while joints with surfaces which were chromate etched all failed outside the joint area.

The following preparation techniques have been applied to the bonding of joints involving two similar high performance thermoplastic composites:

Mechanical abrasion
Peel plies
Corona discharge
Plasma and flame treatment
Chemical etching

3.3.1 Mechanical Abrasion and Peel Plies

Mechanical abrasion, either by hand or using grit blasting, in conjunction with a degreasing treatment, is adequate to achieve good bond strengths with carbon/epoxy composites [16].

Many authors have therefore used this as their baseline treatment in adhesion studies but this is a rather simplified approach. In fact abrasion treatments can vary considerably, Fig. 1, and this may in part explain the rather contradictory results published in the literature. An example is given in Table 1, which tabulates results from various sources for simple lap shear strengths measured on carbon fibre/PEEK joints asembled after grit blast treatments, using the same adhesive. Even accounting for the unsatisfactory nature of the lap shear specimen as a test of bond performance these results are surprising.

The most surprising set of results is that of *Powers* and *Trzaskos* [19], which suggest that a simple mechanical treatment is sufficient to obtain high lap shear strengths for carbon/PEEK composite joints. As will be described below, given the amount of effort that has been expended in developing other more complicated surface treatments this is a most interesting conclusion. The adhesive used by these

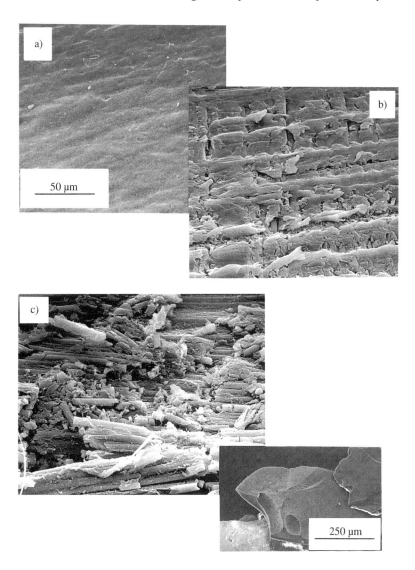

Fig. 11.1 Scanning electron micrographs at same scale of carbon/PEEK surface showing a) as moulded, b) lightly abraded and c) sand-blasted (sharp particles, see insert).

authors was a standard commercial modified epoxy, and the treatment was a grit blast (80 grit aluminium oxide) followed by a solvent wipe (methyl ethyl ketone (MEK)). This gave a mean lap shear strength of over 30 MPa. In contrast to these results, the mechanical abrasion followed by MEK wipe used by *Wu et al.* [17] resulted in bond strengths of less than 5 MPa for the same adhesive film. *Kodokian* and *Kinloch* also measured very low strengths after an abrasive treatment of carbon/PEEK [20]. This therefore suggests that the action of grit blasting is critical and that

Table 11.1 Influence of Mechanical Pretreatments on Lap Shear Strengths of Carbon/ PEEK Joints, all Bonded with the Same 177°C Curing Modified Epoxy Film Adhesive

Treatment	Lap shear strength (MPa)	Reference
Abrasive pad, MEK wipe	4.4	*Wu et al.* [17]
Abrasive pad, "bon ami" cleaner, deionized water cleaning	13.6	
50 mesh alumina grit, deionized water cleaning	20.9	*Silverman* and *Griese* [18]
80 mesh alumina grit, MEK wipe	30.4	*Powers* and *Trzaskos* [19]

the severity of such a treatment (Fig. 1) is necessary to remove a surface layer. The roughening of the surface may also be beneficial.

This is confirmed in a later paper by *Wu*, who obtained high strengths by the use of both release film and Kevlar peel plies during processing [21]. These protected the surfaces from release agent contamination and were removed just before bonding. The same adhesive as that used in the previous studies then yielded lap shear strengths of 28 MPa.

Less data are available on mechanical preparations for carbon reinforced PPS composites. *Krone et al.* showed that the addition of a 50 mesh grit blasting treatment to an MEK solvent wipe resulted in a doubling of the lap shear strength, but values were still low, around 15–20 MPa according to the adhesive used [22]. *Powers* and *Trzaskos* also found values around 20 MPa for PPS composites [19].

Appropriate mechanical treatment which removes release agent and roughens the surfaces to be bonded can therefore significantly improve joint lap shear strengths. The optimization of such treatments is not simple however. When carbon/PEEK joints were prepared using a different adhesive and the sand blast treatment shown in Fig. 1b followed by an alcohol wipe, lap shear strengths of less than 10 MPa were measured [23] only slightly higher than those measured on specimens which had only been cleaned with an alcohol wipe. It may be that weak surface layers are generated by excessively violent sand blasting, but the uncertainty of such methods has led to considerable research into other surface preparations.

Of the methods aiming to increase the reactivity of thermoplastics, which generally show low surface energies, corona discharge treatments have attracted considerable interest. The activation of polymer surfaces before printing is a well-developed industrial process so that equipment allowing a known amount of energy to be applied to a surface is readily available. Typical commercial corona stations include a high frequency generator, a transformer to produce high tension and stainless steel electrodes, either blades or wire, to produce the discharge. The film passes on backing rollers and widths up to 8 metres can be treated continuously [24]. The application of this treatment to carbon reinforced PEEK and PPS has been examined in detail by *Kodokian* [25], although results from other studies have also been published for PEEK [21] and PPS [22]. These studies have shown that corona discharge treatments are extremely effective in increasing bond strengths. Failures at the composite surface could be avoided and the failure locus was moved to the adhesive or the composite above a certain level of treatment [26]. This was attributed to an increase in wettability of the composite surface although chemical modification of the surface was also noted as will be described later.

In parallel with studies of corona discharge treatment, a number of authors have used plasma treatments to improve joint strengths. Plasma modification of surfaces is more a laboratory technique than corona discharge, requiring a treatment chamber and generally being limited to small components, but it allows different activated gases to be used. It has been employed in the treatment of a large number of polymers and generally bond strengths can be rapidly increased for short plasma treatment times [27]. The effects of plasma treatments on polymers are reviewed in several publications [e.g. 28, 29] and results showing the effect of oxygen plasma treatment time on lap shear strength of carbon/PEEK and carbon/PPS joints are shown in Fig. 2.

It is apparent that after a treatment of a few minutes the strength of the carbon/PEEK joint reaches the maximum possible with this adhesive, while the PPS composites failed at the adhesive/composite interface in all specimens even though improved strengths were obtained after longer etching times.

The plasma treatments in Fig. 2 produced an etching effect on the surfaces, as shown in Fig. 3, so that both roughening and chemical activation must be considered in determining the critical parameters for such treatments, as will be discussed below. The damage created on the composite surfaces by these direct plasma treatments is relatively superficial and does not significantly affect short term tensile or compressive properties of the material [21]. However it may be that the surfaces thus created are not stable with time and a preferable means of treating the surfaces would be in a post discharge chamber, such as that described elsewhere for treatment of polypropylene [30] in which the surface treatment only affects a layer of less than 100 Angstroms.

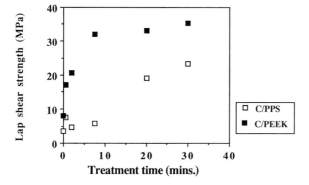

Fig. 11.2 Effect of oxygen plasma treatment time on lap shear strength of carbon/PEEK and carbon/PPS composite/composite joints, same modified epoxy adhesive.

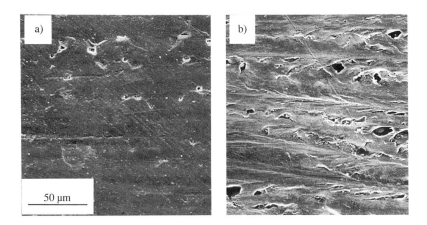

Fig. 11.3 Oxygen plasma etched surfaces carbon/PEEK, same scale. a) 30 s, b) 7.5 mins.

An additional consideration is the stability of these plasma treatments. This has been studied for carbon/PEEK by *Wu* [21], *Kempe et al.* [31], and *Davies et al.* [23]. The results of the latter are summarized in Fig. 4.

The results in Fig. 4 from *Davies et al.* are for specimens plasma treated for 7.5 minutes then stored in air or in a dessicator while their reference specimens were bonded after 4 hours in a dessicator. For specimens assembled after 4 hours in air, mean strengths dropped by over 50%, while after 6 days in a dessicator the values were 30% lower than those of the reference [23]. *Kempe et al.* left tensile shear specimens in air for 72 hours and found values only 70% of those for joints assembled immediately [31]. *Wu* overcame the problem by storing specimens in a neutral paper for up to 15 months before bonding [21]. In this case no drop in lap shear strength was measured in specimens bonded after 6 months while after 15 months values were reduced by 15%. Care is therefore necessary in the storage of plasma treated specimens before bonding.

Flame treatment of composites has not been extensively studied but showed promising results for carbon/PPS on grit blast surfaces [22]. For this material flame treating in a methane burner resulted in an increase of up to 40% compared to strengths measured for the grit blast surfaces.

3.3.2 Chemical Etching

Some of the highest values reported for all the surface treatments have been obtained after acid etching [17, 19]. Chromic-sulfuric acids have been used and can very rapidly eliminate failures at the composite surface, Fig. 5.

However, such treatments are quite aggressive and can remove a surface layer as shown in Fig. 6. This and the hazards in disposal of such products make them less than desirable, and in the near future legislation will no longer allow their use, so the development of alternative treatments is essential.

11. Bonding and Repair of Thermoplastic Composites   345

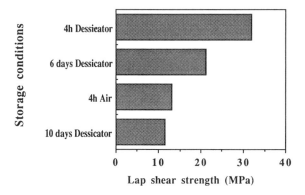

Fig. 11.4 Stability of plasma treatments of carbon/PEEK. Influence of storage conditions.

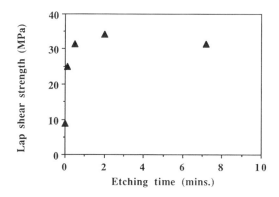

Fig. 11.5 Influence of chromic/sulphuric acid etching time on lap shear strength, carbon/PEEK.

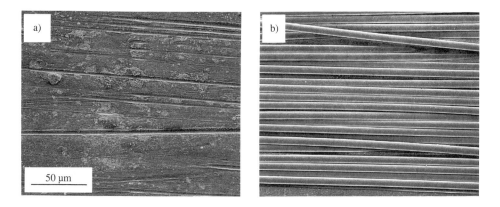

Fig. 11.6 Surfaces of carbon/PEEK specimens after chromic acid etching for a) 5 s and b) 7.5 minutes. Both at same scale, untreated surface in Fig. 11.1.

3.4 Bonding Mechanisms and Surface Characterization

The studies of *Evans et al.* on the adhesion of unreinforced PEEK have included some detailed surface characterization [32], which give an insight into the bonding mechanisms. Both oxygen plasma and chromate etching resulted in high polar components of surface energy, enhanced surface oxygen concentration (measured by XPS, X-ray Photoelectron Spectroscopy) and high joint strengths.

Subsequent correlations of composite joint strengths with surface characteristics have also indicated the high oxygen content of corona discharge [22, 26], acid etched [21] and plasma treated [21, 23] surfaces. At first sight it would therefore appear that this is a requirement for improved joint strength. However, the work by *Wu* on peel plies indicated a low surface oxygen concentration, similar to that of mechanically treated surfaces, but very high strengths [21], and the grit blast surfaces described by *Powers* and *Trzaskos* [19] were presumably similar. In addition, *Davies et al.* showed that leaving plasma treated specimens in air for four hours before bonding caused only small drops in oxygen concentration but dramatic decreases in bond strength [23]. Thus surface oxygen concentration cannot be used as a criterion to ensure good joint strength.

In conclusion, good joints can be obtained with epoxy-based adhesives by both suitable mechanical and discharge treatments but the detailed requirements for avoiding adhesive failure at the composite surface are still not clear.

3.5 Thermoplastic Adhesives

An original approach to bonding of high performance composites is to use a film of a different thermoplastic as an adhesive. This has been developed by ICI under the trade name "Thermabond", in which PEI film, typically 100 microns thick, is included in the joint. Fig. 7 shows a joint prepared using this technique.

This technique is described in reference [33]. It has been tested by several authors [17, 18, 34, 35] and applied to the bonding of stiffeners to large panels [33, 36]. Films of other polymers have also been used: PES did not show good miscibility with PEEK for example, but is effective in bonding PPS composites. When diffusion of the interlayer into the composite adherends occurs good bond strengths can be achieved.

Although this is included as an adhesive technique the addition of films of PEI can also be particularly beneficial to the fusion bonding methods described below, as it allows bonding to be achieved with less energy thus minimizing damage to the composite.

3.6 Summary

From the results described above it is apparent that, contrary to initial expectations, thermoplastic composites can be bonded relatively easily. Thus the application of exactly the same treatment as that used with an epoxy composite may not work, as

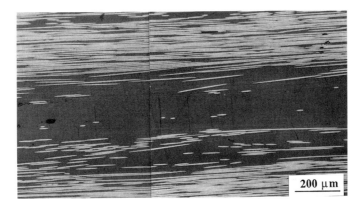

Fig. 11.7 Section through a joint in which carbon/PEEK adherends have been press-bonded at 290°C using a PEI interlayer.

shown in reference [37] for example, but only small modifications may be needed. For comparison, aluminium alloys parts for structural adhesive bonding require solvent cleaning, alkaline cleaning, deoxidizing and phosphoric acid anodizing before the adhesive is applied [38].

The testing of adhesive joints is not straightforward but it is interesting to note that the majority of the work on adhesive thermoplastic composite joints published to date has used the lap shear specimen. A few fracture mechanics specimens have also been tested but virtually no data is available on the long term behaviour of these joints. This is important as the aggresive nature of many of the surface treatments employed and resulting weak surface layers may not be revealed in the short term data. Further testing is required in this area.

4 Fusion Bonding

4.1 Historical

The ability to remelt thermoplastics locally offers the potential to apply to composites the wide range of fusion bonding techniques currently employed to assemble thermoplastic polymers. These techniques have been reviewed by *Watson et al.* [39] and more recently by *Stokes* [40], while *Benatar* and *Gutowski* have reviewed the methods available for fusion bonding of thermoplastic composites [41]. They may be classified according to the type of heat source used:

(i) External
(ii) Friction
(iii) Internal

In this section the requirements for a bonding technique are first discussed. The different methods appropriate to composites will then be briefly outlined. In the following sections some results for thermoplastic composites will be presented and the advantages and disadvantages of each method will be discussed.

4.2 The Bonding Process

In all the bonding methods described here two surfaces are heated, pressed together and held. The parameters involved are therefore temperature, pressure and time. The aim of the operation is to bring the surfaces into intimate contact in order to allow molecular interdiffusion to create bonds across the interface. The interdiffusion process has been reviewed in detail by *Kausch* [42] and for amorphous polymers the intrinsic material properties may be recovered by crack healing processes at temperatures above the glass transition temperature of the polymers. This autohesion process has been applied to composites based on polysulphone, an amorphous polymer with a T_g around 190°C [43]. Around 80% of the parent fracture toughness of the composite could be recovered by healing at temperatues of 213, 225 and 245°C. Results were not strongly time or temperature dependent.

The application of healing to semicrystalline polymers has received less attention. *Wool* has discussed healing of microvoids in polypropylene fibres [44]. In such materials it is in the amorphous regions between crystalline lamellar surfaces that molecular motion can take place. Thus no healing takes place below the T_g but at higher temperatures healing should be possible. However, for unreinforced PEEK no strength recovery has been measured at temperatures below the melting point [45]. This conclusion also holds for carbon fibre reinforced PEEK, for which a series of DCB specimens has been cracked open, healed and then re-cracked, [46] as described in section 6.1 below. Tests to model the consolidation stage of composite forming have reached similar conclusions [47] so the lower temperature bound for the PEEK composite bonding process is around 360°C for all practical purposes.

This places severe constraints on the bonding method to be used, as the interface must reach the melting temperature without distortion of the remainder of the structure. Times must be kept as short as possible and pressure must be controlled to ensure intimate contact between the surfaces without extruding matrix away from the fibres in the joint region. Some of the potential bonding methods will now be considered.

4.3 Fusion Bonding Methods

4.3.1 External Heat Sources

The following methods will be considered:

Hot press bonding
Heated plate welding
Hot gas
Focused Infrared
Laser

These methods are shown schematically in Fig. 8.

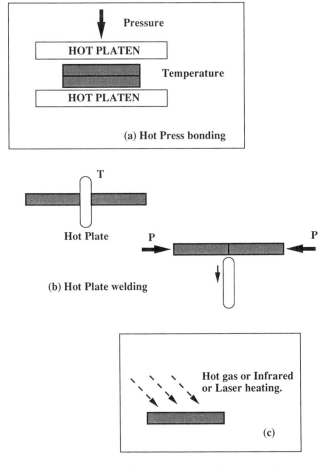

Fig. 11.8 External heat sources for fusion bonding.

4.3.1.1 Hot Press Bonding The hot press bonding technique is essentially a remoulding process. The entire component to be bonded is reheated to the melting temperature and an additional layer of film corresponding to the matrix is usually inserted to ensure adequate polymer at the interface. *Cantwell et al.* have studied the forming parameters associated with this approach in detail, [48, 49], as well as the influence of fibre orientation, specimen dimensions and test conditions. For example, Fig. 9 shows the influence of orientation of the fibre surfaces in contact.

Even at 90° the strength values are very high (compare with Table 1 for example). Very high lap shear strengths and mode I fracture toughnesses have been obtained using this method, both by the authors and elsewhere [50, 51].

4.3.1.2 Hot Plate Welding Hot plate welding of composites has not been widely studied but appears quite promising. It has been very extensively studied as a joining method for plastics and Stokes provides a good reference list [40]. Most commercial hot plate welding equipment is limited to temperatures below 300°C as a PTFE coating is used to prevent the surfaces to be bonded from sticking to the hot plate. Nevertheless, *Taylor* showed promising lap shear results ; up to 84% of the parent material strength could be obtained in a 2 minute welding time at 365°C [52], and more recently *Kempe et al.* showed tensile shear strengths up to 92% of the molded value [31].

4.3.1.3 Hot Gas, Infra-Red and Laser Welding Hot gas welding is a frequently used manual repair method, in which a thermoplastic filler rod is softened between the surfaces to be joined by a hot gas stream [53]. Flow rate and temperature are controlled but the method is very dependent on operator skill.

Finally, two alternative heating sources are infra-red and laser. Infrared heating was described by *Benatar* and *Gutowski* [41], who were not successful in using it to heat surfaces before bonding. *Silverman* and *Griese* [18] report tests with a focused infrared reflector heater, which was robotically controlled and could heat a narrow

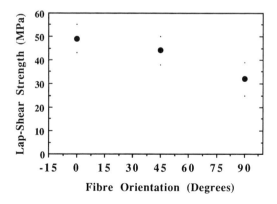

Fig. 11.9 Influence of fibre orientation at interface on lap shear strength of bonds produced in a heated press, carbon/PEEK.

region before bonding. This method produced high lap shear values (32 MPa) in a few minutes, the best of the fusion bonding methods in that study, but with large scatter. Laser heating has been applied in tape winding, to heat thermoplastic composites before applying pressure to consolidate them [54, 55]. It is suitable for very local heating but not well-adapted to heat large areas.

4.3.2 Frictional Heat Sources

Here the two principal methods which will be discussed are:

Ultrasonic welding
Friction welding
Rotation welding

These are shown schematically in Fig. 10.

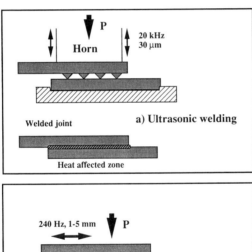
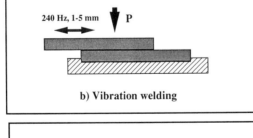
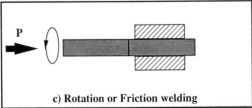

Fig. 11.10 Frictional heating methods.

4.3.2.1 Ultrasonic Welding Ultrasonic welding is widely used for precision assembly of relatively small plastic components. The components to be joined are excited at right angles to the contact area at frequencies around 20 kHz, so that longitudinal waves are transmitted. The process has been extensively studied and *Menges* and *Potente* showed over 20 years ago that for the materials they studied the heating was mainly due to internal friction rather than interface friction between the parts [56]. Thus the damping properties of the material are extremely important. Mechanisms of heating and energy transfer have been discussed by *Wang et al.* [57, 58], while *Habenicht* and *Ritter* also refer to extensive Russian publications in this area in the early 1960's [59]. More practical aspects of the welding process are discussed in references [60, 61]. In particular these underline the problems associated with joining semicrystalline as opposed to amorphous thermoplastics, on account of large energy losses in the temperature range above the T_g before the melting temperature is reached, and high melt viscosities which hinder flow. These necessitate higher amplitudes and longer welding times. In addition, for all materials the inclusion of an energy director at the interface is essential if efficient bonds are to be made.

As far as ultrasonic welding of thermoplastic composites is concerned it is interesting to note that in 1981, in a study of assembly of a polysulfone composite wing for a cruise missile, *Liskay* concluded that ultrasonic welding was not a practical solution [62]. More recently a number of studies have examined the process in more detail, but opinion is still divided over its potential. *Benatar* and *Gutowski* have modelled the welding cycle and identified five sub-processes [63]. These are vibration, viscoelastic heating, heat transfer, flow and wetting and intermolecular diffusion. Experimentally they molded unreinforced PEEK energy directors onto carbon/PEEK composite adherends and obtained excellent bond strengths, measured using the short beam shear test. Dynamic impedance monitoring was shown to be a means of following the welding cycle [64]. *Eveno* and *Gillespie* also moulded energy directors onto their specimens and followed the melt front propagation by using tapered specimens [65, 66]. Fracture toughnesses measured on bonded specimens exceeded those of the parent material. The main parameters which may be varied in ultrasonic welding are the time, pressure and vibration amplitude. A statistical approach was employed by *Strong et al.* to study the influence of these parameters and the combination of pressure and amplitude was shown to have the greatest influence on weld strength [67]. *Todd* varied welding time and achieved lap shear strengths around 35 MPa with both PEI and PEEK energy directors [34]. Other studies have been less successful. *Taylor* obtained very low lap shear strengths without energy directors [52], around 5 MPa, while *Krenkel* measured values up to 28 MPa with energy directors [51]. *Cantwell et al.* obtained similar values when a flat PEI film was included at the interface, but with PEEK films surface damage occurred before full bond strength developed [35], a problem also encountered elsewhere [18].

4.3.2.2 Vibration Welding Vibration welding is a larger scale process, well suited for applications such as automotive assembly. Lower frequency vibrations, typically 240 Hz, with displacements of millimetres rather than microns are used to generate interface friction. Energy directors are not necessary. Maguire obtained lap shear strengths around 40 MPa for carbon/PEEK with this process, with extra PEEK in the form of films placed in the joint [68]. *Nagumo et al.* [50] and *Kempe et al.* [31] were less successful, while *Cantwell et al.* obtained reasonable results with PEI film at the interface [35], up to 30 MPa.

4.3.2.3 Rotation Welding In rotation welding, sometimes known as spin or friction welding, heat is generated by the rapid rotation of one component against the other under an applied pressure [39]. It was first reported in the 1930's and welds can be obtained in a few seconds by the use of rotational speeds of 1 to 20 m/s. A little work on rotation bonding of thermoplastic composite tubes has also been published [31], but few details are available apart from the need to support both the inside and outside of the tubes during welding.

4.3.3 Internal Heat Sources

This heading includes methods where either the properties of the composite itself are employed to generate heat, or where a foreign material is added at the interface to be bonded in order to focus energy there. This includes some of the most promising methods for thermoplastic composites:

Electrical resistance welding
Induction heating
Microwave joining

These processes are shown schematically in Fig. 11.

4.3.3.1 Electrical Resistance Welding Electrical resistance welding has attracted considerable interest in recent years. Some workers have used high resistance wires embedded in resin film as the heating element to obtain good results [34] but this is not as attractive an approach as using the carbon fibres themselves. The idea of using carbon fibres as a heating element is not new, and use of the Joule effect to melt the matrix was described in a patent on impregnation of thermoplastic composites by *de Charentenay* and *Robin* [69]. Some published values of resistance bond strengths show considerable scatter when a carbon fibre tow was used [68], and this is attributed to difficulties with electrical contacts and alignment. *Cantwell et al.* used first a layer of carbon fibres and subsequently a prepreg layer between two PEEK films They also describe contact problems, which were overcome by placing the contacts immediately next to the heating zone and using silver paint to improve contacts [35]. Typical joints obtained are shown in Fig. 12. Reasonably high lap shear strengths (up to 35 MPa) were measured for the prepreg heating element joints, in spite of the 90° orientation of this element. *Silverman* and *Griese* [18] and *Xiao et al.* [70] also reported reasonable lap shear strengths (up to 25 MPa).

The most detailed study of resistance welding to date has been performed by Gillespie and co-workers, who have treated both experimental and theoretical aspects of the process [64, 71–74]. They refer to an MIT thesis which had addressed the contact problem [75] and adapt the solution described there to optimise uniform

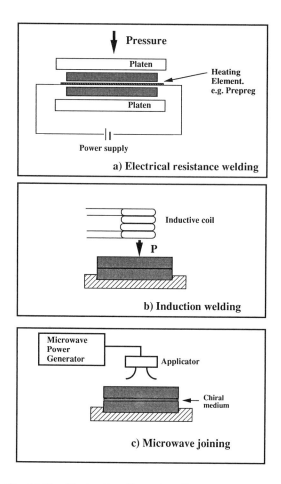

Fig. 11.11 Fusion bonding using "internal" heating.

heating, namely the use of a liquid metal bath. Their modified approach involves burning off the matrix from the ends of a layer of prepreg and then pressing them into a low melting point alloy [71]. Their studies on carbon/PEEK showed that welding initiated at the ends of samples and the melt front then propagated towards the interior. Non-uniform heating is not desirable as it leads to local overheating before complete welding, so that only small areas can be bonded. Longer process times (400 to 800 seconds) at lower power (around 7 amperes rather than 12, for a 208 V a.c. supply) resulted in more uniform heating and better welds.

Maffezzoli et al. used a one-dimensional model for welding processes to predict through thickness temperature and hence crystallinity profiles [76]. *Don et al.* presented a two-dimensional thermal analysis developed specifically for resistance welding, which enabled time to melt to be predicted [72]. They identified variable resistivity of heating elements and heat loss to tools as having the greatest influence on process time. This model was later applied to bonding of co-mingled thermoplastic

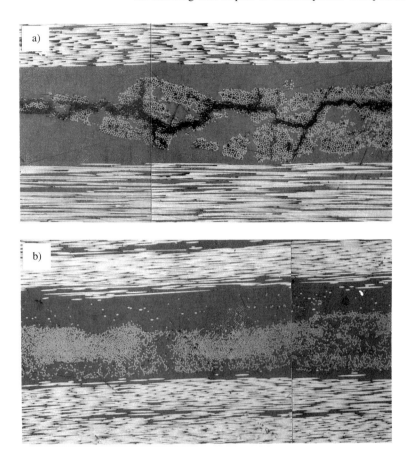

Fig. 11.12 Micrographs showing sections through resistance welded joints, bonded using a) T300 fibres and b) 90° IM6/PEEK prepreg ply as the heating element. Additional PEEK films were added in both cases.

composites with additional PEEK and PES films, and lap shear strengths around 30 MPa were obtained in both cases [74].

4.3.3.2 Induction Welding Induction heating in magnetic materials takes place by hysteresis losses in an applied alternating magnetic field, whereas in electrically conductive non-magnetic materials such as carbon fibres heating is due to eddy currents. The anisotropic conduction paths introduced by the fibres make heating difficult to control however, so metallic materials known as "susceptors", in the form of wire screens or powders are often added at the joint. These encourage local heating by both eddy currents and hysteresis losses but may reduce the mechanical properties of the joint. Examples have been given by *Benatar* and *Gutowski* who used nickel coated graphite fibres to heat and bond J-polymer composites [77], *Taylor* who experimented with silver wire [52], and *Nagumo et al.* who employed stainless steel meshes [50], both with carbon/PEEK. For carbon/PPS *Krone et al.* used the "Emabond" process involving a bonding agent composed of PPS and magnetic particles [22].

Miller et al. have treated induction heating in the context of die-less forming of thermoplastic composites [78], but also discuss induction bonding and coil design. They suggest that by careful selection of tooling material a temperature gradient can be set up which will heat the joint interface more than the rest of the material, and report excellent bond strengths using such an approach. This indicates that metallic susceptors are not essential, a conclusion also reached by *Border* and *Salas* [79]. These authors bonded carbon reinforced PEEK, PPS and PEI composites and obtained lap shear strengths in excess of 45 MPa, by heating the adherends globally. Todd also describes induction heating without metal susceptors but using specially developed carbon fibre mats with high electrical resistance to concentrate heat at the bondline, and achieved reasonable bond strengths, up to 33 MPa [34].

Thus induction bonding can produce very promising results without addition of foreign material at the bondline. The development of continuous welding processes is being actively pursued and in parallel some studies are now concentrating on the mechanisms of heat generation. An example is the work described by *Fink et al.* [80]. These authors discuss experiments which demonstrate that the primary heating mechanism in cross ply laminates is dielectric losses in the polymer region between fibres in adjacent plies. This should enable a simple model based on polymer properties to be established, so that heating parameters can be optimised.

4.3.3.3 Microwave Bonding A third approach which may prove valuable in joining and repair applications is based on microwave heating. *Benatar* and *Gutowski* commented that such a method is difficult in thermoplastic composites, as the fibres either reflect the microwave energy or heat up more quickly than the polymer [77]. However, the localization of heating is possible by placing so-called "chiral" elements at the interface to be heated. These may be in the form of microbubbles or films and a recent study by *Varadan* and *Varadan* suggests that enhanced microwave absorption by the chiral additions could result in a very versatile bonding method [81].

4.4 Structure in Welded Zones

The idea of a heat affected zone (HAZ) is well established in welding of metals and is also appropriate in discussing the fusion bonding of semi-crystalline thermoplastics. Several papers have treated this subject [e.g. 82-86], but so far few studies have extended the idea to thermoplastic matrix composites. The different structural zones which may be present in welds are illustrated in Fig. 13. This figure shows both unreinforced PEEK and glass fibre reinforced nylon joints produced by ultrasonic welding. Distinct variations in structure are apparent. In both cases it is apparent that the heating and cooling cycles undergone by material in and near the joint region vary considerably, producing not only the changes in microstructure visible in Fig. 13 but also internal stresses. More work on heat affected zones is needed as their behaviour is fundamental to all the fusion bonding processes described above.

11. Bonding and Repair of Thermoplastic Composites   357

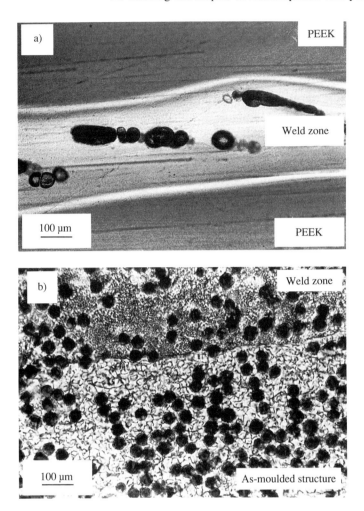

Fig. 11.13 Optical micrographic sections through ultrasonically welded joints, viewed between crossed polars in transmitted light, a) PEEK, b) glass reinforced nylon. In both cases the materials were bonded to similar materials. Voids are apparent in the PEEK joint a), while refined microstructure is developed in the weld zone in b).

5 Comparison of Methods

Having reviewed the large number of fusion bonding methods available a comparison is given in Fig. 14. This is based on several published studies which have examined a number of methods using the same test geometry, the lap shear specimen. The maximum value that can be obtained using this geometry is around 40–50 MPa [49].

This simple comparison only tells part of the story as a number of criteria may be used to judge bonding methods. For example, an important parameter in the selection of a bonding method for some applications will be the toughness of the joints produced. *Cantwell et al.* have compared a number of methods using both lap shear strength and mode I fracture toughness, [87], Fig. 15. Here it is apparent that the toughness of the PEEK film in a heated press joint offers significant advantages.

Finally in this section, Table 2 summarizes some of the principal advantages and disadvantages of the different joining methods.

Table 11.2

Method	Advantages	Disadvantages
Adhesive	Portable	Curing long. May be brittle
Interlayer	Lower temperatures required than hot press	Amorphous interlayers sensitive to solvents
Heated press	Best joints possible	Needs press and tooling. Whole component remelted
Hot plate	Widely studied for plastics	Not portable
Hot gas, IR, Laser	Local heating	Difficult to control, little information available
Ultrasonic	Rapid, automated precision method	Not suited to large areas nor continuous joints
Vibration	Rapid, can be automated	Large machines required, not portable, flash produced
Electrical resistance	Local heating using parent material (prepreg) elements.	Contact, uniformity of heating problems
Induction	Continuous bonding possible	Stress concentrations if susceptors included
Microwave	May be portable	Few studies to date. Foreign material added.

11. Bonding and Repair of Thermoplastic Composites 359

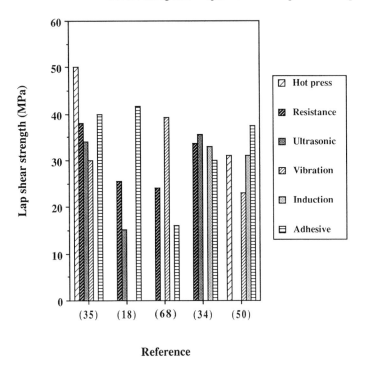

Fig. 11.14 Comparison of methods, lap shear test.

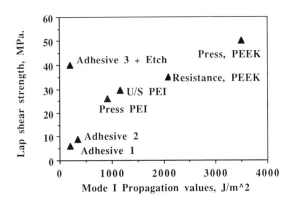

Fig. 11.15 Comparison of methods using both lap shear strength and mode I fracture toughness [87].

6 Repair

As discussed in the introduction to this chapter repair is a vast subject and could not be treated in detail even in a book devoted solely to the subject. Composite structures are repaired in the aircraft industry on an every day basis and much experience has been accumulated (e.g. references [3] and [4] give extensive background in this area). Other recent publications have discussed military [88, 89] and civil [90] composite aircraft repairs, a composite repair methodology [91], preparation and control [92], repair adhesives [93] and bolted repairs [94]. In considering a transition from thermoset to thermoplastic composites and in order to focus on the areas where use of thermoplastic composites might offer distinct advantages, only two types of damage will be considered here; a) delaminations and b) perforations.

6.1 Delaminations

Delaminations may be present in composite structures as a result of poor manufacturing, poor machining, impact damage or environmental effects. Once they are detected in carbon/epoxy parts experience has shown that if they are greater than 30mm in diameter they need to be repaired [4]. Small delaminations may be repaired by resin injection, but this is frequently unsatisfactory. Larger delaminations will require more complicated repair procedures. The thermoplastic nature of polymers such as PEEK allows a heating and pressure cycle to be applied so that delaminations can be completely healed. This was illustrated for compression after impact specimens, Fig. 16.

A second illustration of the healing of delaminations, Fig. 17, is shown below for double cantilever beam specimens [46].

Cantwell et al. have also performed tests which indicate that both flexural and tensile modulus and strength can also be recovered [95] An example is shown in

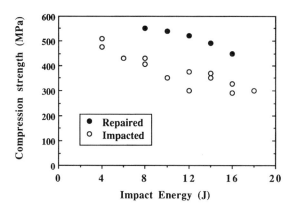

Fig. 11.16 Compression strength after impact, both before and after heated press repair, carbon/PEEK.

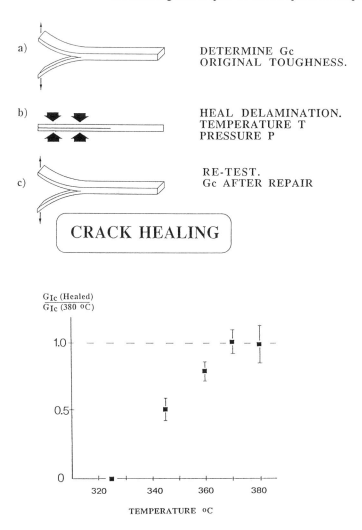

Fig. 11.17 Recovery of toughness by healing in a heated press, carbon/PEEK [46].

Fig. 18 for flexural modulus, which shows that even after impact at relatively high energies stiffness may not be irretrievably lost in these materials. These examples show the potential of fusion bonding processes for delamination repair. Other studies of this type have been published by *Ong et al.* [96]. Those authors tested both PEEK and PPS composites after different levels of impact damage and used both heated press and patch repair to recover compressive strength.

A second type of damage is perforation, which requires the application of a patch.

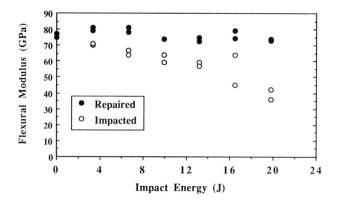

Fig. 11.18 Recovery of flexural modulus by heated press repair after drop weight impact damage at different energy levels.

6.2 Perforation Damage Repair

In this case the simple tensile specmen shown in Fig. 19 was used to examine the effectiveness of different repair techniques. Several of the bonding techniques described in detail in the first part of the chapter were employed, in order to compare their suitability for such repairs. The results of the study are also shown in Fig. 19.

It is clear that the fusion bonding techniques involving a PEI interlayer film are particularly attractive in such applications as they allow high strength to be recovered at relatively low forming temperatures. The adhesive bonding results with a suitable surface treatment are also very promising, and in some cases failure in adhesively bonded specimens occurred outside the patched region.

Conclusions

The last five years have seen considerable activity in the field of joining and repair of thermoplastic composites. These initial studies have allowed the potential of fusion bonding techniques to be demonstrated but have also shown that adhesive bonding remains a serious competitor for both assembly and repair.

If such studies are to be transferred from research groups to application two aspects must be pursued in detail:

1. The creep and fatigue behaviour of adhesive and fusion bonded joints must be studied, in order to generate confidence in their long-term reliability.
2. Bonding processes must be quantitatively modelled, so that optimization of joints for a given application can evolve from the current trial-and-error approach.

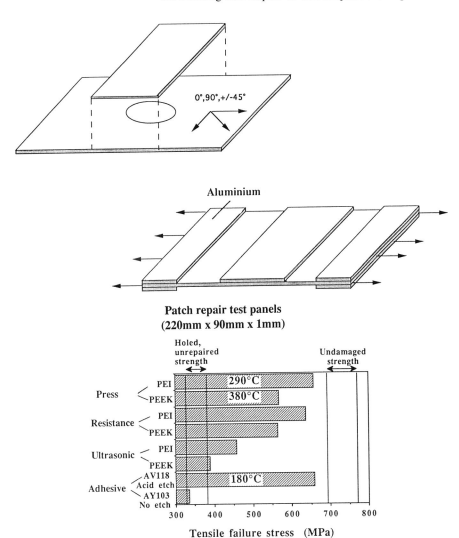

Fig. 11.19 Evaluation of patch repair techniques, carbon/PEEK, using the repair of a holed tensile specimen for comparison.

There are signs that activities in both these directions are underway and with the motivation of the significant advantages offered by thermoplastic composites progress should be rapid.

References

1. Kedward KT, Editor, "Joining of Composite Materials", ASTM STP749, 1981.
2. Matthews FL, "Joining fibre-reinforced plastics", Elsevier, 1986.
3. Brown H, "Composite Repair", SAMPE Monograph No. 1, 1985.
4. Baker AA, "Repair techniques for composite structures", Chapter 13 in "Composite materials in aircraft structures", Ed. DH Middleton, Longman Scientific, 1990.
5. Carlin DM, Proc. SPE ANTEC Conf., 1989, p1447
6. Vinson JR, Poly. Eng & Sci., Mid-Oct., 29, 19, 1989 p1332.
7. Hart-Smith LJ, Chapter 6 in "Joining fibre-reinforced plastics", Matthews FL, Elsevier, 1986.
8. Dastin SJ in "Handbook of Composites", Ed Lubin G, chapter 22, Van Nostrand Reinhold 1982, p602.
9. Phillips JL, Parker RT, ASM "Composites" handbook, 1990, p712.
10. Schwartz C, Ingenieurs et Architectes Suisses, 12, 30 mai 1990, (in French).
11. Dorey G, Ch.6 in "Structural Impact and Crashworthiness", ed Davies GAO, Vol.1, Elsevier, 1984.
12. Carlsson LA, Aronsson C-G, Backlund J, J. Mat. Sci., Vol.24, 1989, p1670.
13. Kinloch AJ, "Adhesion and Adhesives", Chapman & Hall, 1987.
14. Rosty R, Martinelli D, Devine A, Bodnar MJ, Beetle J, SAMPE Journal, July/Aug 1987, p34.
15. Hamdan S. & Evans JRG, J. Adhesion Sci. & Tech., 1, 4, 1987, p281.
16. Lees WA, Int. J. Adhesion & Adhesives, 6, 4, Oct. 1986, p171.
17. Wu S-I Y, Schuler AM, Keane DV, Proc. 19th SAMPE Tech Conf., Oct. 1987, p277.
18. Silverman EM, Griese RA, SAMPE Jnl., 25, 5, Sept/Oct. 1989, p34.
19. Powers JW, Trzaskos WJ, Proc. 34th SAMPE Conf., May 1989, p1987.
20. Kodokian GKA, Kinloch AJ, J. Mat. Sci., 7, 1988, p625.
21. Wu S-I Y, Proc. 35th SAMPE Symp., April 1990, p846.
22. Krone JR, Murtha TP, Stirling JA, Proc. 33rd Int. SAMPE Symp. March 1988, p 829.
23. Davies P, Courty C, Xanthopoulos N, Mathieu H-J, JMSci. Letters 10, 1991 p335.
24. Vetaphone electronik a.s., product data.
25. Kodokian GKA, "Adhesive bonding of thermoplastic fibre composites", PhD thesis Imperial College, London, Sept. 1989.
26. Kodokian GKA & Kinloch AJ, J. Mat. Sci. Letters, 7, 1988, p625.
27. Hall JR, Westerdahl CAL, Bodnar MJ, Levi DW, J. Appl. Polymer Sci., 16, 1972, p1465.
28. Boenig HV, "Plasma Science & Technology", Cornell Univ. Press, 1982.
29. Liston EM, J. Adhesion, 30, 1989, p199.
30. Soulier JP, Chabert B, Garby L, Sage D, Proc. EPF90, (European Polymer Federation) Sorrento Oct. 1990, p455.
31. Kempe G, Krauss H, Korger-Roth G, Proc. ECCM4, Stuttgart 1990, Elsevier Publishers, p105.
32. Evans JRG, Bulpett R, Ghezel M, J. Adhesion Sci. Tech. 1, 4, 1987, p291.
33. Cogswell FN, Meakin PJ, Smiley AJ, Harvey MT, Booth C, Proc. 34th SAMPE Symp., May 1989, p2315.
34. Todd SM, Proc. 22nd SAMPE Tech. Conf., Nov. 1990, p383.
35. Cantwell WJ, Davies P, Jar P-Y, Bourban P-E, Kausch HH, Proc. 11th European SAMPE Conf., Basel, May 1990, p411.
36. Smiley AJ, Proc SPI ANTEC90 meeting, Dallas, May 1990, p1773.
37. Kinloch AJ & Taig CM, J. Adhesion, 21, p291.
38. Scardino WM, "Adhesive Specifications" in ASM handbook of composites, 1990, p659.
39. Watson MN, Rivett RM, Johnson KI, Welding Inst. Research report 301/1986, June 1986.
40. Stokes VK, Poly. Eng & Sci, 29, 19, mid Oct., 1989, p1310.
41. Benatar A, Gutowski TG, SAMPE Jnl., Jan./Feb. 1987, p33.

42. Kausch HH, "Polymer Fracture", 2nd edition, Springer, Chapter 10, 1987.
43. Howes JC & Loos AC, Proc. American Soc. for Composites, Technomic Publishers, 2nd conf., Sept. 1987, p110.
44. Wool RP, in "Adhesion and Adsorption of Polymers" Part A ed. L-H Lee, Plenum publishing, 1980, p341.
45. Davies P, unpublished work, EPFL.
46. Davies P, Cantwell W, Kausch HH, J. Mat. Sci. Letters, 8, 1989, p1247.
47. Lee WI, Springer GS, J. Comp. Matls., 21, Nov. 1987, p1017.
48. Cantwell WJ, Davies P, Jar P-Y Richard H, Kausch HH, J. Mat. Sci. Letters, 8, 1989 p1035.
49. Cantwell WJ, Davies P, Bourban P-E, Jar P-Y Richard H, Kausch HH, Comp. Structures, 16, 1990, p305.
50. Nagumo T, Nakamura H, Yoshida Y, Hiraoka K, Proc 32nd SAMPE Symp., April 1987, p396.
51. Krenkel W, in "Engineering Applications of New Composites", ed. Papetis SA & Papanicolaou GC, Omega Scientific, 1988, p119.
52. Taylor NS, Welding Inst. Res. Bulletin, July 1987, p221.
53. Watson MN, Welding Inst. Bulletin, May/June 1989, p77.
54. Beyeler E, MS thesis, University of Delaware, Dept. of Mech. Eng., 1988.
55. Ferraro F, DiVita G, Marchetti M, Cutolo A, Zeni L, in Proc. ECCM4, Stuttgart 1991, p89.
56. Menges H, Potente H, Welding in the World, 9, 1/2, 1971 p46.
57. Frankel EJ, Wang KK, Poly. Eng. & Sci., 20, 6, April 1980 p396.
58. Tolunay MN, Dawson PR, Wang KK, Poly. Eng. & Sci., 23, 13, Sept. 1983, p726.
59. Habenicht G, Ritter J, Kunststoffe, 78, 6, 1988, p546.
60. Zentralverband der Elektronischen Industrie e.V (ZVEI), "Assemblage par ultrasons de pièces moulées et de produits semi-finis en matière synthetique thermoplastique".
61. Branson Ultrasonics Corp. "Ultrasonic Plastics Assembly", 1979.
62. Liskay GG, Proc. 13th SAMPE Tech. conf., Oct. 1981, p592
63. Benatar A, Gutowski TG, 33rd SAMPE Symp., March 1988, p1787.
64. Benatar A, Gutowski TG, Poly. Eng & Sci., 29, 23, mid-Dec., 1989, p1705.
65. Eveno EC, MS.thesis, Dept. Mech. Eng., University of Delaware, 1988.
66. Eveno EC, Gillespie JW Jr., Proc. 21st SAMPE Tech. Conf., Sept. 1989, p923.
67. Strong AB, Johnson DP, Johnson BA, SAMPE Quart., Jan. 1990, p36.
68. Maguire DM, SAMPE Jnl., 25, 1, Jan/Feb. 1989, p11.
69. de Charentenay FX, Robin J-J, European patent 82.401 904.6, deposed 18th Oct. 1982.
70. Xiao XR, Hoa SV, Street KN, Proc. 35th SAMPE Symp., April 1990 p37.
71. Eveno EC, Gillespie JW Jr., J. Thermoplastic Composites, 1, Oct. 1988, p322.
72. Don RC, Bastien L, Jakobsen TB, Gillespie JW Jr., 21st SAMPE Tech. conf., Sept. 1989, p935.
73. Jakobsen TB, Don RC, Gillespie JW Jr., Poly. Eng & Sci., mid Dec., 29, 23, 1989, p1722.
74. Bastien L, Don RC, Gillespie JW Jr., Proc. 45th SPI Conf., 1990, p20B.
75. Houghton WR, SB Theis, Dept. of Mech. Eng., MIT, 1984.
76. Maffezzoli AM, Kenny JM, Nicolais L, SAMPE Jnl., 25, 1, Jan. 1989 p35.
77. Benatar A, Gutowski TG, SAMPE Quart. 18, 34, 1986 p35.
78. Miller AK, Chang C, Payne A, Gur M, Menzel E, Peled A, SAMPE Jnl., 26, 4, July/Aug 1990, p37.
79. Border J, Salas R, Proc. 34th SAMPE Symp., May 1989, p2569.
80. Fink BK, McCullough RL, Gillespie JW Jr., submitted to Polymer Composites, 1990.
81. Varadan VK, Varadan VV, Poly. Eng. & Sci., 31, 7, Mid-April 1991, p470.
82. Menges G, Zohren J, Plastverarbeiter, 18, 1967 p165.
83. Barber P & Atkinson JR, J. Mat. Sci., 7, 1972. p1131.
84. Barber P & Atkinson JR, J. Mat. Sci., 9, 1974, p1456.
85. de Courcy DR, Atkinson JR, J. Mat. Sci., 12, 1977, p1535.
86. Schlarb AK, Ehrenstein GW, Kunststoffe, 78, 6, 1988 p541.

87. Davies P, Cantwell WJ, Jar P-Y, Bourban P-E, Zysman V, Kausch HH, Composites, Vol. 22, No. 6, Nov. 1991.
88. AGARD Conference on "Repair of Aircraft Structures involving Composite Materials", Oslo, CP402, Jan. 1987.
89. Elkins CA, "In-service use and repair of composite materials", in "Composite materials in aircraft structures", ed. Middleton D, Longman, 1990, p379.
90. Armstrong KB, "British Airways experience with composite repairs", in "Composite materials in aircraft structures", ed. Middleton D, Longman, 1990, p368.
91. Hall SR, Raizenne MD, Simpson DL, Composites 20, 5, Sept 1989 p479.
92. Rogger JS, Composite Structures, 6, 1986, p165.
93. Cichon MJ, Proc. 34th SAMPE Symp., May 1989.
94. Bohlmann RE, Renieri GD, Riley BL, ASTM STP 893, 1986, p34.
95. Cantwell WJ, Davies P, Kausch HH, SAMPE Journal, Nov./Dec., 1991 p30.
96. Ong C-L, Sheu M-F, Liou Y-Y. Proc. 34th SAMPE Symp., May 1989, p458.

Index

19F NMR 48

acoustic emission 263
activation energies for degradation 61
adhesive bonding 339
airy functions 315
amorphous phase, influence of 105
amorphous polymers, microdeformation 129
amorphous specific heat 92
amorphous thermoplastics, crazing 148
APC2 292
-, damage tolerance 186
-, residual stresses 186
aromatic thermoplastic composites 193
autoclave 284

bolted joints 318
bonded joints 323
braiding 285

carbon fiber composite 49
chain ends 73, 74
- characterization 24, 73
- contents 27
characterization 49
co-mingling 279
compaction 283, 297
composites, fibre-matrix interfacial strength 176
compressive fatigue in aromatic thermoplastic composites 221
conformational peak 97
consecutive tape laying 284
consolidation 283
continuous fibre composites 193
cooling rate 291, 294, 296
corona discharge treatment 342
cost effectiveness 333
crack healing 348
crazing in amorphous thermoplastics 148
crazing in PEEK 133

creep in aromatic thermoplastic composites 196
creep modulus in aromatic thermoplastic composites 201
crosslinking reactions 60
crystal growth rate 62, 67
crystalline phase, influence of 105
crystalline specific heat 92
crystallinity 83, 90, 93
- band 94
crystallization kinetics 61, 67, 70, 77
crystallization of PPS, influence of on fibres 159
crystallization temperature 69, 76
cutting and machining 331
cycle time 273

D'Arcy's law 277
deformation in amorphous thermoplastics 147
deformation zones (DZs) in PEEK 132
degradation 70
- mechanisms 63
degree of crystallinity 84, 87, 90
-, determination of 94
-, influence of on PEEK mechanical properties 121
delamination failure, damage zone 188
density determination 84
design data 312
design procedure 306
design structural concepts 310
determination 87
diaphragm forming 288, 327
differential scanning calorimetry (DSC) 61, 87, 89, 90
DMA measurements on TLCPs 261
drapeability 279, 281
dry process 277

effects of impurities 70
elasticity modulus 106

electronical resistance welding 353
elemental analysis of commercial PEEK 68
entanglement 129, 147
extrinsic impurities effects 70

failure criteria 312
fatigue in aromatic thermoplastic composites 208
fibre bridging 182
fibre hybridization 279
fibre-matrix interface in thermoplastic composites 144
filament winding 285, 330
film stacking 279
finite element methods 315
flame treatment of composites 344
fluorine Analysis 32
formability 281
forming pressure 296
fracture toughness, influence of test temperature 183
free-surface crystallization 114
fusion bonding 347

Gegentakt injection moulding of TLCPs 266
glass transition 103, 283

high performance composites, interlaminar properties 174
high performance matrix materials 143
high strain properties of TLCPs 261
hot gas 350
hot plate welding 350
hot press bonding 350
hot-melt impregnation 278
hydroforming 289

impact testing and fracture toughness of TLCPs 255
impregnation 277, 278
- with thermoplastics 143
induction welding 355
inert atmosphere 63

infra-red welding 350
infrared absorption spectroscopy, 94
interfacial region 103
interlaminar fracture, geometrical effects 184
interlaminar shear 282
interlaminar toughness, effect of crystallinity 181
internal stress 292
interply slip 289
intralaminar shear 282
IR absorption bands 99
IR spectroscopy 22

joining thermoplastic composites 317

kinetics 291, 294

laminate 292, 293
lamination of oriented TLCP sheets 165
laser welding 350
light scattering 36
long term mechanical properties 124, 193
- of amourphous thermoplastics 155
- of TLCPs 253

magnetic resonance (NMR) spectroscopy 17
manufacturing 273, 296
- of thermoplastic components 325
- tolerances 331
MARK-HOUWINK-SAKURADA 38, 39
matched die molding 287, 325
material costs 313
matrix morphology of PEEK composites 127
mechanical porperties 227
melting enthalpy 88
microdeformation behaviour in PEEK thin films 129
microscopic interpretations of stiffness in TLCP mouldings 258
microscopy 149
microwave bonding 356

Index 369

mode fracture toughness 177
model compounds 44, 48, 99
modulus-strain function in aromatic thermoplastic composites 198
molecular Mass Determination 35
morphology 291
- of TLCP injection Mouldings 233

nucleating clusters 79
nucleation density 77
nucleation rate 67

orientation distributions in TLCP injection mouldings 246
oxygen diffusion 60

PEEK 38, 49, 63, 66, 74, 83, 84, 87, 89, 90, 92, 94, 99, 100, 141
- composites 111
- crystallization kinetics 69
- degradation 70
- etching 114
- oligomer 66
-, Acylation 6
-, annealing 122
-, characterization 3
-, chemical structure 3
-, crazing in 133
-, crystal growth 112
-, crystalline structure 112
-, deuteration 28
-, impact properties 179, 186
-, in-plane properties 178, 184
-, infrared spectrum 23, 28
-, interlaminar properties 181
-, mechanical properties 111
-, microstructure of 112
-, NMR spectrum 19, 31, 33
-, nucleophilic substitution 7
-, parameters affecting the spherulitic morphology 115
-, physical aging 123
-, physical stucture 111
-, polarized lightmicroscopy 115
-, rate effects 180, 187
-, secondary crystallinity 118

-, solubility 8
-, spectroscopic characterization 15
-, spherulite size 123
-, spherulitic structure 113
-, sulphonation 9, 13
-, thermal degradation 16
--, crystallization kinetics 57
percolation 282
perforation damage repair 362
permeability 278
PES as a composite matrix 156
petrographic techniques 114
plasma treatment 343
ply 282
polarized light microscopy of TLCPs 230
poly (ether ketone), chemical structure 11
polyethersulphone (PES) 141, 146
polyphenylene sulphide (PPS) 141, 157
powder coating 279
PPS composites 158
- , improved 161
pre-preg 278
preconsolidation sheet 282
preform 274, 282
preimpregnated 282
preliminary design 312
preparation of PEEK films 131
press forming 287
processability 276
processing conditions 93
processing window 296
production costs 313
production process 309
property profiles in TLCP injection mouldings 243
pultrusion 286, 329

quality assurance 332

random chain scission 60, 64
recyclability 276
reinforcement 276
relaxation time 276, 295
repair 360
- of delaminations 360
requirements for materials 334

Index

residual stress 293
rigid fraction 100, 101, 103
roll forming 290
room temperature SEC of totally sulphonated PEEK 45
rotation welding 353
rubber-pad press forming 289

scanning electron microscopy (SEM) 114
- of TLCPs 230
SEC of PEEK 41
SEC, calibration 35
self-nucleation 67, 78
service experiences of thermoplastic parts 333
shear banding 155
size exclusion chromatography (SEC) 35, 36, 41, 50, 59
skin-core 293, 296
solidification 283, 293
solubility 5
solution impregnation 279
specific calibration curve 44
specific heat 93, 100, 103
spherulite 292
squeeze flow 282
stress calculation methods 312
structure in welded zones 356
sulphonation of PEEK 5
surface characterization of thermoplastic composites 346
sythesis 5

tape laying 330
tape winding 285
temperature, influence of on compressure fatigue in aromatic thermoplastic composites 221
temperature, influence of on tensile fatigue in in aromatic thermoplastic composites 215
tensile fatigue in aromatic thermoplastic composites 214
-, influence of on cooling rate 223
-, influence of on fibre type 217
-, influence of on matrix 218
tensile properties of TLCPs 238
TGA measurements 66
thermal stability 63, 74
- PEEK, oxidative atmosphere 58

thermogravimetric (TGA) 60
thermoplastic adhesives 346
thermoplastic composites 141
-, surface treatments 340
thermotropic liquid crystalline polymers (TLCPs) 141, 162, 227
- etching 234
- fibre reinforcement in 'in-situ' composites 163
- in conventional continous fibre composites 162
-, influence of on heat treatment time 250
-, influence of on injection moulding conditions 248
-, injection mouldings 231
totally sulphonated PEEK (H-SPEEK) 39
toughness of amorphous thermoplastics 151
tow 279
transcrystalline 292
transmission electron 149
- microscopy (TEM) 131
transverse flow 282
two-phase crystallinity 99
two-phase model 84

ultrasonic welding 352
unit cell 85, 86
universal calibration 43

vacuum bag 284
vibration welding 352
viscoelasticity 196
viscometry 35- 37
viscosity 275, 277, 278
volatile emission 60, 61
volatiles 65

weaving 285
wet process 277
wetting 277
Wide-Angle X-Ray Scattering (WAXS) 85, 86
wrinkling 282